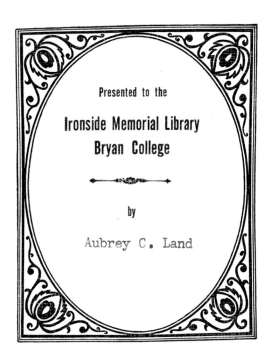

Quicksilver: *Terlingua and the Chisos Mining Company*

Quicksilver

Terlingua and the Chisos Mining Company

by KENNETH BAXTER RAGSDALE

Foreword by JOE B. FRANTZ

Texas A&M University Press
COLLEGE STATION

Library of Congress Cataloging in Publication Data

Ragsdale, Kenneth B   1917–
   Quicksilver: Terlingua and the Chisos Mining Company.

   Bibliography: p.
   Includes index.
   1. Mercury mines and mining—Texas—Terlingua district—
History. 2. Chisos Mining Company—History. I. Title.
TN463.T4R33 1976   338.7'62'2345409764932 75-4081
ISBN 0-89096-013-5

Manufactured in the United States of America
FIRST EDITION

For Janet

Contents

Illustrations

Foreword

THE creosote bush kills any growth within the area over which it drips. It makes the Big Bend country look emptier than its aridity would normally dictate. A wide-lens view of the region recalls those drawings which illustrate how a television screen works—an elliptical blank space with so many dots to the square mile. The creosote bushes provide the dots on a sea of off-white.

Mankind offers the creosote bush little competition. Alpine, the seat of Brewster County, was little more than a section house near Kokernot Springs until the period around World War I. It first appeared as an incorporated town in the national census of 1920, when it had 931 people, hardly a crowd. As recently as 1970 Terlingua boasted an official population of only 25, about the size of a meeting of a bank's board of directors. Metropolitan Terlingua, which would include Study Butte (population 120) and Lajitas (population 6), would run the total all the way up to 151. Urban sprawl is no problem nowadays around Terlingua.

Seventy-five years ago congestion in Brewster County was even less of a problem. The wonder is, as with so many mining towns, that anyone ever walked in there to find minerals in the first place. That someone was enterprising enough to begin a fairly major industrial development in the area borders on the miraculous. It recalls the story of the discovery well which opened the field which in turn has brought the University of Texas and Texas A&M University around seven hundred million dollars in endowment. The wildcatter drilled the well miles from where he had wanted to sink it. When his wagon had broken down and he realized he could go no farther, he made the best of the situation, drilled, and brought in a gusher. Appropriately, the name of the well was Santa Rita No. 1, after the Saint of the Impossible.

Mining quicksilver in Terlingua three-quarters of a century ago seems to have been almost as much of an impossible dream. Even today, with a state university in Alpine and with the Big Bend National Park nearby, Brewster County remains two hundred miles from the nearest commercial airport. Professors at Sul Ross State University at Alpine complain that their graduate students have to make a round trip of four hundred miles to El Paso to find enough materials for theses and that if the students are perceptive enough, they soon realize that they are foolish to make such a round trip and defect to the University of Texas at El Paso. The Sul Ross parking lot has more hitching posts than it has parking spaces for automobiles. Even then, most of the parking spaces are occupied by pickups, probably with a shotgun on a gunrack behind the driver, a deer or catskin lying casually in the pickup bed, and a six-pack of Lone Star beer under the seat. New professors learn quickly to accept an excuse from a student like the following: "Sir, I'm sorry I missed the final examination but my horse was sick." Such reasoning is incontestable.

This is a different world, a faraway world. Appropriately the area's most notable tree is an evergreen called a madrone which produces berries ranging from crimson to red-yellow in color. Equally appropriately, local people refer to the madrone by such popular names as Naked Indian Woman and Lady's Leg. Also, the principal range of mountains, which violates proper American practice by running east-west rather than north-south, is known as the Chisos, which translates as "ghost." In the evenings the ghostliness of grays and blues comes down to dance all around you.

The total effect is one of "geological confusion," a place "where God didn't take much pains." Ludwig Bemelmans wrote that "here I came upon the greatest wonder. The mantle of God touches you; it is what Beethoven reached for in music." H. Allen Smith, who has described the scenery as "rancid," retorted that "old Ludwig must have been boozing it up." A forgotten cowboy said it best in the 1840s when he observed, "You go south from Fort Davis until you come to the place where rainbows wait for rain, and the big river is kept in a stone box, and water runs uphill. And the mountains float in the air, except at night when they go away to play with other mountains."

The problems of procurement of men and materials are magnified. Everything costs more. With so much amplitude surrounding, employers find it difficult to subject men to the discipline of day labor. Money lies at least four hundred miles away, or did when Howard E. Perry ran the Chisos Mining Company at Terlingua. Although in seven decades desiderata have moved two hundred miles closer, distance still presents more inconvenience than enchantment. The land gives nothing, and all the while it is eroding away on the men and beasts who venture into it. Temperatures in the *cañons* and *arroyos* are hotter than Astroturf at sunny high noon, while winters send piercing winds and stinging sleet against which no amounts of clothing and bedding are impervious. It's a land of no-inbetween—either you burn on the plains of Satan, or you freeze in the mountains. Your friends are bobcats and stunted whitetails and diamondback rattlers and horned toads. For a wild night of diversion you can go down to the old Ranger station at Castolon (population 0), with its one store and shaded gallery, and talk about how the government is ruining the cattle business and whatever became of good cigarettes before the government took all the tobacco out of them. Or you can wade the Rio Grande in dry season over to Boquillas, nurse a few *cervezas*, and ponder on how you are going to smuggle a load of Carta Blanca or Bohemia back to your side of the river. A *niño* will take you over on his burro for 25 cents. He charges 50 cents for the return trip, because he's basically a *yanqui* trader who profits off your need.

Looking at Texas as a historian, one is reminded that when western mining states are listed—and most western states owe their primary development to mining—Texas is never included among those states in which mining was a factor. Oh, the hills and valleys from the Sabine to the Rio Grande are sporadically pockmarked by long gentled depressions where some half-demented man imagined he could find Santa Anna's buried gold. Ben Sublett used to pay his bills at Molly Williams' bawdy house in Odessa with gold nuggets he claimed to have mined in the Guadalupe Mountains. East and North Texas enjoyed some coal prosperity before petroleum moved in. But by and large mining is not one of those industries about which Texans boast or exaggerate.

And yet mining contributes to the total picture. Texas' mineral production, exclusive of natural gas and petroleum, ran better than a billion dollars in 1972 and includes such products as talc and soapstone, sulphur, salt, limestone, gypsum, helium, and cement. Though these products are valuable to the entrepreneurs who have developed them, their contributions pale alongside the more than six billion dollars contributed by oil and gas in that same year. When the price of quicksilver hit $775 per flask in 1965, Texas miners thought that the Terlingua area might be revived. But by 1971 the environmentalists had taken over, and the outcry against mercury pollution had become so insistent that the mines in the Terlingua area were again shut down. Permanently? *¿Quien sabe?* Who knows?

In this story of Texas minerals Kenneth Ragsdale has cut and polished a new facet. He has brought new life to our knowledge, hitherto neglected, of the nonpetroleum minerals. For that matter, the story of quicksilver itself, wherever mined, has not been thoroughly told. On both smaller Texas and larger world palettes Ragsdale here contributes new strokes that will further illumine the two portraits. Along the way he gives us an intriguing picture of H. E. Perry, who combined in his mean soul both the stereotyped virtues and blemishes of the Yankee entrepreneur. With equal unconcern he exploited men and minerals. Only profit mattered, but by golly, he saw in a formidable, desolate land a prize that few other men would have had the courage and greed to reach for. While Perry lacked the flamboyance of Jim Fisk or Jay Gould, he was a proper inheritor of a robber tradition. Arguments still rage over whether in an underdeveloped country, such as the United States was less than a century ago, a nation requires such bold buccaneers. While Ragsdale doesn't settle that argument, he gives us new insights and at the same time narrates for us a legitimate piece of Texana that has long been unappreciated and even unknown. When next year and the year after and the year after, the thirsty crowds re-gather in Terlingua to hold a world's championship of chili cook-offs, for the first time they can know why in this God-forsaken place there was ever a Terlingua to begin with.

JOE B. FRANTZ

Preface

COMMERCIAL quicksilver production began in the Big Bend of Texas in the late 1800s, and when the new century dawned, a boom was in the making. Recovery of this rare liquid metal brought a measure of civilization to one of America's last frontiers, yielded considerable wealth to a few mine owners, and became an important factor in the world quicksilver market. From the outset this industrial empire came under the spell of Howard E. Perry, a Chicago industrialist whose Chisos Mining Company became the leading producer in the Terlingua Quicksilver District. This social and economic study focuses on the man, the company, the region, and the influence that Perry wielded.

Today little remains of the Chisos Mining Company installations at Terlingua to indicate its once vital past. The people are all gone. The old company store, the abandoned jail, the slowly decaying Perry mansion, and the vacant adobe ruins are its only testaments to the past. Terlingua is a ghost town. When I first walked down those dusty, silent streets and peered into the darkened mining shafts, I sensed the presence of another age. I wanted to know the people who transformed that barren waste into a prosperous industrial community, where they came from, how they lived, and the man who had occupied that hilltop mansion and guided their lives. The prospect promised high adventure.

After surveying the sparse literature on the Terlingua Quicksilver District, I directed my research toward primary sources and personal interviews and discovered a series of "bonanzas." At the Archives Division of the Texas State Library I examined the extensive Chisos Mining Company Papers. The Federal Records Center at Fort Worth, Texas, made available the voluminous Chisos Mining Company bankruptcy case file. The National Archives in Washington had important National Recovery Administration documents and Big Bend military records relating to Perry and the Chisos Mining Company. San Antonio attorney Wilbur L. Matthews discussed the *Chisos Mining Company–Rainbow Mine* underground invasion case at great length and lent me his files of that landmark litigation. Walter K. Bailey, nephew of Howard E. Perry, provided invaluable data on Perry's childhood and the early Cleveland and Chicago periods.

Next I sought the people who had worked at the Chisos mine. My visits with these friendly and generous people were the highlight and reward of my research. Special thanks go to Harry Carpenter and his daughter, Mrs. Harriet Khrut, who acted alternately as "my man in Alpine" in arranging many of these rewarding interviews. Mrs. Dave Elliot, Mr. and Mrs. Earl Anderau, Mr. and Mrs. William D. Burcham, Mrs. Elijah Bledsoe, Hallie Stillwell, Pablo Sandate, W. D. Smithers, and H. C. Hernandez of Alpine; Hunter Metcalf and Wayne Cartledge of Marfa; Mrs. G. E. Babb of Sanderson; Elmo Johnson of Sonora; Mrs. Petra Benavides of Study Butte; Mack Waters of Panther Junction; and C. T. Armstrong of San Antonio all talked extensively—and frequently with extreme candor—about their Terlingua experience.

The narrative would have lacked much of its color and authenticity without the faithful assistance of the late Robert L. Cartledge. With his vivid recollections of twenty-nine years as store clerk, purchasing agent, and general manager of the Chisos Mining Company, plus justice of the peace and long-time confidant of Howard E. Perry, Cartledge made the greatest single contribution to this undertaking. But in all my adventures in research, the most gratifying surprise occurred when, after many months of interviews, Cartledge presented me a collection of company correspondence exchanged between him and Perry over a ten-year period. This invaluable collection contains Perry's almost daily instructions for conducting every phase of the vast quicksilver operation and Cartledge's comprehensive responses. That Cartledge did not live to see the completion of this history of the company for which he gave so much is the only disappointing aspect of this project.

Quotations from documents and correspondence have been faithfully transcribed, including lapses in spelling and usage, but these slips pass unremarked, to spare the reader a proliferation of *sics*.

I am deeply appreciative of the encouragement and personal guidance of Dr. Jim B. Pearson, who, as a member of the history department of the University of Texas at Austin, supervised the fledgling stages of this research; and the personal thoughtfulness of Dr. William H. Goetzmann, who encouraged me to enter the American Studies doctoral program. My sincere thanks go also to Dr. Ross A. Maxwell, Bureau of Economic Geology; from his firsthand knowledge of the Big Bend region he aided me at the outset of my research and made valuable suggestions for organizing the manuscript. Thanks belong also to Dr. John E. Sunder of the history department for directing me to several high yield research sources, and to Dr. L. Tuffly Ellis for his guidance in seeking a deeper understanding of the Mexicans' perspective on their Terlingua experience.

To Dr. Joe B. Frantz I especially owe a strong debt of gratitude. As friend, employer (Texas State Historical Association), and dissertation chairman (history department, University of Texas at Austin), he made immeasurable contributions. His critical reading of the manuscript gave economy, clarity, and impact to a sometimes casual writing style. And his inimitable marginalia, composed in taxicabs, airplanes, motels, and where else only he knows, added personal humor to a lengthy undertaking that was at best, laborious. Once when I failed to draw a proper analogy, he commented, "I might say the same thing about life on an alligator farm;" and "Whew!" when my thesaurus yielded an overpowering adjective. He inscribed praise, as well as criticism, and "'At's right!" indicated I had composed something to his liking.

To Mrs. Ruth Ragsdale, my mother, personal cheerleader, and grammarian, who read each chapter with mixed praise and criticism and suggested repeatedly the consistent placement of commas, I am also appreciative. But for whatever merit this work achieves, I accord major credit to my wife, Janet, who, during much of this research served concurrently as mother, father, counselor, and breadwinner for my two understanding and admiring children, Keith and Jeffrey. And a very special heartfelt thanks goes to a faithful, fuzzy-faced little dog named Alexander, who maintained a constant vigil beneath my desk throughout the entire preparation of this manuscript.

To all of the above—and many, many more—a very sincere thank you.

Quicksilver: *Terlingua and the Chisos Mining Company*

Introduction

DEEP in the Big Bend of Texas lies a plot of dry, ruggedly barren land known to geologists as the Terlingua Quicksilver District. Located some ninety miles south of Alpine, Texas, in close proximity to Mexico, the region possesses a greater geographic affinity to that country than it does to the remainder of the United States. Geologically a part of trans-Pecos Texas, this is an area of arid, mountainous terrain, characterized by its rugged sierras, high plateaus, and gently sloping plains. The austere beauty of this strange land is delusive and holds many surprises for the newcomer. One visitor reports: "In the clear air of the desert, the mountain masses loom with sharp outlines and clear detail for a distance of many miles, and the plains that surround them are deceptively foreshortened."[1]

From an elevation of 4,500 feet in the Alpine area, the terrain gradually slopes south toward the Rio Grande. Beyond Adobe Walls Mountain the landscape changes abruptly and the smoothly contoured panorama is replaced by rugged, treeless, close-spaced mountains that give an impression of forbidding harshness. Here the thin-soiled, arroyo-scarred land is hardened by searing summer heat, scant rainfall, and high evaporation. This is Terlingua country, the geographic heart of the Big Bend of Texas.

Boundaries of the Big Bend are vague and fluid; it is variously a place, a state of mind, and at times an illusion. Paradoxical views reflect different tastes and different experiences. It is a heaven or a hell, a land of serene beauty or barren ugliness, a place of soothing solitude or haunting loneliness. Hemmed in by the Davis Mountains, the Rio Grande, and an imaginary line drawn somewhere west of the Pecos River, the Big Bend country is a world in itself. The historian J. Evetts Haley paints a dramatic picture of this unique land:

> West and south of Fort Davis, the Rio Grande, that perverse and individualistic stream which gives character to the border of Texas,

[1] Robert G. Yates and George A. Thompson, *Geology and Quicksilver Deposits of the Terlingua District, Texas*, p. 3.

loops leisurely through this desert land, impudently cutting its toughest terrain in two, and then bending back in a neighborly gesture to join the Pecos, before ambling off again to the southeast, eventually to reach the Gulf. The land which lies below the Davis Mountains within the loop is known as the Big Bend, a vast and stubborn and hence challenging and interesting world in itself.[2]

Since this is an overpowering land, everyone who enters—if he plans to stay—involuntarily accepts the challenge of the environment. It is a tough land to settle, never changing. It is an unyielding land, giving up reluctantly whatever man, plant, or animal seeks from its crusty bosom. It is a defensive land, buttressed with steep mountain ranges, deep canyons, trackless deserts, and dry arroyos. It is a land of exaggerated dimensions. Searing summer heat occasionally pushes the mercury above the 120° F. mark, distances seem interminable, and the inhabitants few and solitary. Sometimes nature in the Big Bend reacts as if it had fallen behind schedule and is rushing to catch up. Thunderstorms appear with startling suddenness, emptying their entire contents within a few minutes. Such cloudbursts convert dry water courses into raging torrents and almost anything movable—trees, rocks, earth, and even man and beast—are crushed in its path. Departure is almost as sudden as arrival, and within a few hours the precious water is gone and the arroyos look as dry as usual.

The historian Erna Fergusson assesses the Big Bend's impact on the occupants as follows: "Those who can stand it have had to learn that man does not modify this country; it transforms him deeply."[3] Virginia Madison, another regional historian, remarks: "The vast ranch empires in the Big Bend today are holdings of men who compromised with the country and succeeded despite its drawbacks."[4] The Big Bend allows no winners; there are only survivors.

Such then is Terlingua country, the land the quicksilver prospectors found. Knowing little of its past they were unequipped to cope with the present. Those who compromised remained; those who did not were defeated and left. But the defeated are not part of this story. Those who accepted the challenge succumbed to a lifestyle circumscribed by a condition of scarcity in a land of isolation. Those two factors—scarcity and isolation—touched every facet

---

[2] J. Evetts Haley, *Jeff Milton: A Good Man With a Gun*, p. 74.

[3] Erna Fergusson, *Our Southwest*, p. 18.

[4] Virginia Madison, *The Big Bend Country of Texas*, p. 177.

of life in Terlingua and gave the quicksilver district its uniqueness and individuality.

Settlement came late to the Big Bend. By the time population had penetrated most other regions in the state, the area around Terlingua Creek was still an unknown, uninhabited wasteland. This was the legacy of the red man. The Apaches and Comanches maintained an effective barrier to settlement in the Mexican borderlands, and for over three centuries the Spaniards, and later the Americans, could gain no more than a tenuous hold on the lands along the Rio Grande.

The Spaniards failed because they challenged the region in a manner least likely to succeed. Motivated by an incongruous blend of religion, greed, and an illusionary belief in mythical kingdoms, they fanned out across the Southwest, but instead of colonists, the viceroy dispatched missionaries and soldiers to convert or subdue the savage tribes. They discovered too late, however, that the pacifist Indians along the Rio Grande could not comprehend a sophisticated religion nor adjust to sedentary mission life. Instead the hostile Indians, always a majority, held the balance of power and determined the course of Big Bend history almost until the twentieth century.

As the first Europeans to see the Big Bend the Spanish explorers left journals and reports that describe the land and its occupants. Cabeza de Vaca traversed the region in 1535 searching for the Pánuco. Although de Vaca's route is a matter of scholarly conjecture, the historian Carlos Castañeda traces a southwesterly course across present-day central Brewster County and states that the Cabeza de Vaca party came to the outskirts of the Chisos Mountains. The author adds that "the first geographic description of this vast area, the first information concerning the numerous tribes of Indians that roamed and lived in this immense unexplored territory, the first account of flora and fauna . . . are all to be found in Cabeza de Vaca's inimitable revelation of the American Odyssey."[5] Yet the central theme of this journal—starvation, unpredictable savages, vast unsettled expanses, and unproductive arid wilderness—states unequivocally the conditions for living with this land. For those who followed in Cabeza de Vaca's wake, the message went unheeded.

[5] Carlos E. Castañeda, *Our Catholic Heritage in Texas, 1519–1936*, I: 76, 80.

In 1683–1684 Juan Domínguez de Mendoza made the first documented journey through the Big Bend. This event is symbolic of the Spanish experience in the American Southwest. In 1683 Juan Sabeata, an Indian chief in the Jumano nation, petitioned the governor of New Mexico to send missionaries to his people. Interest in precious metals was still strong, and Sabeata's request revived the "missionary desire to bring into the fold of the church . . . the misguided souls" that roamed the Rio Grande. Spain's response was immediate. On December 15, 1683, the Mendoza party, accompanied by three Franciscans, left Real de San Lorenzo near present-day El Paso, followed a southeasterly course along the Mexican side of the Rio Grande, and "somewhere in the vicinity of Ruidosa in the Chinati Mountains . . . crossed the Rio Grande into present Texas."[6] The party continued to La Junta de los Rios, the Jumano settlements at the confluence of the Rio Grande and Conchos River.

At this point Mendoza paused briefly, probably in the vicinity of the present-day town of Presidio, and then continued eastward. Proceeding along Alamito Creek through Paisano Pass, the party passed just south of the present site of Alpine and continued exploring into Central Texas. Retracing this route, they returned to La Junta, and on June 12–13, 1684, Mendoza "took formal possession of this area in the name of the king as a part of the Province of New Mexico." Mendoza's expedition into the Big Bend led to the establishment of "six missions among the various tribes of Indians" that inhabited that region,[7] four of which were probably on the Texas side of the Rio Grande.

Early in the eighteenth century the Spanish realized the necessity to secure their holdings along the Rio Grande. Nearly five hundred miles separated San Juan Bautista and El Paso del Norte, while midway between lay the more recent establishments at La Junta de los Rios. Castañeda explains that at this period "the chief interest of Spanish officials in this vast region was not gold, or treasure, or mythical kingdoms, but the more practical purpose of putting a stop to the frequent incursions of the numerous savage tribes that lived in this area, who preyed on the frontier settlements of Coahuila, Nuevo Reyno de León, and Chihuahua."[8] The first step in inaugurating this plan was to establish a presidio at La Junta de los Rios.

In 1747, nearly three-quarters of a century following Mendoza's

[6] Ibid., pp. 312, 316.
[7] Ibid., II: 328.
[8] Ibid., p. 311.

visit to La Junta, the viceroy instructed the governor of Coahuila, Pedro de Rebago y Terán, to lead an expedition to La Junta and select a site for the presidio. The Apache and Zuma Indians had forced the abandonment of the six missions established by Mendoza, and the Christian Indians and missionaries had fled to Chihuahua. Although eleven years elapsed before a temporary presidio was built, Terán's expedition was not without rewards. He coursed the heart of the Big Bend and his journals contain detailed descriptions of the region. The Terán party entered Texas on December 5, 1747, "probably a few miles above Boquillas" and on the following day reported seeing recently abandoned Indian *rancherías* ("villages") and "found some pumpkins near the river." They continued their march "upstream between two mountain ranges, one running from west to north and another crosswise to the west. They were now in the Chisoz [i.e., Chisos] Mountains, and very likely following the deep valley of Terlingua Creek." Castañeda concludes that "the expedition had gone fifty-three leagues from Santa Rita crossing, which distance they had traveled on the Texas side of the river through the very heart of the Big Bend."[9]

Construction finally began on the presidio, Presidio del Norte, in December, 1759, and was completed in July, 1760. Although the facility maintained a small garrison, the occupants of the La Junta missions were still vulnerable to Indian attack.[10] Corning cites Indian hostilities as a prime cause of Spain's demise in the American Southwest. He writes: "Spanish withdrawal from the Big Bend area precipitated Indian forays across the Rio Grande into northern Mexico, which, after the 1810 Hidalgo Revolution, ultimately brought about the final dissolution of the Spanish authority in Texas and destroyed the system of presidios and missions."[11]

Spanish settlement in the Big Bend never exceeded a token effort; it remained largely a region to be explored and crossed, rather than a place for permanent occupancy. During the nearly three centuries of Spanish awareness of the area, this policy changed little. Nor did Spain's exit in 1821 begin a new chapter in Big Bend history. It was still the land of the Apache, whose force went unchallenged. However, change was in the offing and two wars plus new ownership of the land north of the Rio Grande led to a new era of exploration and settlement.

The Treaty of Guadalupe Hidalgo, signed on February 2, 1848,

9 Ibid., III: 215–216.
10 Ibid., pp. 229–232.
11 Leavitt Corning, Jr., *Baronial Forts of the Big Bend*, pp. 10–11.

established the Rio Grande as the boundary between the United States and Mexico, giving the United States possession of much unexplored and unsurveyed land lying west of the settled regions of Texas. As a first step in acquiring possession of this territory, the secretary of the interior ordered a survey of the United States and Mexican boundary and appointed Major William H. Emory chief astronomer for the boundary commission. Emory's 1853 report contains the first definitive appraisal of the Big Bend region in terms of its geologic past, its potential for permanent occupancy, and the wonders of its natural beauty. He saw the Big Bend as potential ranching country, but he tempered his enthusiasm with practical logic: "The grass is rich and luxuriant; low, scrubby bushes are found, but no growth of timber. No water, except what collects in the gullies during heavy rains." As Emory's team moved farther south toward the Terlingua country, his journal entries reflect the changed terrain: "The nearer we approach the river, the more rough the country becomes; deep ravines and gullies constantly impede the progress of the wagons, and the whole surface is covered with sharp angular stones and a growth of underbrush armed with thorns."[12]

Although Emory's party made a careful appraisal of the region's geology, his failure to mention the exposed cinnabar deposits was not an oversight. Keenly aware of the panic that followed discovery of California gold in 1849, Emory explains: "I kept the search for gold and other precious metals as much out of view as possible, scarcely allowing it to be the subject of conversation, much less actual search; for I well knew if this mania was once to seize my party, it would be attended with the worst consequences."[13]

Unlike the Spaniards, Emory viewed the region realistically and cited two deterrents to the "development of the mineral wealth of the country, no matter how rich it may prove. One is the hostility of the Indians, which makes it unsafe for parties of less than fifteen or twenty to traverse the country; another is its remoteness from navigation and the scarcity of water."[14] These were prophetic words.

While Emory's assignment was to gather data, his journal en-

[12] Major William H. Emory, *Report of the United States and Mexican Boundary Survey*, United States House of Representatives, 34th Cong., 1st Sess., pp. 74, 75–76.

[13] Ibid., p. 94.

[14] Ibid., p. 95.

tries contain glowing descriptions of the region's physical grandeur. The survey party explored the Mesa de Aguila and saw the Santa Elena Canyon. He writes: "From this dizzy height the stream below looks like a mere thread, passing whirling eddies, or foaming over broken rapids. . . . From the point formed by its last projecting ledges the view is grand beyond all conception." From his perspective overlooking the eroded Rio Grande basin "set off by the rugged volcanic mountains of the Chisos, we trace the winding stream in the basin below, to which distance gives a softening character of fertility."[15] Emory's reconnaissance identified and described two natural wonders—the Grand Canyon of the Rio Grande and the Chisos Mountains—that one day would become regional landmarks of the Big Bend National Park.

In the interim between Spain's exit from the Rio Grande missions and the appearance of Emory's party, another event ushered in a new phase of Big Bend history. In 1839 a group of Chihuahua traders seeking a shorter route to the United States avoided the established Santa Fe Trail and struck out on a northeasterly course from Chihuahua City. They crossed the Rio Grande at Presidio del Norte, continued north along Alamito Creek, turned east through Paisano Pass (the same route followed by Mendoza party), passed just north of the present site of Alpine, and continued north via Comanche Springs at present-day Fort Stockton.

By the time Emory reached this area in 1853 the route of the Chihuahua traders was etched deeply in the Big Bend soil. He notes: "Numerous trails from the Pecos and the Escondido here unite to form a large broad one, running south to the Rio Grande; there are unmistakeable signs of their constant use."[16] While the trail bypassed the Terlingua area, this commercial artery had a profound effect on the settlement of the Big Bend. Following the Treaty of Hidalgo in 1848 several of the trail's wagoners took up land "on the outside edge of that little wave of Mexican settlement that washed up from Chihuahua and played out on the north bank of the Rio Grande."[17] By mid-century John Burgess, Ben Leaton, and John W. Spencer had established permanent trading houses at Presidio del Norte as the nucleus of the first Anglo settlement in the Big Bend region.[18]

---

[15] Ibid., p. 56.
[16] Ibid., p. 75.
[17] Haley, *Jeff Milton*, p. 74.
[18] Corning, *Baronial Forts*, p. 16.

Until late in the nineteenth century the Indian remained the pivotal factor in the settlement of the Big Bend. However, with the increased traffic to the California gold fields in the early 1850s, the balance of power began to shift. The military escorts dispatched to assure the emigrants' safe passage through this unsettled region marked the beginning of the end of Indian autonomy in the Big Bend. Fort Davis was established at the far southeast projection of the Davis Mountains in 1854, and Fort Stockton, located approximately eighty-five miles to the northeast, was opened five years later. The latter facility had the specific assignment of protecting the San Antonio–San Diego mail route. Indian encounters began immediately and continued until the outbreak of the Civil War. When the Union troops departed in 1861 the Indians resumed unrestrained control of Big Bend and began sweeping east to the frontier settlements in Central Texas.

Following the Civil War, troops returned in greater strength, which had a two-fold impact on Big Bend settlement. First, after June, 1867, Fort Davis became the hub of Indian warfare in trans-Pecos Texas, and by 1881 the cavalry had ended the Apache menace. Second, military demands for supplies and services stimulated farming and ranching in the Big Bend–Davis Mountain area and attracted new settlers. The historian Mrs. O. L. Shipman writes that "settlers along the river began raising corn, wheat, and other forage for the soldiers stationed west of the Pecos. The first big ranches and stores were established in this section to trade with the post commanders. Don Milton Faver was enabled to prosper and amass his great fortune by selling his cattle and produce to soldiers."[19]

Thus the occupancy of the land passed from red man to white, the longhorn replaced the buffalo, line camps replaced the Apache *rancherías*, wheatfields and corn crops now grew where Indian squaws once tended pumpkins, maize, squash, and melons. And even greater changes were imminent, according to western historian Robert Utley: "In 1881 and 1882, surveyors and construction workers, guarded by detachments of Texas Rangers, slowly pushed railroad tracks across the Trans-Pecos. . . . Binding the Big Bend for the first time to the settled part of Texas. . . ."[20] Meanwhile, in the early 1880s Marfa, Alpine, and Marathon sprang up in the wake of

[19] Alice Jack (Mrs. O. L.) Shipman, *Taming the Big Bend*, p. 51.
[20] Robert M. Utley, "The Range Cattle Industry in the Big Bend of Texas," *The Southwestern Historical Quarterly* 69 (April 1966): 429.

the railroad, and with Presidio, they became the settlement nuclei for the Big Bend.

The heyday of ranching did not reach Terlingua country until early in the twentieth century, after the passing of the longhorns.[21] From 1882 to 1885 cattlemen moved into the Big Bend and rapidly claimed the springs, creeks, and isolated waterholes that dot the area. Early in 1885 General Richard M. Gano established the Estado Land and Cattle Company on 55,000 acres in Block G-4, a survey number destined to play a key role in the history of Terlingua quicksilver operations. Cattle bearing Gano's G-4 brand grazed on an open range "bounded by the Agua Frio, the Rio Grande, Terlingua Creek, and the Chisos Mountains." Gano's neighbors were James P. Wilson and Clyde and Louis Buttrill, who ranched in the Rosillos range from 1884 to 1894.[22]

Some of the last Texas range land to be taken up was in the Terlingua area. Scant water and short grass were obvious factors in this delay. Stuart T. Penick, a member of a United States Geological Survey team, camped in the Terlingua area in 1902 and reported the region was still "pretty primitive. . . . There were temporary cow camps in the area where the spring and fall roundups were made, but . . . only a few cowboys were in the area between roundups."[23] C. A. Hawley, who moved to Terlingua in 1910, remarked that "there was but one ranch house between Terlingua and Alpine and that belonged to W. M. Pulliam and his son, Roselle."[24]

Although still sparsely settled the remote lands bordering Terlingua Creek had passed into the hands of the cattleman. The new tenant's hold on the land would be brief, however, as the next step in the settlement process was at hand. Settlement begets settlement, and through three and one-half centuries of exploration, occupation, and reoccupation—however tenuous—the settlement pattern of the Big Bend emerges as a microcosm of the classic frontier experience. A series of eight sequential frontiers are definable and each progresses in logical order—the Indian, the explorer, the trader, the soldier, the rancher, and the farmer. Only two remained, the mining frontier and the urban frontier, and the way was now

[21] Ibid., p. 42.

[22] Shipman, *Taming the Big Bend*, p. 100.

[23] Virginia Madison and Hallie Stillwell, *How Come It's Called That?*, p. 10.

[24] C. A. Hawley, "Life Along the Border," *Sul Ross State College Bulletin* 44 (September 1964): 60.

clear for these to be drawn into the settlement orbit. The Indians were gone, travel routes defined, and at the threshold of the new century stood the next recipients of the land, who would soon enter, inventory its economic potential, and re-enact the inevitable processes of occupancy and reoccupancy. A new drama was about to begin.

CHAPTER 1

Discovery and Early Development

AS the nineteenth century drew to a close reports of a rare mineral discovery in the Terlingua area began to circulate in the Marfa-Alpine region of Texas. It was neither gold nor silver—popular subjects in the American West—as these reports mentioned a strange red formation called cinnabar from which quicksilver, or mercury, is extracted. The mood was one of interest, rather than excitement, as few persons could comprehend the economic potential these rumors portended. Farmers, ranchers, and storekeepers were more likely to be preoccupied with the weather, range conditions, and completion of the railroad.

Quicksilver is many things to many people, and its multiple terminology used synonymously by the lay public is confusing. Properly stated, "mercury" is the technological term used by the chemist and pharmacist to identify one of the basic chemical elements; "quicksilver," on the other hand, designates the silver fluid metal as an item of commerce. Although both words apply to the same material their application varies with different fields of endeavor.

The vast usages of the rare liquid metal for which no substitutes have been developed account for quicksilver's distinct value (see appendix A). Mercury, the natural chemical element, is one of the basic precious metals found in the ore-bearing rock. Like other metals it is mined from the earth, and most of it is the refinement of the crude cinnabar ore (native quicksilver is extremely rare). However, it differs from all other metals in that it alone remains liquid at normal temperatures.[1]

In addition to chemical, industrial, and pharmaceutical uses of mercury, the development of mercury fulminate gave the metal a

---

[1] Mercury freezes at a temperature of minus 39.5° F., and boils at plus 357.25° F. Volatilization occurs at plus 360 degrees, i.e., it is transformed into a gas or fume. The property has a specific gravity of 14, i.e., fourteen times as heavy as rain water, and of all metals, only gold and platinum are heavier.

strategic importance. The experiments of the scientist Edward Charles Howard in 1799 resulted in the compounding of fulminate of mercury crystals. Howard discovered that by dissolving mercury in a solution of nitric acid and alcohol, this compound, when dry, formed a crystalline substance—essentially the same process used today in the commercial manufacture of mercury fulminate. The economic potential of this compound became evident when it was discovered that striking the crystals lightly caused an explosion. Used as primer to detonate gunpowder in cartridges and shells, mercury became a critical material in war as well as in peace, and its economic potential multiplied manyfold.

Thus no one in Marfa or Alpine realized that far to the south an economic giant lay sleeping. Nor could they foretell that beneath those barren hills lay deposits of cinnabar so vast that within the foreseeable future Terlingua country would become one of the nation's leading producers of quicksilver. At this point Terlingua cinnabar, or quicksilver, or mercury, was still rumor, gossip, and supposition. But a boom was in the making.

Facts concerning the discovery of cinnabar in the Terlingua area are so shrouded in legend and fabrication that it is impossible to cite the date and location of the first quicksilver recovery. Early inhabitants reported seeing aborigine pictographs executed with brilliant red pigment on the limestone cliffs and cave walls in the Terlingua Creek area. Although there is no record of these pictographic treatments ever having been chemically analyzed, students of the Plains Indians state that the aborigines used cinnabar for this purpose, as well as for personal adornment.[2]

While the Indians along the Rio Grande had long been aware of the cinnabar, its development awaited the coming of Anglo settlement. Prior to 1880 Texas had virtually no mining, as the potentially productive areas were located in the far western portions of the state. Neither had Texas sponsored a geological survey. Geologist W. H. von Streeruwitz cites a lethargic legislature as one reason for this neglect: "In 1883 the Legislature of the State passed a mining law, but its contents and ruling were not very tempting. . . . It was quite natural that no mineral resources were expected in a State which did not deem it worthwhile to pass sensible mining laws."[3]

The first geological survey in the trans-Pecos area began in the

[2] Interview with W. W. Newcomb, director, Texas Memorial Museum, Austin, Texas, April 17, 1965.

[3] W. H. von Streeruwitz, *Geology of Trans-Pecos Texas*, pp. 228–230.

late 1880s after rail transportation made the region accessible. The complete lack of knowledge of that area is explained by geologist E. T. Dumble: "The geography and topography of the Trans-Pecos region was almost entirely unknown. The location of the different mountains, and even the names by which they are designated, are differently laid down by the various cartographers. . . . [therefore] I organized a topographic survey there with Mr. W. H. Streeruwitz in charge."[4]

The Streeruwitz party found promising evidence of mineral deposits but attributed the underdeveloped state of these resources to the lack of water and the danger of Indians. Streeruwitz described the perils of that country with great feeling: ". . . travelers hurried through parts of the country, as the Sierra de los Dolores, ("the Mountains of Misery,") . . . with its Puerta de los Lamentaciones ("Gate of Lamentations"), and no body stopped long enough to examine the mountains for their mineral resources; or if perchance some one did stop, he did so at the peril of his life, as is proved by the numerous graves which are found in the mountains."[5]

Although the region was remote, the native occupants along the Rio Grande were aware of the mineral deposit and had become curious about its worth. Charles Laurence Baker, former geologist with the Bureau of Economic Geology at the University of Texas, recalls that "Graf (Count) W. H. von Streeruwitz told me that previous to 1890 Mexicans were bringing him samples of quicksilver ore [cinnabar] from the Terlingua area though that country was nearly inaccessible."[6] This growing interest in minerals stimulated others to seek gold. Lock Campbell, a Big Bend settler, engaged a group of prospectors to search for the precious metal on the Mexican side of the Rio Grande, but, fearing Mexican bandits, they prospected the Texas side of the river instead and found cinnabar in the Terlingua area. Campbell reportedly scoffed at their "worthless" find and instructed them to continue their search for gold. Virginia Madison credits Campbell's mercenaries with the discovery of the Terlingua cinnabar; however, she cites no date for this event, nor does she relate it to the period of discovery that led to quicksilver production.[7]

[4] E. T. Dumble, *First Annual Report of the Geological Survey of Texas, 1889*, p. xxiv.

[5] Streeruwitz, *Geology*, pp. 228–230.

[6] Charles Laurence Baker, Cardova, Illinois, to Kenneth B. Ragsdale, November 16, 1964, in possession of author.

[7] Virginia Madison, *The Big Bend Country of Texas*, p. 117.

In 1884 Juan Acosta, another Big Bend resident, brought a specimen of cinnabar to an Alpine merchant named Klein, who identified the ore and knew its potential worth. Klein negotiated an agreement with Acosta to develop the prospect, but before beginning operations they sold their claim to a group of Californians who began the first real mining operations in the Big Bend. They named the site of their claim "California Hill." While Jack Dawson is reported to have produced the first flask of quicksilver in the Terlingua district in 1888, official United States government publications cite the initial Texas production in 1899.[8]

Although the Acosta episode is well documented and represents the first discovery that resulted in quicksilver recovery, there was no "rush" to the Big Bend. More than a decade elapsed before the mining journals reported the event. Late in 1894 the Los Angeles *Journal*, the *Manufacturers Record* of Baltimore, and several El Paso newspapers published accounts of the find.[9] Concurrent with these reports, mining engineer William P. Blake visited the area, inspected the deposits, and reported his findings at a meeting of the American Institute of Mining Engineers in March, 1895. His paper, published in the same year, appears to be the first scientific report on the Terlingua quicksilver deposits.[10]

Blake stated that early in 1894 George W. Wanless, an employee of the Rio Grande Smelting Works of Jiménez, Mexico, learned that some Mexicans had found rich deposits of cinnabar in the Terlingua area. In association with Charles Allen of Socorro, New Mexico, he explored the region, "finding the cinnabar-deposits and locating them for development."[11] While Wanless and Allen left no record of the disposition of their holdings, Madison states they "sold their interests in the claim to some Big Bend citizens, who developed it and sold it to the Marfa and Mariposa Mining Company, owned by the Normand Brothers, Tom Golby, and John Sharp, for a price reported to be over $125,000."[12]

Golby, however, later claimed priority for the discovery. In a

8 J. Harlan Johnson, "A History of Mercury Mining in the Terlingua District of Texas," *The Mines Magazine* 36 (September 1946): 390; Donald A. Brobst and Walden P. Pratt, eds., *United States Mineral Resources, Geological Survey Professional Paper 820*, p. 403.

9 William P. Blake, "Cinnabar in Texas," in *Transactions of the American Institute of Mining Engineers*, XXV: 69.

10 Ibid., p. 68.

11 Ibid., p. 69.

12 Madison, *Big Bend Country*, pp. 178–179.

letter to the *Engineering and Mining Journal-Press* in 1924, he stated: "I was one of the first men to know anything about the Terlingua quicksilver. I think I first saw a piece of ore which was brought to me by a Mexican in 1893–1894, and, although I made several attempts to locate the ore in place, it was not until 1896 that I was able to do so."[13]

He added that the land had not been properly surveyed. Not until the summer of 1898 were the owners sure of the extent of their holdings. Golby added that "we owned about 2,500 acres of cinnabar land, four miles along the strike of the belt." The following year the Marfa and Mariposa Mining Company erected the first Scott furnace in the district, and by 1901 had doubled the furnace capacity and "were running out more 'quick' than the New Almaden [in California]."[14]

By the close of the century the word was out—quicksilver had been discovered in the Big Bend of Texas! By the spring of 1900 about 1,000 flasks of the liquid metal had been recovered and four major producers—Lindheim and Dewees, Marfa and Mariposa, the California, and the Excelsior companies—were established in the district.[15]

From the beginning the name "Terlingua" identified the quicksilver producing area, and by 1904 developments in the Terlingua Quicksilver District were reported in the professional engineering and mining journals. "Terlingua" is a corruption of two words, *tres lenguas*, which in Spanish, means "three tongues." Folklorists differ on the connotation of the word "tongue." Some believe it refers to the three languages spoken in that area: Indian, Spanish, and English. Others maintain it is the three tongues, or forks, of Terlingua Creek.[16] However, long before the discovery of quicksilver, Terlingua identified a Mexican village on a small stream near the Rio Grande that bore the same name. Later when the name "Terlingua" was used to identify the mining community, the former village became known as Terlingua Abaja, or lower Terlingua.

[13] Thomas Golby, "The Story of a Quicksilver Mine," *Engineering and Mining Journal-Press* 118 (October 1924): 579–580. Golby makes no mention of a partner named John Sharp. He notes, "my partner, Montroyd Sharpe, of Santa Cruz, Calif., and I . . . were lucky enough to get Normand Bros., of Marfa, Tex., interested with us, and the four of us controlled the Marfa & Mariposa mine."

[14] Ibid., p. 579.

[15] Johnson, "Mercury Mining," 36 (September 1946): p. 391.

[16] Virginia Madison and Hallie Stillwell, *How Come It's Called That?*, p. 54.

Since the growth and location of Terlingua followed the development of the mining industry, the site of the town shifted with the fortunes of recovery. During the life of the district, Terlingua was identified with three different locations.[17] Stuart T. Penick, a member of a United States Geological Survey party, provided one of the earliest descriptions of the village.

> When I was there in the summer of 1902 Terlingua was a sprawling camp of temporary sheds and shelters composed of various kinds of material, such as tin, canvas, old sacks, sticks, and adobe bricks. The only permanent buildings were the commissary and smelter. There were from 200 to 300 laborers of the lowest class of Mexicans there. They seemed to be temporary, for very few of them had families. They lived in a very primitive way. Their principal diet consisting of beans (frijolies), chilli, made of goat meat, and tortillas, cooked on an open fire in skillets and Dutch ovens. They partook quite freely of the Mexican liquors called tequilla and sotole in their off times, and at these times were sometimes somewhat quarrelsome. Otherwise they seemed very quiet and peaceful.[18]

Three years later when C. A. Hawley joined the Marfa and Mariposa Mining Company, the village had grown substantially. He recalled that "in 1905, five years after the mines were opened . . . the adobe village had a population of close to a thousand." This same year geologist M. P. Kirk estimated the population of Terlingua at 1,200 and steadily increasing.[19]

To determine the population of Terlingua at any given time is impossible, as the name applied synonymously to two different mining camps and the entire quicksilver mining district. Also, the boundaries of the district were as fluid as the population, and estimates of the number of people centered around five or six principal camps or villages vary from 1,200 to 2,500.

The growth of the mining district is best shown in a chronological sampling of published descriptions of the productive area. The mining engineer J. Harlan Johnson states that by the middle of 1901 prospectors, working along a six-mile strip of land, reported

[17] Madison, *Big Bend Country*, p. 181.

[18] Stuart T. Penick, La Grange, Texas, to Kenneth B. Ragsdale, November 22, 1964, in possession of author. The Terlingua referred to in Penick's letter was the village centered around the Marfa and Mariposa Mining Company in the California Hill area, approximately six miles west of the Chisos mine.

[19] C. A. Hawley, "Life Along the Border," *Sul Ross State College Bulletin* 44 (September 1964): 31; Johnson, "Mercury Mining," 36 (September 1946): 393.

finding large quantities of surface ore.[20] A year later Benjamin F. Hill, a geologist for the University of Texas Mineral Survey, visited the region and reported the expanding productive area as "a rectangular strip, approximately fifteen miles long and four miles wide. . . . Fresno canyon is the western boundary. . . . The eastern end of the rectangle would be several miles east of Terlingua Creek."[21] High quicksilver prices during World War II stimulated prospecting and expanded the district to its greatest proportion; by 1946, the total area that had been prospected and developed was forty miles long and twenty miles wide.[22]

Surface outcroppings of the cinnabar deposit yielded the first recovery, and individual samplings assayed "from forty-eight to seventy-four per cent quicksilver. Single pieces of residual ore were reported which were eighty-two per cent cinnabar."[23] After a thorough appraisal of the district the geologists reported that the cinnabar occurred mainly "in calcite veins, in the limestones and the Comanchean Cretaceous, and in the flags and marls of the Upper Cretaceous. Some andesite dikes and sills have also been found to contain the ore."[24] "Nuggets" and "boulders" that carried an even greater metal content were discovered, and the Excelsior Mine reported masses of almost pure cinnabar weighing several hundred pounds. Early prospectors claimed finding small amounts of native mercury near California Hill.[25]

Cinnabar assaying this high metal content can be refined by one of the simplest methods of recovery employed in any mining process—retorting. Geologist C. N. Schuette explained the economic position of the small independent operator: "A . . . small furnace is all that is needed to make a finished product that has a ready cash

[20] Johnson, "Mercury Mining," 36 (September 1946): p. 391.

[21] Benj. F. Hill, "The Terlingua Quicksilver Deposits of Brewster County," in *The University of Texas Mineral Survey, Bulletin Number 4*, p. 9.

[22] James M. Day, "The Chisos Quicksilver Bonanza in the Big Bend of Texas," *The Southwestern Historical Quarterly* 64 (April 1961): 428.

[23] Johnson, "Mercury Mining," 36 (September 1946): 391.

[24] Day, "Chisos Quicksilver Bonanza," p. 428.

[25] The geological phenomenon creating the quicksilver ore was the subject of a distinguished paper by Dr. J. A. Udden, a former director of the University of Texas Bureau of Economic Geology who later was engaged as a consulting geologist in the Terlingua district. The substance of Udden's paper, "The Anticlinal Theory as Applied to Some Quicksilver Deposits," briefly stated, is that "the mercury sulphide ascended as a hot gas, was trapped in permeable limestones capped by impervious clays in the rocks and thereupon by cooling condensed as a solid, usually the cochineal-red mineral, cinnabar" (Charles Laurence Baker, Cardova, Illinois, to Kenneth B. Ragsdale, November 16, 1964).

market. The cost of concentration, shipping expense to a smelter, and smelting charges usual in . . . other prospects are absent, together with the delay in obtaining payment."[26]

After depleting the surface deposits, it became necessary to prospect at deeper levels. "Follow the ore" became the watchword of the quicksilver district, and surface panning began the precarious process of locating the ore-bearing deposits. After discovering the stringers—the minute concentrations of the cinnabar ore, sometimes no larger than the thickness of a straight pin—the prospectors sank a shaft along the stringer until they found ore in sufficient amounts to warrant furnacing. Schuette, who inspected the Mariposa Mine, reported these stringers were "followed up and down in the exploratory drifts, resulting in an amazing series of passages resembling the burrows of a rabbit warren."[27]

In 1902 the mining methods employed were still primitive, and picks, shovels, hand drills, and sledges made up the usual complement of equipment. By August eighty feet marked the deepest penetration, and the operators employed a crude windlass to remove the ore. However, in shallower workings Mexican laborers surfaced the ore-bearing material in rawhide buckets attached to their backs. These vertical shafts were equipped with notched poles, which the natives negotiated with great agility.[28]

Burro-drawn wagons or carts hauled the ore to the retort where hand-sorting usually preceded the smelting process. Since widely fluctuating assays required different periods of smelting, sorting became increasingly important. During this early phase of development Mexicans performed most of this work under Anglo supervision. Labor was fairly abundant at $0.90 to $1.25 per day.

As the operations grew more extensive and results yielded greater financial reward, the miners discovered the magnitude of their undertaking greatly increasing; however, their biggest obstacle still lay ahead. With the penetration of the deeper formations, especially the Del Rio clay, all work was conducted below water level. The mining engineer Johnson described the difficulty of this maneuver: "Exploration . . . from which future ore supply must come, is difficult. . . . Grouting against a water pressure of 50 pounds per square inch is no small problem."[29] When pumping and grouting techniques failed, the mine operators were faced with one of the

26 C. N. Schuette, *Quicksilver: Bureau of Mines Bulletin Number 335*, p. 3.
27 Ibid., p. 7.
28 Johnson, "Mercury Mining," 36 (September 1946): 392.
29 Ibid.

most frustrating ironies of the Terlingua district: the constant lack of surface water necessary for human consumption, and the ever-present subsurface water that constantly threatened to inundate the rich cinnabar ore.

In a few years the industry grew from the status of the small operation to a level where quicksilver mining demanded new techniques that transcended individual effort. Since the use of the retort proved wasteful and inefficient, conversion to the large and expensive condensing furnace appeared inevitable to achieve maximum recovery. This metamorphic change which occurred in the Terlingua district is a fundamental segment of the history of mining in the United States: the application of mechanical equipment, engineering skill, and investment capital to an industry that had outlived the age of individual effort.

By 1903 the "rush" was on. To the Terlingua country had come the Normand brothers, prosperous ranchers from Marfa, to establish the Marfa and Mariposa Mining Company; the Sanger brothers, mercantile giants from Dallas, invested heavily in the Texas Almaden Mining Company (known locally as the Dallas Company); the Gleim family, operators of the flourishing Shafter silver mine, established the Big Bend Cinnabar Mining Company; from Del Rio came Louis Lindheim and Del Dewees to erect an installation bearing their names; and the Colquitt-Tigner combine constructed a major facility for the recovery of quicksilver. The combined investment of these major installations is estimated to have exceeded $2 million. Thus one of the nation's most primitive and isolated regions witnessed the arrival of the age of industry.

This is the story of mining—the story of people attracted by precious metal. Beneath that barren desert lay a mineral deposit of magnetic allure, and the prospect of profit attracted the bold and the adventurous to the Big Bend of Texas—investors and financiers, geologists and engineers, and some just plain mining men, all in search of a bonanza. While many failed, some became wealthy. A few are remembered, but most are forgotten. Yet from this human throng only Howard E. Perry was destined to make millions, dominate life at Terlingua, and eventually own most of the producing mines in the district. He was not forgotten.

CHAPTER 2

Perry Opens His Mine

HOW Howard E. Perry acquired his Big Bend property is legend-
ary; the truth remains one of the quicksilver district's best-kept
secrets. One version states he received the land in lieu of payment
of a debt. Another less likely version is that Perry bought the land
from a Big Bend rancher, General Richard Montgomery Gano,
"who had a predilection to preach the gospel and a couple of remunerative hobbies—cattle and land." Madison reported that Perry,
traveling in Kentucky, attended a revival meeting conducted by
Gano and was exposed to his liberal religious doctrines: "Saved
thirty-two souls and sold sixteen sections of land." The version to
which many Big Benders give credence was told by Ed Nevill,
Brewster County rancher, who maintained he saw Perry purchase
two sections of land at auction in front of the court house in Alpine
for $150.[1] It is, however, a matter of record that in 1942 Perry
owned 7,440.8 acres in the Big Bend of Texas, valued at $1,137,500,[2]
of which four sections were acquired as early as 1887 for a consideration of $5,760.[3]

Of the three versions the first is probably the most authentic.
Perry told both Wayne and Robert Cartledge, former administrative
employees at the Chisos mine, he lent a friend $300, accepting as
security some Brewster County land. When the note went unpaid,
he gained possession by default. Perry was more explicit in the ver-

[1] Virginia Madison, *The Big Bend Country of Texas*, p. 184.
owned 7,440.8 acres in the Big Bend of Texas, valued at $1,137,500,[2]

[2] Petition and Schedules, *Cause No. 688, Watson-Anderson Grocery Company et al. vs. The Chisos Mining Company*, Federal Records Center, Fort
Worth, Texas.

[3] Both Day and Madison report the amount of this transaction as
$5,780.00, "but the amount stated on the deed is $5,760.00" (Sara Pugh,
county and district clerk, Brewster County, Alpine, Texas, to Kenneth B. Ragsdale, December 29, 1965, in possession of author). "The sections or surveys
listed are Surveys 69 and 67 in Block G-12, and Surveys 297, 295, 320, 321,
and 319—Block G-4—but the deed refers to '4 ½ Sections or Surveys' yet it
lists all the above but it does state 'excepting the undivided ½ of Survey of No.
69'" (Pugh to Ragsdale, January 5, 1966).

sion he told his nephew, Walter K. Bailey. Bailey states that "the story which I believe is reasonably correct is that he, and another person [name unknown], in Chicago accepted ownership of large acreage in the Big Bend." Bailey believes that the two men had no connection except a common debtor and they took this land as his only asset. Bailey further explains that Perry's "having learned that there was mining on the property, . . . did not tell his fellow creditor, and bought the land at a tax sale."[4]

If, according to Bailey, Perry did purchase the property at a tax sale, then this substantiates Nevill's version of the Perry land legend. The date of the transaction (1887), however, predates organized mining in Brewster County, and it is highly unlikely that at this time Perry was aware of the mineral deposits. And even if Perry had been to Alpine, it is also apparent that he had not seen the property in question. Writing to S. A. Thompson (position unknown) at Fort Davis, Texas, on January 3, 1888, Perry states:

> You say you expect to leave in a few days for the country in Foley Co. [now abolished and the land incorporated in Brewster County] in which my land is situated. If it would not be asking too much, if you will drop me a line *how the land in question looks* [emphasis added] I shall be greatly pleased and indebted to you.
>
> Herewith memo of description:
>
> 640 acres survey #321, Block G-4
> 640 acres survey #323, Block G-4
> 640 acres survey #295, Block G-4
> 640 acres survey #67, Block G-12
> 640 acres survey #69, Block G-12[5]

Many unanswered questions concerning Perry's land transactions remain. Whether the Wilbur A. Reeves, from who Perry acquired the four original sections of land, was the former creditor, and whether the Cartledges have long since forgotten the amount of the transaction are matters of conjecture. By whatever means Perry acquired the property, it included the controversial Survey 295 (or 296) that became the subject of a legal battle and proved to be the most valuable property in the quicksilver district.

Former Big Bend employees believe that when Perry acquired the Terlingua property, he thought it had little or no value; however, after paying taxes on the land for several years and receiving

---

[4] Walter K. Bailey, Cleveland, Ohio, to Kenneth B. Ragsdale, May 24, 1968, in possession of author.

[5] Clifford B. Casey, *Mirages, Mysteries and Reality, Brewster County, Texas, The Big Bend of the Rio Grande*, p. 396.

no income, he was surprised to receive a purchase offer of $0.50 per acre. When Perry ignored this proposal it was soon followed by others, each for a larger amount. Later he reportedly rejected one offer of $50,000.[6]

The sources of this sudden interest were two Big Bend cattlemen, T. D. (Devine) McKinney of Marathon and his father-in-law, Jess M. Parker. In the late summer of 1900, while branding cattle in the Long Draw, they noticed the heat from their branding fire caused globules of quicksilver to form on the limestone rocks.[7] With the Marfa and Mariposa mine already producing quicksilver about four miles west of this location, they knew the importance of their discovery.

Before attempting to develop the property, they added to their partnership R. C. McKinney and J. J. Dawson, an experienced quicksilver miner. Dawson appears to have acted as the leader of the group, as on September 10, 1900, he filed two applications for mining claims on what was assumed to be state land, designated on the Brewster County land map as Survey 296 (or 295). On October 3, Brewster County surveyor William M. Harmon surveyed the property and located two claims for Dawson, "General Long," and "May." Both were identified as being "on the waters of TasLinguas Creek, a tributary of the Rio Grande River, and 80 miles south of Alpine in Brewster County."[8] After Dawson and his partners posted notices that mining claims had been legally filed on the land, they began developing their claim.

How Perry was drawn into the land controversy is not entirely clear. Wayne Cartledge, long-time Perry employee and confidant, maintains that the line established by Harmon between Survey 70, Block G-12, and Survey 295 (or 296), Block G-4, bisected the ore deposit. The Brewster County land map shows Survey 70 as school land and subject for filing mineral claims. Survey 295 (or 296) is listed as the property of Howard E. Perry. Cartledge believed it was at this point that Dawson's group made Perry the first purchase offer for the property, which he rejected. Cartledge reported that they later added John R. Holland of Alpine to the partnership, and

[6] Interview with Wayne Cartledge, Marfa, Texas, December 29, 1964.

[7] Ibid.

[8] Petition for Preliminary Injunction, *Cause No. 133, Equity, Howard E. Perry vs. W. M. Harmon, et al., Defendants*, Federal Records Center, Fort Worth, Texas. A standard mining claim is an area 600 feet wide and 1,500 feet long containing 20.66 acres.

with his financial backing, engaged a Judge Reagan to go to Chicago to present Perry their largest and final offer.[9]

Although Perry again turned them down, their persistence intrigued him. He asked the Chicago law firm of Shortall and Morrison to investigate. In turn, it contacted the Austin firm, Cochran and West, which assigned a young attorney, Eugene Cartledge, to represent Perry.[10] Cartledge knew of the quicksilver discovery in Brewster County and felt this recent development had prompted the offers for Perry's holdings. An investigation at the General Land Office in Austin confirmed Cartledge's suspicions; several mineral claims had recently been filed on land adjoining Perry's. He transmitted this information to Perry, who requested that Cartledge go to Brewster County and make a personal inspection of the property.

Arriving at the quicksilver district Cartledge found exploratory work in progress, rich ore deposits exposed, and a crude retort in operation on land he believed belonged to his client. Geologist Robert T. Hill, who visited the district in 1902, confirmed Cartledge's findings. He reported, "Parker and McKinney produced some metal in a small furnace [retort] on their claims near the eastern boundary of the district."[11] After discussing the matter with the supposed trespassers Cartledge returned to Alpine to inspect the surveyor's field notes before beginning litigation. He discovered in the Brewster County land records the basis for a possible error in Harmon's calculations.

In establishing the boundaries of railroad land grants—which these were—the law requires that these surveys must be located by first establishing their relationship to a prior or known grant. In following this stipulation Cartledge discovered that Harmon had been remiss. In this case Survey 1, Block-17, located at the mouth of Santa Elena Canyon, approximately fifteen miles south of the disputed property, constituted the prior grant. Instead of beginning at this point, Harmon originated his survey from the corner of the

[9] Wayne Cartledge interview. This is Wayne Cartledge's version of the episode. His brother, Robert, also a Perry employee, believes their Chicago emissary was a brother of Big Bend miner Del Dewees, then living in Chicago. Wayne and Robert Cartledge were later employed by Perry and played key roles in the development of the Chisos property. They were sons of Eugene Cartledge, member of the Austin legal firm, Cochran and West, that first represented Perry in Texas.

[10] Ibid.

[11] J. Harlan Johnson, "A History of Mercury Mining in the Terlingua District of Texas," *The Mines Magazine* 36 (September 1946): 392.

adjoining survey. Cartledge, therefore, believed that Harmon's boundary deviated from the true property line.

At this point Cartledge requested Harmon to resurvey the land, this time basing his calculations on measurements originating from Santa Elena Canyon. Harmon refused and Cartledge countered with a threat that if he did not provide him with a corrected set of field notes, he would employ a licensed surveyor to rerun the lines and mandamus him to accept the survey in court. Harmon remained adamant.

Returning to Austin, Cartledge engaged a licensed surveyor, M. V. Homeyer, to reestablish the boundary between Survey 70 and Survey 295 (or 296), this time following the procedures defined by the state legal code. Cartledge's suspicions were well founded. The legal boundary established by Homeyer cut heavily into the claims originally thought to be on the adjoining property. When apprised of the change, Holland chose not to develop the claims and sold Perry his interest for a reported $15,000.[12] Dawson and his other partners were less conciliatory and Perry's attorney instituted court action to compel them to abandon the property.

On November 8, 1900, Perry's attorneys filed a motion for a temporary injunction prohibiting the defendants—J. J. Dawson, R. C. McKinney, T. D. McKinney, and J. M. Parker, all of Brewster County—from mining property believed to belong to the plaintiff, Howard E. Perry. The court granted the injunction on November 26, 1900, and further restrained William M. Harmon, county surveyor of Brewster County, Texas, "from surveying any mining claims upon the claimant's land."[13]

Records of the ensuing litigation contain no mention of Survey 70, Block G-12, nor is there any reference to the authenticity of Harmon's survey, key issues in Wayne Cartledge's version of the episode. Instead, the litigation focused on some still unexplained and unqualified changes made in the numerical listing of several surveys in Block G-4, which include Perry's Survey 295 (or 296; see appendix B).

The legal history of this property began on March 7, 1882, when John T. Gano, deputy surveyor of Presidio County, surveyed two sections of land for R. M. Gano and Sons, the land and cattle company operating in the Big Bend area. (Prior to 1887 Presidio County included present-day Brewster County.) Gano identified

[12] Wayne Cartledge interview.
[13] Petition for Preliminary Injunction, *Cause No. 133, Equity.*

1,280 acres of land in two contiguous surveys and "did temporarily number the more northern survey as No. 295 and the more southerly survey as No. 296,'[14] *the reverse of the present designations.* The General Land Office in Austin filed John T. Gano's certified field notes on April 2, 1882; however, eight months later, for reasons never divulged, Land Commissioner W. C. Walsh reversed four of Gano's entries. His directive, dated December 7, 1882, instructed Presidio County Surveyor E. G. Gleim as follows: "I have this day changed numbers of surveys 295 to 296 and 296 to 295 also surveys 297 to 298 and 298 to 297, these surveys are in Presidio County made by virtue of Certificate No. 3165 and 3166. You will please change your records to correspond."[15]

The northern survey, temporarily numbered 295, was changed to 296; and the southern survey, temporarily numbered 296, was changed to 295. Presidio County Surveyor Gleim, however, failed to effect these changes on his county land records. On May 8, 1884, the commissioner granted to Augustus Norton, assignee for the Gulf, Colorado, and Santa Fe Railway Company, title to the more southerly survey, "temporarily numbered 296 but changed to 295."[16] In the ensuing land transactions this error was compounded.

Perry's attorneys succeeded in holding valid Commissioner Walsh's changes, and on May 13, 1901, in *Cause No. 133, Equity, Howard E. Perry vs. W. M. Harmon, et al., Defendants,* the court granted Perry legal possession of Survey 295, 640 acres that richly rewarded the owner.

Perry's victory is best recorded in the flamboyant rhetoric of Wigfall W. Van Sickle, sage of the Big Bend, Perry's adversary in this proceeding who later became his legal aide and personal confidant. He later wrote that "a string of United States Marshalls came out and put to flight all the defendants and section number of the map 296 was really 295 and so held by the court. Seeing the intrepid quality of the owner of the mine and finding out he would rather have a fight than a frollic the writer went to his side of the fence where he has been stationed on the watch tower as trumpter of the territory."[17]

---

[14] Ibid.

[15] Records of Surveys, General Land Office, Austin, Texas.

[16] Petition for Preliminary Injunction, *Cause No. 133, Equity.*

[17] Wigfall W. Van Sickle to Howard E. Perry, December 8, 1937, Cartledge-Perry Chisos Mining Company Correspondence, in possession of the author. All further reference to company correspondence will be to this collection unless otherwise indicated.

Though this matter was settled in court, it raised more questions than it resolved. How Perry originally acquired the property does not appear in the court records. Why Walsh reversed these particular survey numbers and why Gleim failed to respond to his handwritten note was never explained. Although the commissioner is empowered to "number said surveys from one upwards,"[18] there appears no consistent method of survey numbering on the Big Bend land maps. Significantly, though in each land block some logical sequence appears, the only exceptions are sections 295, 296, 297, and 298. Also significantly the General Land Office's Presidio County (and later Brewster County) maps were altered immediately to conform to the commissioner's directive. However, the field notes were not corrected until September 29, 1900, less than three weeks after Dawson filed his two mining claims, and about one month prior to Perry's filing for temporary injunction.[19]

This is the legacy of the Big Bend, a part of the aura of mystery that envelops the region. It touches all who enter. Howard E. Perry had arrived and he was now part of the complex of forces that would write a new chapter in Texas history, and in the process would add his measure to the lore of the Big Bend.

Although Perry's arrival in the quicksilver district drew slight attention, his inconspicuous appearance belied the events of the not-too-distant future. Once cleared of legal entanglements, Perry made preparations to begin mining operations. First he needed local financing, and Eugene Cartledge accompanied him to Austin and introduced him to Dr. E. P. Wilmot, C. M. Bartholomew, and Morris Hirshfeld, directors of the Austin National Bank, who arranged a $50,000 loan.[20] Although living in Chicago, Perry incorporated the Chisos Mining Company under the laws of the State of Maine on May 8, 1903, and listed Augusta, Maine, as the principal office.[21] On July 10, 1903, he registered the company with the secretary of state's office as a foreign corporation doing business in Texas.[22] When actual developmental work began is not known, but the company reported its first recovery in 1903.[23] In 1904 the mining

---

[18] Petition for Preliminary Injunction, *Cause No. 133, Equity.*

[19] Records of Surveys, General Land Office, Austin, Texas.

[20] Wayne Cartledge interview.

[21] State of Maine, Office of the Secretary of State, Records of Incorporation, Augusta, Maine.

[22] State of Texas, Office of the Secretary of State, Records of Incorporation, Austin, Texas.

[23] Wayne Cartledge interview.

journals describe the Chisos Mining Company as a major operation, and by 1905 the geologists were predicting that "the mine may become one of the principal producers in the district."[24] History proved them correct.

The first two decades of the twentieth century were halcyon years for Perry and his company. After establishing the mine and beginning production, the problems of supply, maintenance, and operation were solved, and the increased demands for quicksilver created by World War I richly rewarded the owner. Near the end of this period, the Chisos mine achieved the distinction of being the largest producer of the precious liquid metal in the United States. This was the highpoint in the mine's history and a personal triumph for its owner, who, though seldom seen, became a living symbol of the company and the village that mushroomed around it. "Chisos" and "Terlingua" would become synonymous in the regional lore that grew in Perry's wake.

Who was this strange, secretive, and eccentric little man who defied the obstacles of both man and nature to carve out an empire in a remote corner of the Big Bend of Texas? How did he establish an industry in a region devoid of the basic necessities of life? The answers add an enlightening dimension to the economic history of Texas and the Southwest.

Perry entered the Big Bend as a Yankee embroiled in a legal battle over property claimed and made productive by local citizens. Although Perry gained title to the property, the victory remained suspect. This event cast a lengthening shadow of doubt over Perry and led inevitably to fabricated stories of bribery, fraud, and theft. Frontier people harbor an innate distrust for strangers, and when, like Perry, they attempt to conceal their intent and actions with secrecy, suspicion reaches inordinate proportions. Perry did not consciously seek this identity but could not avoid it, as no area of compatibility ever existed between the man and the people of the region. Two entirely different standards of business acumen separated them: what one accepted as legally acceptable, the other rejected as morally wrong. Thus Perry was destined to become the distrusted enigma of the mining district.

Perry's business principles were nurtured in a climate altogether foreign to the simple and uncomplicated rural lifestyle of the Texas frontier. During the last three decades of the nineteenth century, northern commercial centers seethed with business activity

---

[24] Johnson, "Mercury Mining," 36 (September 1946): 393.

stimulated by unprecedented industrial expansion. Free competition drove men to obtain their share of the rewards, with many indifferent to how their wealth was achieved. As the economic historian Edward C. Kirkland charges, "Piracy, no longer found on the high seas, was domesticated on land; it invaded business" and became part of the competitive technique of fortune expansion.[25] A statement by James B. Burke, a member of a New York law firm that represented Perry, supports the Kirkland thesis. He saw Perry as probably "rather typical of some early twentieth century businessmen who developed industries, but with what we now regard as a rather cold attitude as far as his employees, and those with whom he dealt, were concerned. . . . I don't think he inspired affection, to say the least."[26] Walter K. Bailey, while candid, was more judicious in his appraisal of his uncle. Examining him within the context of his age Bailey equates Perry's methods with the benefits the system produced. Bailey envisioned Perry as "one of the last of that breed of business pirates who were, in many respects, very ruthless in their business operation, but who had a great deal to do with the building of this country."[27]

His Big Bend neighbors were less generous. With uncharitable consistency their verbal impressions of Perry ran the gamut of negative human qualities—selfish, cocky, conniving, thieving, boastful, and damned overbearing.[28] The only point at which all reached accord was that Perry was a man of mystery. Friends and former business associates still know little of his past prior to his appearance in Texas. After he established the Chisos mine, Perry confided in few people, revealing little of his private life. Former associates in Texas cite Cleveland, Chicago, and Portland, Maine, as Perry's home. All three are correct. Nor is there any uniformity about his other business interests. One source close to Perry understood he worked for the C. M. Henderson Shoe Company, a manufacturing firm in Chicago,[29] while others claim real estate and lumber as his other occupations.[30] In his correspondence references to ventures in printing and manufacturing appear: "As a boy. . . . I got my father

[25] Edward C. Kirkland, *Business in the Gilded Age*, p. 15.

[26] James B. Burke, New York, New York, to Kenneth B. Ragsdale, September 28, 1965, in possession of author.

[27] Bailey to Ragsdale, May 24, 1968.

[28] Adjectives gleaned from various Big Bend area interviews. Some subjects requested that they remain anonymous.

[29] Wayne Cartledge interview.

[30] Interview with Robert L. Cartledge, Austin, Texas, July 8, 1965.

to stake me on a printing office," and "in my iron concern in Chicago which employed over 500, . . . we built heavy machinery."[31]

Although Perry lived more than thirty years in Portland, little is known of him there. Portland newspaper files contain nothing about him, nor do local historical publications. Recollections of older men who had business dealings with him confirm the impression formed by his contemporaries in the Big Bend—Perry was a man of mystery. They recall that he mentioned the real estate business in Chicago, and that while there he lost considerable money in a Canadian mining venture, but that he "told of making a bundle in early aviation."[32] J. E. Colcord, a former Chisos mine employee and later a neighbor at Portland, writes: "As for what he did in the business line, I don't know. He was always very busy about something—I am inclined to think *only he knew what!*"[33] Such is the aura of confusion and wonder that surrounded Perry. Actually the facts are far less romantic.

Howard Everett Perry, the son of Lansford Perry and Nancy Wilson, was born in Cleveland, Ohio, on November 2, 1858. Both families came to the Western Reserve in the early 1800s; the Perrys emigrated from Batavia, New York, around 1800, and the Wilsons arrived from New Hampshire about 1815. Lansford Perry attended Chester Academy at the same time as James A. Garfield and later entered the lumber business in Cleveland. At the time of Howard's birth he was the managing and operating partner in the Woods-Perry Lumber Company, one of the largest in the Middle West. The family lived on Pearl Street (now West 25th Street) on the west side of the Cuyahoga River, overlooking the valley where the lumber yard was located.[34]

At an early age Howard Perry exhibited personal characteristics that remained dominant throughout his life. One was his love for boats and sailing. As a youth he spent his summers traveling on the lake schooners—"lumber hookers"—that brought lumber down from the Michigan peninsula. His mother recalled that when the schooners approached the Cleveland docks, she could see Howard perched atop the highest mast watching the docking process. Perry also

[31] Perry to Robert L. Cartledge, Austin, Texas, September 18, 1937, and February 17, 1938.

[32] Clipping from unidentified newspaper in possession of Robert L. Cartledge, Austin, Texas.

[33] J. E. Colcord, South Portland, Maine, to Kenneth B. Ragsdale, September 15, 1965, in possession of author.

[34] Bailey to Ragsdale, May 24, 1968.

exhibited evidences of stubbornness and independence as a youth. "A difficult child," a nephew, Walter K. Bailey, recalls. He adds: "I do not know how much schooling he had; but I do know that he attended Oberlin College (probably the Academy), and that he left 'by direction' shortly after having been enrolled. I asked him once what happened, and he said, 'Oh, nothing much. It just had something to do with tarring and feathering somebody, and riding him on a rail.'" According to Bailey, Oberlin College records indicate only that Howard Perry left that institution rather abruptly.[35]

Perry worked for his father at the Woods-Perry Lumber Company until his twenty-first birthday, when his father announced that the son should seek other employment in Chicago. The elder Perry explained that if he remained in his employment he would always be known as "Old Man Perry's Boy." Although reluctant, Howard nevertheless heeded his father's advice. Years later he learned that his father had arranged with friends in the Chicago lumber business that the third place he applied for work "he would be hired, and he was." Bailey further explained that, "I know he was in the lumber business [in Chicago] for a considerable period of time, and I think part of the time as a salesman. . . . I have reason to believe that he was in a number of business enterprises, and successfully so."[36]

Perry lived in Chicago from 1881 to 1914, thirty-three important years in his business career. He held several apparently responsible positions with the C. M. Henderson firm, was closely associated with some of the city's top industrialists, and gained valuable business experience. When he departed for Portland, Maine, in 1914, he was a wealthy man managing two family fortunes, his and the Hendersons'. Entries in the Chicago city directories reflect Perry's changing status. The 1881 edition lists him as a clerk, boarding at the Matteson House (no address). The following year he had advanced to the rank of salesman and had moved to a more prestigious address at 1808 Prairie Avenue, where he remained until 1889, when he moved to 3140 Calumet Avenue for ten years. In 1899 he returned to Prairie Avenue, this time at 1619, and listed his occupation as director. Five years later, still at the same address, he was listed as secretary of C. M. Henderson & Company.[37]

The last entry suggests another important event in Perry's life.

[35] Ibid.
[36] Ibid.
[37] Joseph C. Wolf, custodian, local history and genealogy, The Newberry Library, Chicago, Illinois, to Kenneth B. Ragsdale, March 15, 1968, in possession of author.

On February 2, 1897, Howard E. Perry married Grace Henderson, daughter of Charles Mather Henderson, senior partner of the firm, C. M. Henderson & Company.[38] On her one-hundredth birthday Mrs. Grace Henderson Perry, Howard Perry's widow, recalled the details of their courtship and marriage: "I was among the guests of a wedding party using a private railroad car to travel from Chicago to St. Louis and Howard Perry was aboard." She explained that he courted her "in the fashion of the day" and they were wed and went on a two-year honeymoon.[39]

Apparently Perry was successful almost from the time he arrived in Chicago, as his 1882 address, 1808 Prairie Avenue, places his residence in the same block with the Hendersons', a prominent Chicago family. The Henderson firm, located at the corner of Adams and Market streets, employed five hundred people in the manufacture of boots and shoes, "the largest combined manufacturing and jobbing house in the United States, [whose] facilities place them beyond the competition of Eastern establishments." C. M. Henderson was also a vigorous Chicago civic leader, active in reform movements to improve city government. In 1874 he contributed large sums of money to a campaign that removed Mayor Harvey D. Colvin from office, adopted a new city charter, and reorganized the fire department.[40]

According to family records, Perry's marriage to Grace Henderson "was somewhat disturbing to Mrs. Henderson," although Henderson and Perry were mutually compatible, shared common business and political views, and Henderson expressed great confidence in Perry's business ability. This respect was confirmed by Henderson's will which, according to Walter K. Bailey, provided that at his death "Howard was to settle the estate, and I know that having done this, he invested most of the Henderson fortune in United Shoe Machinery Corporation [stock]. This took care of the family quite well, and also created considerable wealth for his wife, Grace Henderson Perry."[41]

After establishing the Chisos Mining Company, Perry directed the company's operations, first from an office at 1619 Prairie Avenue; in 1914 his office was moved to Room 304 in the Sharpless Building

[38] Chicago Record, February 3, 1897.
[39] Portland (Maine) Evening Express, December 17, 1965.
[40] A. T. Andreas, History of Chicago, p. 729.
[41] Bailey to Ragsdale, May 24, 1968. Mrs. Grace Perry's estate inventory, filed November 14, 1968, in Cumberland County, Maine, probate court, listed the appraised inventory at $333,284.40.

at 565 Washington Boulevard, his last Chicago address.[42] The following year the Perrys moved to Portland, Maine. The Perry estate
at Cumberland Foreside, a Portland suburb, included a twenty-
three room mansion, three smaller homes, and a greenhouse overlooking Casco Bay. Perry wrote a business associate that this property "cost well over $500,000 and has been called the finest in the
state between Beverly, Mass. and Bar Harbor."[43] Prior to moving
there Perry spent summers sailing in New England, and it was while
cruising Casco Bay that he first saw the home. Mrs. Perry recalled
that day with fond memories: "We tried to buy it then but found
the owner didn't want to sell. . . . Later we were in Texas and heard
from his family that he had died and they were ready to sell. Mr.
Perry returned to Maine and bought the house the same day. . . . It
was a clear, cold day and the snow covered everything. . . . It
couldn't have been a more beautiful sight."[44]

When asked why he moved to Maine, Perry explained that the
answer was perfectly clear: "You could not join the New York Yacht
Club if you had a Chicago address."[45] At that time he was a member of the Portland, Eastern, and New York yacht clubs and maintained a sailing yacht large enough to have a year-round captain
and a sailing crew of five or six people.[46] When asked how he could
afford such a yacht, "he again said the answer was very simple. If
he had not wanted the yachts, he would not have made so much
money, which he had to do in order to have them."[47]

Although Perry's achievements in business and financial circles
identified him as a person of considerable stature, his physical appearance was not a sight to inspire confidence. Harris Smith, a Big
Bend miner, remembers him as "a little-bitty short feller, couldn't
have been over five feet. I'm five feet, seven and one-half inches and
I looked 'way down on him. Walked like Charlie Chaplin, his feet
sticking out at about forty-five degree angles. Every step he took

[42] Wolf to Ragsdale, March 15, 1968.

[43] Perry to R. V. Rinehart, June 25, 1941, Chisos Mining Company Papers,
Archives of the Texas State Library, Austin.

[44] Portland (Maine) *Evening Express*, December 17, 1965.

[45] Bailey to Ragsdale, May 24, 1968.

[46] "Howard E. Perry was a member of the New York Yacht Club from
1920 to 1944. Two yachts that he owned during most of this period were the
59-foot power yacht *Lassie*, and the 80-foot schooner *Mistral*. He also owned
for a few years a 74-foot power yacht, *San Toy II*" (Sohei Hohri, librarian, New
York Yacht Club, New York, New York, to Kenneth B. Ragsdale, April 15,
1968, in possession of author).

[47] Bailey to Ragsdale, May 24, 1968.

you could hear his britches legs flop against each other."[48] Because of his diminutive size, his associates reported he purchased boy's clothes, as the smaller men's sizes were much too large.

If Perry was self-conscious about his slight stature—and undoubtedly he was—it was never manifest in his relations with his physical superiors. He always spoke with confidence and argued with no one. If someone disagreed—which was rare—he ignored them. If this positive manner compensated for his unimposing physical appearance, then this is a clue to the inner character of Howard E. Perry.

This self-assured attitude assumed a form of optimism that bore the indelible mark of a true mining man, whether owner or lonely prospector. He approached each new venture with unquestioned confidence; success was imminent and failure inconceivable. If the price of quicksilver was down, the market was bound to improve; if recovery was off, the next assay would be richer; and if new ore deposits were not forthcoming, then the next drift would reveal the greatest bonanza yet discovered. Optimism grew into fanaticism, and Perry approached the world as if guided by divine Providence. In Perry's own words: "I learned many years ago that my enemies all die before long—Mrs. Perry always says 'Don't worry he will die after a little.' "[49]

Perry categorized the people he encountered as those who either accepted or rejected his omnipotence. The former were fondly cajoled with "dear old Dahl" or "will you ask our good sheriff"; the latter impatiently rebuffed with "that skunk Johnson," "Dr. Wright is a genuine rat!" or "I wrote you not to pay old stick in the mud Jacobson for that cylinder."

Perry's egomania inheres in the self-appraisal of his success at Terlingua. "There have not been very many mistakes made at Chisos Mine . . . no other man in the world could have created Chisos like I had done. And scarcely any man in the world but that would have lost it many years ago." Not only did he feel endowed with qualities of natural leadership, but he also possessed the capacity to mold other men in his own image. An obscure mine employee named Fulton became the object of this transformation. "I like Fulton exceedingly," wrote Perry, "in fact, very fond of him. I think he has possibilities . . . if he can 'conform.' I can build him up into something truly big and fine if he listens to me and willing

---

[48] Interview with Harris Smith, Austin, Texas, January 21, 1966.
[49] Perry to Cartledge, May 31, 1939.

to learn."[50] As Perry's wealth grew and his investments became more diversified, business turned into a game. As he became consumed in his own self-importance, the problems of others became insignificant. Nowhere is this facet of his personality more evident than in his decision not to attend his sister's funeral. "I had first thought that I would go to Cleveland on the same train that my sister's body was shipped, but so many things came up for immediate attention that I gave it up as quite impossible. Anyway, the poor little sister was dead and there was nothing more I could do for her."[51]

This was the same strange little man who seldom bathed, but daubed his body with swatches of cotton soaked in alcohol,[52] who forbade guests in his Portland mansion to converse with the domestic employees, and who rebuked inefficiency in public employees with fits of temper accompanied with eruptions of profanity and threats of violence.[53] This was also the same little man who, with blatant boldness, engaged the elements to establish the Chisos Mining Company in the inhospitable vastness of the Big Bend of Texas. Success was imminent and failure inconceivable!

The facts concerning Perry's earliest efforts in establishing the Chisos mine are obscure. But once the mine became productive he directed the operation from Chicago through a regular exchange of correspondence. None of these records are preserved and the details of early explorations are sketchy and incomplete. However, two retorts yielded the first recovery in 1903 and the following year two more were added.[54] Local journalists, recognizing the mine's potential, reported Perry's early success: "With the four retorts they will take out 200 flasks per month, as they have plenty of good ore in sight."[55] If projected on an annual basis, this figure indicates that the Chisos mine probably recovered about half of the 1903 Texas production, reported as 5,029 flasks valued at $211,218. Total United States production for that year was 35,620 flasks of quicksilver, which yielded $1,544,934.[56] Although Perry's initial production was substantial, it was destined to increase.

The Chisos mine was apparently surfacing high grade ore, as

[50] Perry to Cartledge, February 15, 1938.
[51] Perry to Cartledge, October 13, 1939.
[52] Interview with W. D. Burcham, Alpine, Texas, December 29, 1964.
[53] Source wishes to remain anonymous.
[54] Wayne Cartledge interview.
[55] Alpine *Times*, Supplement, June 29, 1904.
[56] Department of the Interior, *Mineral Resources of the United States*, 1903, p. 281.

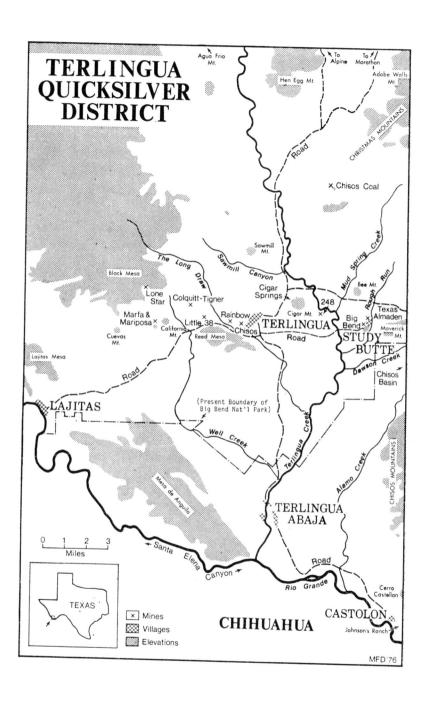

TERLINGUA QUICKSILVER DISTRICT

Agua Frio Mt.

To Alpine

To Marathon

Hen Egg Mt.

Adobe Walls Mt.

CHRISTMAS MOUNTAINS

× Chisos Coal

Road

Mud Spring Creek

Sawmill Mt.

Sawmill Canyon

The Long Draw

Black Mesa

Bee Mt.

Rough Run

× Lone Star

Colquitt-Tigner ×

Cigar Springs ×

248

Texas Almaden ×

Marfa & Mariposa ×

Little 38 ×

Rainbow × × Chisos

Cigar Mt. ×

Big Bend ×

× Maverick

California Mt. ×

Reed Mesa

TERLINGUA

STUDY BUTTE

Cuevas Mt.

Road

Dawson Creek

Lajitas Mesa

Road

Chisos Basin

LAJITAS

(Present Boundary of Big Bend Nat'l Park)

Well Creek

Terlingua Creek

CHISOS MOUNTAINS

Mesa de Anguila

Alamo Creek

TERLINGUA ABAJA

0 1 2 3
Miles

← Santa Elena Canyon →

Road

Rio Grande

Cerro Castellan

TEXAS

× Mines

⊠ Villages

▨ Elevations

CHIHUAHUA

CASTOLON

Johnson's Ranch

MFD '76

it is not considered profitable to retort material with less than 4 percent metal. When the geologist Kirk visited the district in 1905, he reported some of the ore surfaced assayed as high as 25 percent quicksilver. Although little work had been done on the property, the Kirk study predicted the presence of vast ore reserves.[57]

When Perry opened the quicksilver properties, he knew little about mining; but he possessed business experience, had the capacity to finance the needed expansion, and chose wisely the men to supervise the operations. Although he seldom visited the mine, he was not content to remain uninformed about a business that yielded such handsome profits. He inspected the New Almaden mine in California and went to Almaden, Spain, to study methods of exploration and recovery employed in the oldest and largest quicksilver mine in the world.[58]

From this experience Perry realized that the potential of the Chisos mine far exceeded the recovery facilities afforded by the four retorts, and in 1905, when the Colquitt-Tigner mine suspended operations, he leased that company's Scott furnace to treat the Chisos ore. This installation had a 10-ton daily capacity, "but Devine McKinney [Perry's former adversary], who has charge, is putting through 15 tons per 24 hours with excellent results."[59] With this installation located on the west side of the Long Draw, the Chisos hauled its ore about five miles to the Colquitt-Tigner facility for conversion.

Perry's search for an experienced mining engineer in 1905 coincided with the Texas legislature's failure to appropriate funds to continue the Texas Mineral Survey. Dr. William Battle Phillips, director of the survey and a competent and widely experienced mining engineer and geologist, accepted Perry's offer and became the first superintendent of the Chisos mine. Phillips possessed an intimate knowledge of the geology of the quicksilver district, having first visited there in 1902 and returning the following year with Dr. J. A. Udden, a brilliant geologist also with the Texas Mineral Survey.[60] During these field surveys Perry probably became acquainted with the two geologists. Phillips joined the Chisos staff sometime in 1905 and Udden returned to the district some months

57 Johnson, "Mercury Mining," 36 (September 1946): 393.
58 Interview with Robert L. Cartledge, Austin, Texas, December 27, 1964.
59 Alpine *Times*, Supplement, June 6, 1906.
60 Monica Heiman, *A Pioneer Geologist: Biography of Johan August Udden*, pp. 22–28.

later as Perry's special consultant.[61] Both men exerted profound influence on the eccentric mine owner, and Charles Laurence Baker,
who knew both professionally at the Bureau of Economic Geology
(at the University of Texas at Austin), credits them for the early
success of the Chisos.[62]

The task facing Phillips was formidable. With minimum planning and administration, the Chisos Mining Company had grown
in three years from an obscure beginning to a highly profitable
operation employing more than one hundred men. Phillips' job was
to transform this underdeveloped property into a functioning industrial organization. The commercial market stimulated by an
expanding economy advanced the price of quicksilver to $36.22 per
flask and the growing military market left orders unfilled. The
economy of the quicksilver district responded to the national prosperity as the Alpine *Times* reports "much good ore being taken out
of the different mines. At the Chisos Co.'s camp, where Dr. Phillips
is in charge about 100 men are employed and much ore is in sight.
. . . The Dallas people have had their smelter fired up about two
weeks. . . . At nearly 200 feet they are in a fine body of ore. The
Marfa and Mariposa Co. are . . . taking out several thousand dollars
worth of ore each week.[63] Bustling with activity the district produced 4,723 flasks of quicksilver in 1905, approximately 20 percent
of the national production, and the prospects of employment had
attracted more than 1,200 persons to the region.[64]

Drury M. Phillips, Dr. Phillips' son who accompanied him to
the Chisos, speculates why his father accepted the position:

> He [Dr. Phillips] was convinced there were large deposits of various
> quicksilver ores in the Big Bend Region, and he had the opportunity
> to prove some of the theories he and others had developed as to
> origin, occurrence and utilization. Perry had money, and saw in
> Phillips an opportunity to make more. . . . Mining and refining
> equipment and methods were crude, and only partially effective

[61] Heiman, ibid., cites the year 1909 as the beginning of Udden's association with the Chisos mine. However, the Bureau of Economic Geology in Austin, Texas, has on file map work performed by Udden for Perry dated August,
1906.

[62] Charles Laurence Baker, Cardova, Illinois, to Kenneth B. Ragsdale,
November 16, 1964, in possession of author.

[63] Alpine *Times*, Supplement, May 9, 1906.

[64] In 1905 California produced 24,734 flasks of quicksilver, or 81 percent
of the United States total production of 30,451 flasks (U.S. Department of
Interior, *Mineral Resources of the United States, 1905*, pp. 393–397).

. . . . labor was plentiful, but painfully unskilled, and a close control practically unknown.[65]

Undoubtedly language constituted a barrier, as practically all labor was native Mexican, who worked under Anglo supervision. The low wage scale was another factor contributing to the low efficiency of labor. Of the six states producing quicksilver, Texas ranked lowest in wages paid employees; miners who were paid $2.00 per day in Texas would have received $6.00 for the same skills in Washington. The Indian roustabout in the Arizona mines who received $3.00 per day ranked ahead of the Mexican at Terlingua who received only $1.25.[66] In 1905 Phillips reported his lowest paid employees were the muckers who received $1.00 per day, while those who performed the more skilled tasks of blacksmith and enginemen received from $3.00 to $4.00 per day.[67]

One year later, in March, 1906, the miner's daily wages varied from $0.50 to $1.25, while the highest paid Mexicans were the furnace employees who received $1.50 per day. Lupe Moscoro, for example, worked twenty-nine days for $0.50 per day. From his month's income of $14.50, $12.00 was deducted for the store account and $0.50 for the doctor's fee; he received the remaining $2.00 in cash. The highest paid employee was Frank Dryden, who worked twenty days at the furnace at $1.50 per day. Dryden received a gross salary of $30.00, and after $5.25 was deducted for the store account and $1.25 for the doctor's account, he received $23.50 for the month's work.[68]

Despite the labor deficiencies Phillips established the Chisos Mining Company on a sound operational basis, developmental work continued, and by the end of the year, the company was "in good bodies of ore at the depth of over 200 feet . . . and running about 15 tons per day through the [10 ton] smelter."[69] In February, 1906, Phillips moved his family from North Carolina to Terlingua.[70] Laurance V. Phillips, another son, recalled the three-day trip from

[65] Drury M. Phillips, Huntsville, Texas, to Kenneth B. Ragsdale, January 17, 1966, in possession of author.

[66] C. W. Schuette, *Quicksilver: Bureau of Mines Bulletin Number 335,* p. 26.

[67] Johnson, "Mercury Mining," 36 (September 1946): p. 393.

[68] Chisos Mining Papers. The company store is discussed in detail in chapter 8.

[69] Alpine *Times,* January 7, 1907.

[70] Ibid., February 21, 1906. Dr. Phillips had three sons, but only Laurance V. Phillips and Drury M. Phillips went to the mine.

Alpine to the mine in a covered wagon as a delightful outing, and "the Doctor," who loved to travel, found this trek with his family a pleasant respite from his duties at Terlingua.[71]

Phillips established his family—his wife and two sons—in two tents on a hill overlooking the mine. In one tent used for cooking, the food was prepared by a Mexican woman and Mrs. Phillips, the only American woman in the camp. Drury Phillips recalls, "It was not an easy life—but it was a *good* life—I really learned the meaning of 'Amigo.' "[72]

Early in March, 1906, Perry arrived at the mine to inspect the recent changes effected under Phillips' reorganization. The ore was still being treated in the four Chisos retorts and the leased Colquitt-Tigner furnace, whose combined productive capacity did not equal the volume of ore being surfaced. Since both men contemplated expansion this subject probably dominated their conversation. Before Perry left for Chicago he announced signing a contract for a 20-ton furnace whose designer, Robert Scott, would personally supervise the installation (see appendix C).[73] This acquisition placed the Chisos mine in the category of a major quicksilver producer. With vast ore reserves and this new condensing facility, the Chisos mine would help write a new chapter in Texas' economic history.

The Phillips-Perry relationship, though mutually advantageous, was destined to end. Late in the summer of 1906 Mrs. Phillips was seriously injured in a fall from a burro. Drury Phillips, recalling the tragic event, states: "Our mother and [Laurance] were riding bareback on a burro thru the camp—a pig ran squealing from a hut—they were thrown—she struck her head on a rock—suffered a concussion from which she later died." Phillips left the mine immediately and moved his wife to Alpine, "three days in two covered wagons with four mules each."[74] Drury Phillips and A. P. Stramler, a mining engineer and former student of Phillips at the University of Texas, remained at the mine temporarily to continue Phillips' work program. After his wife's accident Phillips returned to the Chisos only briefly prior to terminating his employment with Perry.[75]

[71] Interview with Laurance V. Phillips, Austin, Texas, December 14, 1964.

[72] Drury M. Phillips to Ragsdale, January 17, 1966.

[73] James M. Day, "The Chisos Quicksilver Bonanza in the Big Bend of Texas," *The Southwestern Historical Quarterly* 64 (April 1961): 436.

[74] Phillips to Ragsdale, January 17, 1966.

[75] Following his wife's death, Phillips went to Birmingham, Alabama, as a geological consultant and in 1909 returned to the University of Texas as direc-

Although Mrs. Phillips' accident no doubt hastened her husband's departure from the mine, in all probability the Perry personality was a contributing factor. Perry later described the conflict that existed between them: "I used to rave and tear my hair over things which were going wrong and many a time I have given Dr. Phillips . . . an awful going over. Phillips used to say, 'You will kill yourself if you think too much about these things.' After a time I learned that I could scarcely ever remake anyone down there and of course, Mexicans never."[76]

Conflict was inevitable, since Phillips was primarily a scholar, highly regarded in his profession, whose services and advice were widely sought. Although economic considerations were present when he entered Perry's employment, his prime objective in going to the Chisos mine was professional betterment through firsthand study of the Terlingua quicksilver deposits. Drury Phillips also cites Perry's interference as the probable turning point in their association. He believes that

> so long as Perry allowed Phillips to "run the business," they got along quite well, but when Perry began to dictate mining and refining procedures, the two strong willed men disagreed sharply. . . . I am of the opinion that Phillips' efforts to improve operations ran directly counter to Perry's insistence on high profits, and that this not uncommon conflict had much to do with their parting company.[77]

Those who knew both men recall they parted friendly. "I do not recall," Drury Phillips wrote, "ever hearing 'The Doctor' make any derogatory remarks about Perry,"[78] while Perry was quoted by former employees as crediting Phillips with his early success with the Chisos mine. Geologist Charles Laurence Baker, who knew both men, expressed Phillips' lasting contribution far more succinctly: "He taught Perry how to mine."[79]

The first developmental phase of the Chisos Mining Company was at an end. With a well-equipped plant furnacing what appeared

---

tor of the newly formed Bureau of Economic Geology and Technology. In the spring of 1915, Udden became director of the bureau when Phillips accepted the presidency of the Colorado School of Mines. During Phillips' career, he acted as geological consultant to the Lucas interests at Spindletop and to the Hogg family at West Columbia, Texas. Phillips died June 7, 1918.

[76] Perry to Cartledge, July 17, 1935.
[77] Phillips to Ragsdale, January 17, 1966.
[78] Ibid.
[79] Baker to Ragsdale, November 16, 1964.

to be an endless ore supply, the first decade of the company's history ended on an optimistic note. However, many problems still lay ahead for Perry—transportation, fuel and water supply, education, and all facilities of civilization—the solution to which would accord Terlingua a distinctiveness among frontier mining settlements of the American Southwest.

CHAPTER **3**

The Basic Necessities: Wagons, Wood, and Water

TERLINGUA, a village born in isolation, was spawned by the discovery of cinnabar, grew in response to mining needs, and died when the lengthening shafts no longer yielded the cochineal-red ore. While quicksilver was the prime economic base of Terlingua, other factors affected life far more profoundly. Surrounded by desert and cut off from civilization by near-impassable mountain ranges, the area evolved a cloistered lifestyle that was doubly conditioned by scarcity and isolation. These two factors became the controlling forces in the mining village and were reflected in all patterns of human conduct; travel and transportation, fuel, and water supply—each developed in proportion to the geography, topography, and climate of the region. In meeting the challenges of scarcity and isolation the people developed a society, not in harmony with, but in opposition to these factors that fashioned life in the Big Bend of Texas.

Terlingua's first lifeline was approximately one hundred miles of wagon ruts that connected the mining district with Marfa, the Big Bend's first supply center. C. A. Hawley, an early mine employee, emphasized the importance of this link with the outside world, remarking that practically everything required to sustain the village and operate the mine had to be shipped in.[1] The terrain, as well as the distance, was a prime factor in the region's isolation. The mining engineer W. P. Blake visited the area in 1895 and reported that the cinnabar deposits are best reached from Marfa by team through the open country that descends to the Rio Grande. "The last six miles of the route," according to Blake, "is impassable for wagons, and the cinnabar camp is reached by a pack-trail which

---

[1] C. A. Hawley, "Life Along the Border," *Sul Ross State College Bulletin* 44 (September 1964): 17.

turns westward from the wagon road and leads across a country much broken and intersected by dry 'washes' or creek beds."[2]

By 1902 some improvement in transportation was noted as the district had once-a-week mail and passenger stage service from Marfa. Although Alpine was a few miles nearer, a better road to Marfa gave that village a distinct trade advantage.[3] The immigrant bound for the mines usually arrive in Marfa by train and made his stage reservations at the St. George's Hotel, located on the south side of the railroad tracks. A one-way ticket cost $9.00[4] and the adventure began when the Mexican wranglers attempted to hitch the mules to the stage. Since these wild little beasts resisted every attempt at restraint, "it took some time to get them hitched to the hack, and after they were in place . . . it took three or four strong men to hold them until we were off." Hawley, who first went to the mines on the Terlingua stage, recreated the excitement of the departure. "Finally the word was given, the buckaroos jumped out of the way, [the driver] shouted to the mules in Spanish, and away we went, the mules on a dead run and we passengers gripping the seats and holding on for dear life."[5]

Isolation, a constant factor throughout the life of the mining district, not only delayed discovery of the ore deposits but also retarded development once the cinnabar was located. Journalist W. D. Hornaday reported in 1910 that while there were many promising claims in the Terlingua area, "it is difficult to get the properties financed owing to the long and tiresome journey" required to reach the strike zone. Hornaday further explained it was not the personal discomforts of travel, "so much as it is the heavy cost of bringing in machinery and supplies, that makes the long overland trip a serious barrier to the development of the quicksilver."[6]

A source of personal discomfort was the Terlingua stage, a simple two-seated buggy with a canopy top that followed the wagon ruts between Marfa and Terlingua. Mail sacks and express packages were carried in the rear and the storage of suitcases and personal belongings was left to the discretion of the passengers. Stuart T.

---

[2] William P. Blake, "Cinnabar in Texas," in *Transactions of the American Institute of Mining Engineers,* XXV: 69–70.

[3] W. D. Hornaday, "The Cinnabar Deposits of Terlingua, Texas," *The Mining World* (December 17, 1910), p. 1133.

[4] Chisos Mining Company Papers, Archives of the Texas State Library, Austin.

[5] Hawley, "Life Along the Border," p. 13.

[6] Hornaday, "Cinnabar Deposits," p. 1133.

Penick, a United States Geological Survey member, rode the stage from Terlingua to Marfa in June, 1903, and his recollections of this experience dramatize one of the hardships of Big Bend travel.

> On the first day nothing unusual happened, but on the 2nd day, shortly after we got on the way, a brisk north wind with rain came up. The coach was not entirely water proof, and the passengers became wet and uncomfortable. There were two other men and a woman with two small children aboard, and as the day wore on the rain came harder and the going got slower, for we were now traveling over the dobie flat that stretches for some distance south out of Marfa. When that dobie soil gets wet it tends to ball up on the wheels, so that the horses could only proceed at a slow walk. Therefore, it was almost dark when we reached the stage-stop where we were to eat dinner and change horses. The passengers were all tired and especially the woman with the children, for they were cold, hungry and fretful. . . . it was decided to spend the night there and proceed to Marfa the next day. Instead of the five or six hours it ordinarily took to cover the distance, it was late afternoon when we got to Marfa.[7]

Midway between Marfa and Terlingua was a halfway house, a key point on the stage route. As separate drivers began simultaneous runs from both terminals, they met at the halfway house, spent the night, exchanged passengers and mail, and returned the following day to their original departure points. Hornaday recalls that this en route facility rivaled the torturing stage ride for personal discomforts.

> [During the summer] the heat is terrific [and] the region is rough and uninviting in appearance. . . . Human inhabitations are few and far between in the whole 110 miles. . . . There are a few ranch homes below Marfa for about 40 miles, but from there on the country is barren of homes, with the exception of the dilapidated adobe "halfway" house. It is a primitive hostelry that is kept by a lone Mexican. Near the house he has provided tents for the occupancy of any overflow of guests that may happen to pass that way.[8]

Jeanette Dow went to Terlingua in 1909 to teach school and remained one year. Her departure was also on the Terlingua-Marfa stage, and while she did not object to the hardship of a two-day stage trip, the overnight experience at the halfway house was fright-

---

[7] Stuart T. Penick, La Grange, Texas, to Kenneth B. Ragsdale, December 6, 1964, in possession of author.

[8] Hornaday, "Cinnabar Deposits," p. 1133.

ening. She recalled, "I was so afraid to spend the night at this half-way house that the Doctor's wife and I arranged to go together, so that solved that situation."[9]

The power source of all Terlingua transportation was a breed of Spanish mule that seldom weighed more than six hundred pounds. Smaller than the regular mule, it endured heat better, required less feed, and could go longer without water than a horse. Moreover, it was larger and possessed greater strength and mobility than its burro progenitor. Evidently no horses were used for transportation purposes in the Terlingua district.[10]

Mexican contractors handled the movement of equipment and supplies, as well as passengers and mail. They owned their own equipment and were available to the mine operators as the need arose. Eight contractors operating twenty wagons served the Chisos mine, and Feliz Valenzuela, its largest freighter, maintained four wagons and a herd of between thirty and forty mules.[11]

The Studebaker wagon, the brand used exclusively in the mining district, was a heavily built vehicle designed to carry three to four tons, but the freighters frequently exceeded its capacity by several thousand pounds. The front wheels (forty-eight inches in diameter) were slightly smaller than those in the rear (fifty-two inches), which gave the vehicle an awkward and ungainly appearance, but provided greater maneuverability on rough terrain. Normally eight to twelve mules pulled each wagon and the driver rode in a saddle on the rear, left-hand, wheel mule from where he directed his charges with leather guide lines, the ever-present *malacate*, or whip, and a colorful flow of Mexican profanity.

Chisos freighters were the elite of the Mexican community, as they were its most prosperous citizens and the only ones self-employed. The owners seldom drove the wagons, but usually employed relatives in that capacity; their primary duties were to schedule the equipment and see that the freight was properly dispatched. As the freighting industry progressed, the drivers, most of whom had previously been farmers, developed a characteristic mode of dress and acquired a complement of artifacts incident only to the Mexican freighter of the Terlingua Quicksilver District.

[9] Jeanette Dow Stephens, Shreveport, Louisiana, to Kenneth B. Ragsdale, June 25, 1968, in possession of author.

[10] Interview with Robert L. Cartledge, Austin, Texas, December 26, 1964.

[11] Other freighters employed by the Chisos mine and the number of wagons each operated were Raffel Carrasco, two; Cleofas Acosta, two; Valentine Rodríguez, three; Paz Molinar, one; and Antonio Franco, two.

Khaki pants, blue chambray shirts, and a white broad-brimmed Mexican straw hat became the freighters' standard dress. The hat afforded maximum protection against both the sun and the infrequent rainsqualls; to increase its utility, the drivers impregnated the hat with melted bee's wax, making it impervious to water. Instead of shoes they wore *huaraches*, or sandals, first fashioned from rawhide, but after the appearance of automobiles in the Big Bend, made from sections of tires and tethered to the feet with rawhide thongs.[12]

Harris Smith, a Big Bend miner, pointed out that rawhide was basic in the Mexican's economy. He explained that "when an animal died—sheep, goat, mule, or cow—it was skinned and the hide converted into some item needed in their struggle for a livelihood. These were just poor laborin' boys. Had a lot of ingenuity. Lived off the land all they could."[13] The most characteristic item of their trade was their *malacate*, or whip, also referred to as an *isador de caballos*, or horse pusher. This was also fashioned from rawhide, and plaited to the length of their particular hitch of mules.

The only purchased items they owned were an axe and a capstan jack. The axe was essential for making emergency repairs to the wooden portions of the wagon and for chopping firewood. The capstan jack, a screw-type device, served to elevate the wagon for greasing the "hounds," or axles, which, left unattended, emitted a loud growling noise and gave rise to the expression, "Your dogs [hounds] are growling."[14]

Travel time to the railroad depended on the weight of the load. Wagons carrying quicksilver required six to eight days; lighter loads required less time, but the drivers always spent several days on the road. The life of a freighter was rugged; they lived in the open, were always in transit, prepared their meals by campfire, and slept inside a small tarpaulin on the ground near their wagons. Water was always a major concern, as it was impossible to carry a large supply from the mine. When conflicts arose with the local ranchers over the use of water holes, Perry purchased one section of land about twenty miles north of the mine that provided permanent water, and later acquired a watered section north of the 02 Ranch, approximately midway between Terlingua and Alpine, for the same purpose.[15]

12 Interview with Harris Smith, Austin, Texas, February 2, 1966.
13 Ibid.
14 Ibid.
15 Cartledge interview, April 30, 1965.

The increased flow of freight and supplies between the quicksilver district and the supply centers reflects the industrial activity spawned by the quicksilver development. In July, 1904, for example, the Alpine *Times* reported that "eighteen wagons loaded with machinery and lumber came in last week from Alpine and Marfa for the Dallas Mining Co."[16] Two years later Marfa still maintained a monopoly of the Terlingua trade and the stage service had increased to two scheduled trips a week. With improved roads, the distance was spanned in two days. Marfa dominated the quicksilver district trade until about 1909, when most of the mine activity shifted to the east side of the district and Alpine became the new supply center. Operators in the Terlingua–Study Butte area felt that being nearer the railroad at Alpine and Marathon would give them a bargaining advantage over the Mexican freighters. They soon discovered, however, that this was an illusion. As the geologist J. Harlan Johnson explained, "[they] charge just as much for hauling from Marathon, 90 miles, as from Alpine, 100 miles, or from Marfa by way of Alpine, 130 miles. It is fifty cents per 100 pounds, and they do not regard distance at all. Whether they are five days on the road or ten does not disturb their serenity or enter in their calculations."[17]

Alpine held the Chisos trade until World War I, when a misunderstanding developed between Perry and that town's merchants. No one recalls the nature of the dispute, but Perry retaliated by establishing two warehouses and an oil storage tank in Marathon and dispatched all future shipments of quicksilver from that point.

Quicksilver constituted the principal cargo on the freighter's northbound journey, while equipment and supplies made up the load on the return trip. The mine superintendent, who originated orders dispatching a load of quicksilver, sent shipping instructions to the furnace operator, the person responsible for recovering and storing the reduced metal.

> Mr. Dahlgren, Chisos Furnace Operator
> Please let Feliz Valenzuela have one hundred-eighty (180) and Rafarl Carrasco one hundred-eighty-one (181) total three hundred-sixty-one (361) flasks quicksilver for shipment.
> > The Chisos Mining Company
> > W. R. Cartledge[18]

[16] Alpine *Times*, July 20, 1904.

[17] J. Harlan Johnson, "A History of Mercury Mining in the Terlingua District of Texas," *The Mines Magazine* 36 (September 1946): 394.

[18] Chisos Mining Papers.

This shipment of 361 flasks weighed approximately 31,768 pounds,[19] and since weight, not volume, determined the number of wagons used, probably Valenzuela and Carrasco each used two wagons on this trip.

After the freight wagons departed, loading instructions for the return trip were dispatched by mail. Carried on the stage, this information reached the supply centers several days before the slower quicksilver wagons.

> Mr. L. T. Votaw
> Alpine, Texas
> Dear Votaw:--
> Two of Paz Molinar's wagons should arrive there by Friday 19th. Please load them back as follows.
>   10 bbl. Gasoline
> From Watson-Anderson
>   15 Sks. Sugar
>   4 Cases 6/8 Crusteene lard
> Complete load out of bookings of July 11th at Watson & Anderson.
> > Robert L. Cartledge[20]

As Terlingua's population grew the freighter's responsibility increased in proportion to the demand for more supplies. Traffic along the Terlingua-Alpine road increased steadily and during the early years of Chisos operation the company had as many as twenty-six wagons on the road at one time.[21]

The specialized problem of moving heavy mining machinery over mountainous terrain provides part of the romance of mining in the American West. The Terlingua quicksilver miners, for instance, faced both distance and mountainous terrain that challenged them as long as the mines remained productive. The magnitude of this problem was first reported in 1905 when T. P. Barry, manager of the Dallas Company, attempted to haul an 18,000-pound boiler from Marfa to Study Butte. As the Alpine *Avalanche* reported, "It took two wagons to load the boiler on, and 26 mules were employed to pull it . . . the outfit got bogged in the mud before they got across

---

[19] A flask of quicksilver was established at seventy-five pounds net weight. In June, 1927, it was changed to seventy-six pounds to conform to the new weight of the European flask. The empty container weighed approximately thirteen pounds, making the filled flask total eighty-eight pounds gross weight.

[20] Chisos Mining Papers.

[21] Hawley, "Life Along the Border," p. 20.

the arroyo just in the edge of town, and they were several hours digging it out."[22]

This was not an isolated incident; as late as 1930 the Chisos mine experienced a similar problem, when apparently the roads linking the mines with the rail centers had been little improved and the mainstay of heavy transportation remained the mule-drawn Studebaker wagon. In this latter year Perry purchased a 425-horse-power engine from the Shawnee Milling Company as part of a pro-ducer gas system. Although the engine arrived at Marathon in the last week of May, it was still in transit to Terlingua on September 2. Perry, who had anticipated a problem, wrote Robert Cartledge on May 25, "It may be that you will ask Feliz [Valenzuela] to buy a very heavy strong wagon for such extra heavy loads."[23] Cartledge responded and the dismantled engine, estimated by Chisos mechan-ic J. E. Anderau to weigh 14,000 pounds, was loaded on a new Studebaker wagon designed to carry only 7,000 pounds. Valenzuela hitched fourteen mules to the overloaded vehicle and the ill-fated trip to Terlingua began. The mules experienced so much difficulty pulling the load that he increased the team to twenty-two mules to negotiate Del Norte Gap. The drama of this experience is inherent in a letter Cartledge wrote Perry:

> . . . in the middle of the road it bogged down and the 22 mules were unable to pull it. . . . Up thru Del Norte Gap in the bed part of the creek, he would have to make a pull of about ten feet at a time. . . . He first broke an exles about 15 miles out of Marathon. . . . He made it to the top of Del Norte, but there had to over haul a rear Axle. From there he has made it to the steel tank in 02 pasture where he broke a rear wheel. . . . I hardly know what we are going to do when we get to Adobe Walls and below. Guess, we have put on about 40 mules. . . . I am also afraid that at Terlingua Creek we will have to build wood floor road, as bridge will not hold up wagon.[24]

On reaching Adobe Walls Mountain, Cartledge inspected the load and had the shaft and bearings removed from the engine to reduce weight. After making these changes, he was able finally to get the engine to the mine in late September, almost five months after delivery in Marathon. When the engine was removed from the wagon and weighed, instead of the originally estimated 14,000

[22] Alpine *Times*, September 20, 1905.
[23] Perry to Cartledge, May 25, 1930.
[24] Cartledge to Perry, September 2, 1930, Chisos Mining Papers.

pounds, it actually weighed 22,500 pounds! And it had been hauled one hundred miles in a Studebaker wagon designed to carry only 7,000 pounds.[25]

Increased mechanization during World War I made the nation conscious of automobile travel, so that by 1917 nearly 200,000 cars were operating in Texas, 342 of them licensed in Brewster County.[26] As the automobile ceased to be a novelty even Terlingua became accustomed to the sight and sound of this new mode of transportation. The first automobile appeared there in 1911 driven by United States revenue officers from Eagle Pass.[27] The following year Wayne Cartledge purchased a Model-T Ford, replacing it around 1915 with a Buick D-16. Shortly thereafter, F. Harry Fovargue, another Chisos employee, purchased an air-cooled Franklin.[28] As wider public acceptance brought the price of automobiles within reach of more people, ads like the following appeared regularly in the Alpine Avalanche: "BUY NOW! $295. F. O. B. Detroit. Starter and Demountable Rims, $85. Extra. Every spring the demand for Ford Cars is several hundred thousand greater than the available supply. Place your order immediately, to avoid delay in delivery."[29]

The first commercial use of the automobile in the Big Bend occurred around 1917 when a mail truck replaced the mule-drawn stage between Terlingua and Alpine. This vehicle transported both mail and passengers, but the twice-weekly schedule failed to meet the district's needs. Perry is reported as paying Julius Bird, Alpine Buick dealer, $25 to drive him to Terlingua, while Robert Cartledge said he paid $40 for the same service.[30]

Isolated uses of mechanized equipment inspired the editor of the Avalanche to report in 1918 that "wagons as freighters to the mines and other river points are being relegated [to] the scrapheap. . . . Six big motor trucks have taken the place of the slower team drawn vehicles."[31] The newspaper was more optimistic than accurate as the 1918 equipment proved inadequate for the industry's load requirements, as well as the region's road conditions.

Gradually improved county roads brought Terlingua in closer contact with the outside world, until "going outside" had a very

25 Cartledge interview, December 26, 1964.
26 Alpine Avalanche, March 1, 1917.
27 Hawley, "Life Along the Border," p. 81.
28 Cartledge interview, December 26, 1964.
29 Alpine Avalanche, March 6, 1924.
30 Cartledge interview, April 30, 1965.
31 Alpine Avalanche, July 18, 1918.

special meaning in the quicksilver district. By the late 1920s automobile trips to Alpine and Marathon were common, but travel was still slow and unpredictable. Edith Hopson, Brewster County court stenographer, explained the hazards of this experience: "In 1930 we traveled by car to Terlingua. If the weather was dry it took all day to get there, sometimes longer. . . . We carried picks and shovels in the car and if the banks of a creek had been cut back . . . we built it back as the occasion arose. . . . I can recall on two occasions sitting in the car all night waiting for water in creeks to run down so we could cross."[32]

The mule-drawn freight wagon, however, was still the Chisos mine's transportation mainstay in the early 1930s, carrying large volumes of supplies and equipment. On April 29, 1932, the Marathon State Bank billed the Chisos Mining Company $17.20 for the month's pasturage on 129 mules.[33]

Perry first witnessed the performance of the tandem transport truck in April, 1935, while at Sierra Blanca supervising the removal of some mining machinery. He enthusiastically wrote Robert Cartledge that he was "much impressed with the performance . . . of those big twin trucks. . . . Before long it will probably lead to hauling all of our store freight in this way and we will make a great saving."[34] Cartledge concurred and Feliz Valenzuela acquired a truck to haul the Chisos freight. On September 28, 1936, Cartledge apprised Perry of the great change that had occurred in one of the critical phases of life at Terlingua. "There are no more freighters down here. . . . The only freight wagons left in the country are Tiburcio Ramirez's. Then one of these is a burro team and the other is only a six mule team. All old freighter people with mules have sold [them]. . . . Feliz got $1,000.00 for his mules and I paid $1,278.00 for his truck, I putting up the balance that he needed."[35]

As the cycle of progress begot its inevitable changes one phase of frontier life disappeared and a symbol of modern technology took its place. The color and romance of the freight caravan had passed from the Terlingua country, and now existed only in the minds of the few who could recall that "the strung-out teams of small Mexican mules pulling his [Perry's] freight wagons were fully

[32] Edith Hopson, Alpine, Texas, to Kenneth B. Ragsdale, March 25, 1968, in possession of author.
[33] Chisos Mining Papers.
[34] Perry to Cartledge, April 5, 1935.
[35] Cartledge to Perry, September 28, 1936.

as picturesque as the much advertised 20-mule teams hauling borax out of Death Valley."[36]

The moving panorama of the slowly rising plumes of alkaline dust following the heavily loaded wagons across the desert, the rumbling of the steel-tired wagon wheels accompanied by the staccato crack of the *malacate*, and long after the freighter's campfire ceased to glow, the fragrance of exotic food mixed incongruently with the odors of perspiring human bodies and the dried sweat of the sleeping work animals—these were the recollections of a way of life that had faded from the industrial scene in the Big Bend of Texas.

One era was at an end, another was beginning. As improved travel increased the contact frequency with the urban centers, life in Terlingua gradually grew more nearly cosmopolitan as its citizens responded to new social and cultural forces. The process was slow and its impact was greatest in the Mexican community. But the pattern was established and this transformation would gain momentum with the approaching decade. Life in the quicksilver district would never be the same. The automobile made it so.

That this desert-imbedded ore deposit required no water directly for conversion was an auspicious circumstance. It did require heat—intense and prolonged heat—and wood for the furnaces and retorts was almost as rare as water in the Big Bend. As long as the Terlingua mines produced quicksilver, wood and water were prime concerns of the mine operators.

When the mines first opened, limited supplies of both cottonwood and mesquite were found along the Rio Grande. Mesquite, however, was preferred, as it is heavier and makes excellent fuel. It burns longer and hotter, is almost smokeless, and deposits minimum ash. In the early years of development, Mexicans brought loads of this wood to the mining camp tied to the backs of burros, which the vendors "sold to families for domestic use. The heavier pieces were brought in on wagons and sold to the company [Chisos] for fuel at the smelter."[37] Constant demand for wood soon depleted these sources, and Johnson noted the impact on the district's economy. "As supplies dwindled the price increased to $6 per cord, or higher if the seller was adept at merchandising."[38] By 1905 the price

[36] Charles Laurence Baker, Cardova, Illinois, to Kenneth B. Ragsdale, November 16, 1964, in possession of author.

[37] Hawley, "Life Along the Border," p. 19.

[38] Johnson, "Mercury Mining," 36 (September 1946): 393.

of wood advanced to $7 per cord and the Alpine *Times* reported the developing fuel crisis: "J. J. Dawson could not start up the Terlingua Mining Company furnace as he intended on account of failing to get wood, and other companies also had men out riding day and night to get wood hauled."[39]

William B. Phillips and the geologist, Johan August Udden, had prospected the Terlingua area a year earlier for coal-bearing formations. Probably on Phillips' advice the Chisos mine began purchasing coal as a heat source. This fuel proved acceptable under the retorts, and in June, 1905, D. W. Gourley delivered about four tons a day to the mine.[40] However, when Perry leased the Colquitt-Tigner furnace his problem of wood supply increased manyfold.

By June, 1908, the wood shortage had grown so critical that Perry contemplated importing large quantities of wood from Mexico. Although he knew such wood was stolen, the exigency demanded drastic measures. He shared this problem with the mine management. "I really don't know whether it is best to make arrangements with Acosta for two or three hundred cords of wood or not. Much would depend in this matter upon how the first Mexican wood trade turned out. Did it work out satisfactorily and at about the price which we figured, or was same increased by wood being stolen?"[41]

The following day Perry wrote his department heads, again soliciting their opinions before making a final decision. "If we are going to use cottonwood at Tigners . . . we will have to canvass the cottonwood situation pretty thoroughly. I suppose there is some cottonwood to be picked up on the U.S. side of the Rio Grande . . . but it is just possible that it would not be much cheaper than the Mexican cottonwood. . . . please write me how the wood matter appears to you all along the line."[42]

In Perry's business transactions price alone took precedence over all other considerations. As a former Brewster County associate recalled, "He always bargained for anything he was going to have done to secure the very lowest price."[43] Purchasing wood was no exception. When mine superintendent Marcus Hulings was unable to increase the furnace heat to operating level, Perry advised him to use larger wood. This, however, involved purchasing wood from a

[39] Alpine *Times*, June 28, 1904.
[40] Ibid.
[41] Perry to Chisos Mining Company, June 22, 1908, Chisos Mining Papers.
[42] Ibid., June 23, 1908.
[43] Hopson to Ragsdale, March 25, 1968.

man named Watters, which Perry hoped to avoid. Since Perry feared any interest exhibited by the company would lessen his bargaining advantage, he advised Hulings to proceed with caution. "You could buy a few loads of Watters' wood . . . without agreeing on any stated amount. . . . I have much hope that you won't have to do this for anything like this on our part will mean the giving up hope of Watters finally naming us a lower price."[44]

Failing wood reserves led to considerable experimentation to develop a substitute heat source, and in 1911 the Texas Geological Survey recommended producer gas, using the local coal deposits as the gas source. This process converts coal into gas by blowing steam and air through a bed of burning coal, forcing the condensed gas into a reserve container, from which it is directed to burners in the quicksilver furnace.[45] An oil-compressed air method was tried, as well as a mixture of wood and compressed creosote wood brush, commonly called greasewood. Why this method failed was never fully known; however, one local theory held that the creosote wood contained some chemical element that combined with the mercury to prevent condensation.[46]

The producer gas method appealed to Perry's frugality. This not only would solve the Chisos' fuel needs, but with coal deposits located on the company property, it might also prove more economical. The Chisos management began erecting a small producer gas plant in 1923 and ordered two Nash engines to provide the power. The engines arrived at the Marathon terminal in July and the Perry correspondence for the month focused on the search for coal. Economy, as usual, was a concurrent theme. "Marcus wrote me that Octaviano had not found any coal on our section of 248. . . . Am not surprised for one thing Oct is no prospector. . . . Inefficient bad mining and prospecting might mean a higher price for coal. I think you ought to get Delores to make a definite effort to find that other man, even going himself to Glen Springs to locate him."[47]

Who the other man was and whether he was ever found is not known, but a permanent coal supply was not located until 1926, when Perry summoned Udden from Austin to direct the prospecting.[48] Robert Cartledge recalled that Udden studied the area where

[44] Perry to Marcus Hulings, undated, Chisos Mining Papers.

[45] Johnson, "Mercury Mining," 36 (October 1946): 445.

[46] Ibid., 37 (March 1947): 28.

[47] Perry to Cartledge, July 13, 1923, Chisos Mining Papers.

[48] Monica Heiman in A Pioneer Geologist: Biography of Johan August Udden, pp. 36–37, erroneously states that "when the perceptive owner of the

he had previously found surface indications of coal and made several locations for test drilling. At each site he told Cartledge the depth at which the coal would be encountered, which varied from 40 feet to 100 feet. Test cores verified Udden's exact predictions except at the 100-foot level, which he missed by 2 feet![49]

After the coal mine was opened, producer gas became the principal heat source and the problem of a fuel supply was solved; however, the Chisos continued purchasing wood. In the late 1930s both wood and gas were still in use, and an Alpine source provided an adequate supply of wood at $12 per cord, double the price paid when the mine was first opened. On January 18, 1938, Cartledge wrote Perry: "[I] . . . have been buying all the wood that comes in here. . . . I can stop buying wood . . . but I am afraid that we will not be able to get wood later on and we will need all . . . we can get to run the furnace. . . . [Also] we had gotten the coal mine in shape to where our costs would have gradually have grown less."[50] During that year (1938) 179.75 cords of wood were charged to the furnace and the coal mine remained productive.

When quicksilver production began around the turn of the century, the Terlingua country was primarily ranching country and most ranges were state-owned land under lease to the stockmen. While this method afforded cheap grazing, the ranchers acquired only those tracts that gave access to water. Hawley explained their reasoning. "I am told that those who held leases on the tracts where springs and water holes were located did not bother about the rest of it, because whoever had control of the water controlled the range."[51] A close parallel existed between ranching and quicksilver mining, because of all the necessities, water was basic; and with the increasing population converging on the mining district, this problem remained acute until the 1930s.

While substitution partially solved the fuel problem, no simple solution existed for water. After a thorough investigation the mine owners concluded that there were four possible solutions to the problem: (1) to haul water from the Rio Grande; (2) to conserve

Chisos Mining Company, Howard E. Perry, asked his Austin lawyer to recommend a geologist, it was not more cinnabar (there was plenty of that), it was coal that Perry wanted." Heiman cites their first association as about 1909. Udden first made a geological survey of the Chisos property in 1905 or 1906.

[49] Cartledge interview, December 26, 1964.
[50] Cartledge to Perry, January 18, 1938.
[51] Hawley, "Life Along the Border," p. 69.

the limited amounts of rainfall; (3) to utilize the natural water sources—*tinajas* or springs and creeks located near the mines; and (4) when water was encountered in the deeper workings of the mines, to pump it to the surface for domestic usage. All methods were tried and each contributed to the needs of the district.

The Rio Grande was the first water source for the mining district. The geologist Robert T. Hill visited the area in 1902 and reported that most of the water was being hauled from the Rio Grande, a distance of twelve miles.[52] When Phillips joined the Chisos mine in 1905 that camp received its water from Mexican freighters, operating 350-gallon burro-drawn tank wagons from the river. The circuitous road negotiated many arroyos and washes and reached the mining camp after traveling nearly twice the actual distance and climbing a grade of nearly one thousand feet. The Mexicans received one cent per gallon. Phillips reported that water was in such short supply that "the current system of supplying the men with water is to allow one bucket for each single man per day and from two to five buckets for a family."[53]

The Big Bend has an annual rainfall of about twelve inches, most of which falls between mid-June and late September. This paucity of moisture, plus the fact that it occurs mainly during the hot summer months, contributes little to the district's water supply. However, early in the century some operators constructed earthen and rock dams across the canyons and arroyos to impound surface water. The Marfa and Mariposa mine built a dam across a narrow gorge adjacent to the company store, and the late summer rains filled "the reservoir behind the dam . . . [which] constituted the only water supply for the entire camp." Although this source was adequate, the water quality had specific shortcomings. Hawley recalled that during one particularly dry summer the water level dropped to a point that enabled the workmen to "remove silt and settlings which were a foot or more in depth." As the workmen dredged the reservoir they discovered a carcass of a burro "that had been carried down in the freshet two years before." Hawley added, "Of course everyone was shocked, and I think there was a sudden increase in the number of cases of illness. However, no death occurred, and in July the rains came, the reservoir was filled."[54]

Terlingua Creek traverses the heart of the eastern district, about four miles east of the Chisos mine. In 1905 Phillips investi-

[52] Johnson, "Mercury Mining," 36 (September 1946): 392.
[53] Ibid., p. 393.
[54] Hawley, "Life Along the Border," pp. 34–35.

gated this stream as a possible water source and found that while it was not a continuously flowing stream, a temporary dam might provide a more convenient water supply than the Rio Grande. This was tried with moderate success.

Apparently much actual suffering resulted from the scarcity of water in a region afflicted by such intense heat and high evaporation. During periods of scarcity water was sold at ten cents per bucketful. A former employee recalls the touching sight of Ursalo Olguin, the water man, rationing water to the Mexican women and children at the Chisos camp. He would "always assist the girls and women," according to Hawley, "by placing the filled pails carefully on top of their heads. . . . Ursalo was provided with a punch with which to punch out a dime for every pailful supplied to any Mexican family in the excess of the allotted two for any one day."[55]

Because of the scarcity, Perry's stinginess became inordinate. Former employees of the Chisos tell that when one time "the postman asked for a drink when he reached Terlingua, Perry refused him. He had to travel fifteen miles farther before he was able to quench his thirst."[56]

Although the foregoing incident probably was exaggerated, the water shortage remained critical and there appeared little prospect of relief. In 1907 the Terlingua correspondent to the Alpine *Times* reported:

> This section of the country is as dry as the desert of Sahara. The water supply . . . is about exhausted. . . . It is rather a sad sight to witness . . . women and children presenting their checks at the tanks for small quantities of God's free drink. Mr. Jimmie Laurie, the affable and genial storekeeper . . . says that dispensing water by the check [rationing] system is very repulsive to him. If it was beer instead of aqua, he avers, on affidavit, that he would have no kick coming.[57]

Prolonged drought and a steadily increasing population aggravated the water shortage until mine operators began looking beyond the boundaries of the mining district for other water sources. Subsequently, when a series of springs were discovered in the Christmas Mountains about ten miles northeast of the district, the Study Butte miners proposed laying a pipe to that point. The Chisos management discovered other springs about two miles from the

[55] Ibid., p. 35.
[56] Ibid., p. 87.
[57] Alpine *Times*, April 3, 1907.

mine at the foot of Cigar Mountain, and these became Terlingua's first permanent water source.

From these springs a Chisos attendant hauled water to the mine in a tank wagon and made daily deliveries to the homes of the Anglo employees. Hornaday reports that "several wagons, pulled by six to ten burro teams, are constantly employed hauling water to the mine and camp."[58] Later the company-owned houses were piped for water, which was then deposited in large galvanized storage tanks outside each house. The Mexican employees' water supply came from two central tanks near the Chisos Store; these tanks were kept locked and a company employee supervised the distribution. Although the Chisos management used a system of allocation among the Mexicans, former employees recall no restrictions were ever imposed on the Anglo families.[59] Hornaday, however, reported in 1910 that a regular system of water allowance was in operation. Each single man was allowed "one bucket of water per day for his household needs. The allowance for married men runs from two to four buckets per day, depending on how many there are in the family."[60]

The Mexicans carried water to their homes and stored it in barrels. This was the duty of young children—usually girls—and to simplify their task, they balanced a three-foot pole across their shoulders and attached a wire to each end that extended downward to the bail of their water buckets. To avoid waste, the children nailed two sticks together to form an X. These were slightly shorter than the diameter of their water bucket, and after filling the bucket, these crosses were floated on the water to prevent spilling.[61] Many carried an additional bucketful on their heads. Beginning as small girls, they would carry ten-pound lard pails filled with water, and as they became more adept, they increased their capacity with twelve- or fourteen-quart galvanized buckets. Former Chisos employees recall that the more skilled "would sometimes engage in needlework as they made their way home with a bucketful on their heads, never spilling a drop."[62]

Cigar Springs also served the Chisos camp for a number of years until Perry purchased some property a few miles east of the mine on Terlingua Creek and drilled for water. When this well be-

[58] Hornaday, "Cinnabar Deposits," p. 1133.
[59] Interview with Earl Anderau, Alpine, Texas, December 27, 1964.
[60] Hornaday, "Cinnabar Deposits," p. 1133.
[61] Source wishes to remain anonymous.
[62] Hawley, "Life Along the Border," pp. 36–37.

came productive, the company pumped the water about seven miles to a large storage tank located on the summit of a mesa. Cigar Springs was then abandoned and water was hauled to the village from this tank.

In the mid-1920s when the Chisos shafts penetrated below the 700-foot level, much water was encountered; however, these formations yielded a quality of water unsatisfactory for household use; they were therefore sealed off to permit further exploration. Useable water, later discovered near the 800-foot level, was pumped to the surface and stored in a tank on a hill above the village. From there it was piped to the Anglo homes and a new tank was erected for the Mexicans east of the store nearer their homes. This water was used without restrictions.[63]

No one recalls when this water system became operative, but Johnson reported that in 1927 "water was still being pumped about 7 miles to within 1½ miles of Terlingua, from where it was delivered by tank wagon."[64] As long as quicksilver was mined by the Chisos, the same source of water that prevented surfacing the deepest ore deposits also provided a never-ending supply of one of life's basic necessities.

Perry's success at Terlingua was contingent upon solving the economic problems created by scarcity and isolation. That the Chisos Mining Company operated for nearly four decades deep in the inhospitable wasteland of the Big Bend attests to the success of those who served the man and his company.

[63] Anderau interview.
[64] Johnson, "Mercury Mining," 37 (March 1947): 38.

CHAPTER 4

Perry Assumes Control

WITH Phillips' departure from Terlingua in the autumn of 1906 following his wife's accident, the Chisos mine entered a new phase of administrative control. This event marked the beginning of Perry's direct involvement in the mine's operation, and from that date a new management policy bore the indelible imprint of the Perry personality. The most unique aspect of this policy was Perry's prolonged absence from Terlingua. Nevertheless, from his Chicago office he exercised firm control over every detail of company business and community life and was in return apprised of on-site happenings. Exploration, development, recovery, sales, local politics, personnel, and domestic problems were prime subjects of the Chicago-Terlingua correspondence.

The success of this policy depended upon the United States mail, the telegraph, and Perry's careful selection of administrative personnel. Despite his lofty self-appraisal, Perry respected ability in others when company business was involved, and he consistently engaged men of outstanding capacity who served him faithfully with extended tenure. Without apparent reason, most of them remained blindly devoted to him long after the Chisos mine closed. This was one of the ironies of the Terlingua experience.

Phillips headed a long line of highly skilled geologists, mining engineers, and company administrators who contributed immeasurably to the Chisos' success. Together they pursued well-calculated development programs that greatly increased the mine's productivity and richly rewarded its owner.

During the early period of exploration the eminent geologist, Johan August Udden, was one of Perry's greatest assets. The Swedish-born scientist first visited the Big Bend in 1903 as Phillips' associate in the University of Texas mineral survey. Shortly after Phillips joined the Chisos in 1905, Udden returned to the district as Perry's special consultant. Monica Heiman, Udden's biographer, defining the nature of the Perry-Udden relationship, wrote, "J. A. Udden was a professional geological consultant, not H. E. Perry's employee, and

Perry always treated him with great respect, as did other mine owners and operators."[1]

Although Perry held Udden in high esteem, the mine owner's ultraconservative persuasion frequently ran counter to Udden's insistence on expanded exploration. While Perry never admitted his folly, "the richest ore in the mine would never have been reached if Dr. Udden had not been able to persuade Perry . . . to spend considerable money in penetrating barren ground in order to reach it." Udden theorized correctly that geologic faulting occurred "later than the ore deposition; hence ore bodies, supposedly exhausted," were discovered well beyond fault displacements.[2]

Charles Laurence Baker, Udden's colleague at the University of Texas Bureau of Economic Geology, explained further Udden's contribution. "When Perry got into trouble with displacements and pinchouts of ore," Baker wrote, "Dr. Udden was able to find him more ore. Eventually Udden persuaded him to sink deeper through the nearly barren Del Rio Clay and in the underlying limestone he found the greatest ore body in the entire district."[3]

Adding a second dimension to Udden's contribution to the Chisos, Heiman states that he "indicated how to avoid drilling into fault lines in the substructure," which contained the water-bearing formations.[4] Columnist W. D. Hornaday observed Udden's apparent success when he visited the Terlingua district in 1910. He noted that "the scarcity of water in the Chisos mine is all the more astonishing to geologists, when it is considered that [the Study Butte mines] 6 to 10 miles away . . . had to be abandoned a few years ago" because their workings were flooded.[5] Following Udden's death in 1932, however, the Chisos drillers pierced a fault below the 800-foot level, which inundated those workings. Jack Dawson, Perry's adversary in the Section 295 land litigation, paid Udden the supreme compliment while predicting Perry's ultimate defeat at Terlingua. Dawson maintained that "the only man who could read

[1] Monica Heiman, A Pioneer Geologist: Biography of Johan August Udden, p. 38.

[2] E. H. Sellards, "Economic Geology of Texas," in The Geology of Texas, vol. 2, Structural and Economic Geology, by E. H. Sellards and C. L. Baker, University of Texas Bulletin No. 3401 (January 1, 1934), p. 552.

[3] Charles Laurence Baker, Cardova, Illinois, to Kenneth B. Ragsdale, November 16, 1964, in possession of author.

[4] Heiman, Pioneer Geologist, p. 38.

[5] W. D. Hornaday, "The Cinnabar Deposits of Terlingua, Texas," The Mining World (December 17, 1910), p. 1133.

the Terlingua area was Udden. When he died, the area was finished."[6]

Although Udden was unique among the Chisos staff, his colleagues at Terlingua were also well-schooled, experienced men of high professional caliber. That Perry succeeded in maintaining a competent staff so remote from the labor market is a tribute to his persistent recruiting and careful selection. He personally chose each staff employee.

In staffing the Chisos mine, Perry used three techniques: to fill administrative and lesser staff positions, he advertised widely in newspapers and frequently relied on personal contacts and the recommendations of friends and colleagues; to fill professional staff positions, he solicited colleges, universities, and technical schools that offered degrees in mining and metallurgy. All three sources yielded results.

Graduates of the Texas and Colorado schools of mines appeared on the Chisos payroll. Phillips and Udden, who came to the Chisos via the Texas Mineral Survey, were graduates of the University of North Carolina and Augustana College (Minnesota), respectively. Marcus Hulings, who joined the Chisos staff as an engineer and advanced to mine superintendent, was a 1903 mining and engineering graduate of Rensselaer Polytechnic Institute. He came to Terlingua from an assignment in the Mexican silver mines. Perry's inquiry to the Case Institute of Technology yielded another superintendent, F. Harry Fovargue, a 1906 graduate who came to the Chisos from the West Virginia coal mines.[7] Fovargue remained at the Chisos until he resigned in 1918. Perry, however, persuaded him to return to his former position in 1939, which he held until the mine closed in 1942.[8]

Anglos held the more responsible positions at the Chisos: mine superintendent, general manager, furnace superintendent, assayer, store manager, purchasing agent, store clerk, doctor, master mechanic, and blacksmith. James M. Day's research reveals that "white persons were employed for furnace work for a time, but the turnover was so high that Mexicans were placed on that job."[9] To fill many less skilled positions, the company advertised in El Paso, San

[6] Heiman, *Pioneer Geologist*, p. 38.

[7] Ruth H. Fovargue, Painesville, Ohio, to Kenneth B. Ragsdale, August 13, 1966, in possession of author.

[8] San Angelo *Standard-Times*, June 8, 1958.

[9] James M. Day, "The Chisos Quicksilver Bonanza in the Big Bend of Texas," *The Southwestern Historical Quarterly* 64 (April 1961): 439.

Antonio, Houston, and Cleveland, Ohio, newspapers. Leo Moore of Bay City, Texas, seeking a clerical position in 1910, was offered a furnace position instead. His rejection reads as follows: "Your letter of 8th instant was forwarded me from Houston. In answer I will say that your offer is very liberal, but as I have had no experience in furnace work will say that I think best that had better not except the position as I am a bookkeeper and Stenographer by trade but, as I stated in my application I have had long experience in store work."[10]

After 1910 responses to the company's newspaper advertisements were mostly from family men over forty seeking positions that offered advancement. A representative reply came from E. W. Gold of San Antonio, who wrote: "I am 40 years of age, married, three children. I have had a good education, and practical business experience covering many years. . . . Regarding salary. . . . It should not be less than $75. to begin with. I am now employed but my work gives me no chance for enriching my abilities. I prefer therefore to start at a reasonable salary with some concern where I can have a future."[11]

Whether Gold found employment is not known, but the Chisos management succeeded in attracting men of his apparent caliber who were stable, well qualified, and when once engaged, usually remained in the company's service for a number of years.

Although their productivity was sometimes restricted through administrative blunders, Perry gave his employees wide latitude to exercise their skills. Explaining his philosophy of a man's capacity, Perry once remarked, "If we dissect men we destroy them. We have to strike a balance and see if on the whole they are not good."[12]

The employees who served the Chisos longest and were most responsible for the company's longevity—possibly excepting Udden —occupied nontechnical administrative positions. The general manager held the key administrative post at the Chisos camp, as it was his responsibility to implement the mine owner's directives. Perry's long absence from Terlingua, plus his personal idiosyncrasies and fluctuating policies—if they were policies—made the general manager's position increasingly difficult. Much credit for Perry's success at Terlingua is accorded two brothers, Wayne and Robert Cart-

[10] Leo E. Moore to Chisos Mining Company, December 13, 1910, Chisos Mining Company Papers, Archives of the Texas State Library, Austin.
[11] E. W. Gold to Chisos Mining Company, November 20, 1910, Chisos Mining Papers.
[12] Perry to Cartledge, August 4, 1936.

ledge, who successively occupied that post for almost three decades. The son of Eugene Cartledge, Perry's Austin attorney, Wayne joined the Chisos mine in 1909 as store clerk, subsequently advanced to store manager and purchasing agent, and became general manager in 1913. When he resigned to become Perry's partner in a general store at Castolon, Perry appointed Robert, his younger brother, as general manager.

The relationship that developed between Perry and Robert Cartledge remains an unexplained phenomenon in the history of the Chisos Mining Company. Perry exercised a strange hold over his employees, who accorded him unquestioned loyalty and dedication in return. None was more dedicated than Robert Cartledge; for over two decades he formed a human bulwark between Perry and a gathering host of agitators. This strange devotion Cartledge held for Perry is one of the most perplexing facets of the later history of the company. Their correspondence, exchanged daily—"please write me each mail day—so I will know about things"[13]—was voluminous, factual, and filled with graphic insight into every aspect of life and business in the Terlingua district. Cartledge, as Perry's most vigilant observer in the Big Bend, apprised the mine owner of everything from production problems to local gossip, including a case of parental neglect in the Chisos village.

> I am satisfied that we are going to have to get another cook. . . . I talked to a few of the boarders . . . and asked them not to discourage him . . . but they have all ready started kicking.[14]

> Weller and Brown of Rainbow arrived here this morning . . . as soon as I learn there business I will write you.[15]

> Fred owes it to his children to give them the proper invirments. . . . If we try to and show him that he is not doing the best by his children, he seems to think you are butting into his business. . . . Several times I have felt like going over to Fred's and giving him a good whipping, tho I may have had to kill him.[16]

Perry's relationship with his subordinates was neither personal nor cordial; he remained always aloof, distant, and formal. While many of his Big Bend contemporaries referred to him as "Old Man Perry," "that little bastard," or "Yankee thief," the Chisos employees

---

13 Ibid., June 19, 1939.
14 Cartledge to Perry, August 16, 1921, Chisos Mining Papers.
15 Cartledge to Perry, September 6, 1932, Chisos Mining Papers.
16 Cartledge to Perry, undated, Chisos Mining Papers.

at every level respectfully referred to him only as "Mr. Perry." This was true long after the mining company was dissolved, and even after Perry had callously dismissed his oldest and most faithful employees. Robert Cartledge served as Chisos general manager for thirteen years and repeatedly risked his life for Perry in that capacity as well as justice of the peace. He once wrote Perry, "My life has been threatened now on account of the work I have been doing. Of course, if I get killed that will be my misfortune."[17] His loyalty knew no bounds. He injured his health, alienated his family, and even spent his personal savings to satisfy creditors when the mine was in financial difficulty. Yet in 1940 Perry fired him without notice. Nevertheless, Cartledge's loyalty remained steadfast. Until he died thirty-two years later, he categorically defended Perry whenever his integrity was challenged and was never heard to refer to his former employer as anything but "Mr. Perry."

Ed Babb had served the Chisos almost twenty years when Perry fired him for not accepting the position vacated as a result of Cartledge's dismissal. Although Babb later described Perry as being "damned overbearing, dominated everything and everybody he came in contact with,"[18] he also retained a portion of his allegiance to his former employer. When Babb failed to respond to this researcher's questions about Perry's policies at Terlingua, a mutual friend explained: "Perry brainwashed those people at the mine. I knew most of them and I still can't explain it. He treated them like dirt and they loved him. And after what he did to Bob Cartledge, Bob never said a disrespectful thing about Perry. Don't blame Babb for not answering your questions. Perry told him to keep his mouth shut thirty years ago and he hasn't talked yet."[19]

Almost four decades separate Perry's assuming administrative control of the Chisos mine and Cartledge and Babb's dismissal in 1940. These events symbolize the beginning and the end of the Perry era at Terlingua. During the intervening years he slowly and persistently gained control of most of the quicksilver properties, and in turn, became the dominant factor in the political, social, and economic life in the Terlingua Quicksilver District. The Chisos mine held the key to Perry's success, and Phillips' departure from Terlingua in 1906 placed in Perry's hands the power that he began to wield with facile expertise.

[17] Cartledge to Perry, September 7, 1936.
[18] Interview with Mrs. G. E. Babb, Sanderson, Texas, December 27, 1964.
[19] Interview with Dr. Ross A. Maxwell, Austin, Texas, January 8, 1965.

Perry first perfected the technique of management in absentia; the careful selection of administrative personnel assured its success. As Phillips' replacement, Perry chose Jim Lafarelle, an old Big Bend prospector and a person totally lacking the professor's knowledge, experience, and insight into the problems of operating the quicksilver complex. Lafarelle's apparent deficiencies fit the Perry plan, as he could now articulate company policy from Chicago and the witty Irishman Lafaralle would translate his orders into action at the mine.

With the knowledge of mining gained under Phillips' tutorage and the vast ore reserves blocked out in Section 295, Perry looked to the future with the optimism of a true mining man. With characteristic verbosity, he explained his plan:

> I believe that there is no method for my handling our correspondence which is at one and the same time so easy for me and so useful and direct all along the line, as to send letters intended for Mr. Ferguson [purchasing agent and store manager] and Mr. Lafarelle addressed to the Company same as this, and of course you understand that all communications which I address to The Chisos Mining Co. are directly intended for the above named gentlemen, and letters addressed to the Chisos Mining Co. with name of some individual following, such as Mr. Johnson [furnace operator], Mr. Hamilton [position unknown], Mr. Quayle [assistant superintendent], Mr. Knight [position unknown], etc. are intended first to be read by [the party named] . . . and then sent to the further parties named.[20]

Perry thought this multipurpose correspondence would result in a more comprehensive understanding of his directives and would in turn provide him with a complete report on the company's operation. However, the apparent strength of the system was its basic weakness; with various department heads reporting directly to Perry lines of authority became vague, and misunderstandings resulted. Although this method remained in effect as long as the Chisos mine produced quicksilver, it was never satisfactory, and during the latter periods of declining profits its inherent flaccidity hastened the final collapse.

Robert Cartledge believed this complex organizational structure facilitated the Perry method of handling men. As Cartledge explained, "Perry seemed to try to develop friction between men at the administrative level." By having the department heads report

[20] Perry to Chisos Mining Company, May 25, 1908, Chisos Mining Papers.

directly to him, abrasive competition would develop among the men on his staff. Apparently Perry believed that by creating tension, he would cause his employees to be more prone to confide information to him about their fellow employees' deficiencies and shortcomings. Further, Cartledge felt Perry believed that by creating fear and insecurity in his employees, they would strive harder to gain Perry's acceptance.[21] Unknowingly Cartledge himself was a victim of the plan's efficiency.

Earl Anderau, a former Chisos employee, recalls how Perry established personal loyalties at the expense of company morale. While Perry rewarded some employees with periodic pay raises, the company conformed to no salary schedule, and these individual advances were administered with characteristic Perry formality and secrecy. Anderau states that Perry told his father, Chisos mechanic J. E. Anderau, that he was going to raise his salary "but definitely didn't want it to show up on the company payroll, as he did not want Bob Cartledge to know about the raise." Anderau also recalled that despite all the secrecy, Perry prepared a written agreement and the additional income arrived in three-months' installments from the Portland, Maine, office.[22]

As Perry assumed greater administrative control, the position of mine superintendent became marginal. Apprised of every facet of the operation, he followed each detail of the mining operation with great interest and insight. He wrote: "I think that Mr. Knight must have finished the boarding up or shade on the west side of No. 1 condenser at Tigners by Friday or Saturday of last week, and we should be receiving the benefit of this cooling this present week. I shall await with much interest to see whether any improvement becomes noticeable."[23]

Supervision-in-detail became the distinguishing feature of the Chicago-Terlingua correspondence, and Perry's directives reflected the change in company policy. Each new business transaction became a confrontation of two forces—only one was immovable. The result bore the Perry image. Vicious competitiveness lay at the crux of all negotiations with suppliers, and when a new cable for the No. 9 shaft was ordered, he explained the delay as follows: "I have been taking time to rub the cable people together and still at it. I

[21] Cartledge interview, October 27, 1965.
[22] Interview with Earl Anderau, Alpine, Texas, December 27, 1964.
[23] Perry to Chisos Mining Company, May 25, 1908, Chisos Mining Papers.

have already gotten Roebling down 10%, but wrote them the other day that I didn't know even yet whether I could send them the order, so I am letting them sweat on it."[24]

By June, 1907, with construction of the Scott furnace well underway, brick procurement became a major issue. Word reached Perry that the Texas Almaden Mining Company at Study Butte had contracted with local brick maker Harry Dryden for 300,000 bricks and had an unused surplus. Hoping to avoid paying Dryden's price, Perry negotiated with Texas Almaden for 35,000 surplus bricks. However, after consummating the deal, he learned that the Marfa and Mariposa mine planned to suspend operations; therefore, on June 8, 1907, he wrote the Chisos staff, "[this] puts the brick proposition in a more favorable light, as they [Marfa and Mariposa] will have a large amount of brick for sale, and at [a] lower price than what we paid Dallas, so that I presume it will be just as well not to take very many more."[25]

The lower-priced brick apparently failed to supply the Chisos' needs, since in November the company took delivery of the remaining furnace materials from the Evans and Howard Company of St. Louis, Missouri. To reimburse the Chisos management for some damaged merchandise in an earlier shipment, the St. Louis firm included extra items without cost. Perry immediately wrote H. A. Ferguson, Chisos purchasing agent, to disregard such conciliatory acts; "Since we will deduct in cash . . . leave out $50 or less . . . you can just simply send them check on account without comment."[26] Perry's methods obviously never endeared him to people.

Once in control of the operation, Perry experimented with several innovations. The "investigative" Perry, as Robert Cartledge described him, conceived an idea to spray the interior walls of the condensers in the Tigner furnace with a fine mist of water, which he believed would force quicker condensation and result in greater metal recovery. Although Perry later conceded failure, he upheld the veracity of his idea, allocating failure to Providence. "I cannot feel otherwise than that our use of the spray in condensers is a failure . . . so far as visible results are concerned. . . . Some great controlling factor must have got in its deadly work last week to cause so tremendous a falling off."[27]

Perry's resourcefulness assumed many forms. In 1908 he wrote

24 Perry to Cartledge, December 28, 1931, Chisos Mining Papers.
25 Perry to Chisos Mining Company, June 8, 1907, Chisos Mining Papers.
26 Perry to H. A. Ferguson, November 28, 1907, Chisos Mining Papers.
27 Perry to Chisos Mining Company, May 25, 1908, Chisos Mining Papers.

the mine management, "I have got on to a small device for making ice in a cheap and quick way without the use of ice machines. . . . This would certainly be a great thing for us."[28] When the columnist Hornaday visited the mining camp, he reported the success of this Perry experiment:

> Mr. Perry has shown a very liberal spirit in making improvements to the property. He is now installing an electric light and power plant which will lessen considerably the cost of operations. He is also putting up an ice factory, the chief purpose of which is to provide ice for the condensers in order that a greater precipitation of quicksilver may be obtained. By the adoption of improved methods . . . the former loss of 30% of the quicksilver in the fumes and ore had been reduced to 10%. A still further saving of the metal will be made by the use of ice in cooling the condensers to a greater degree.[29]

With the completion of the Scott furnace, Perry foresaw even greater production increases by adding another smokestack. After the Chisos staff effected the modification, Robert Cartledge recalls that this experiment resulted in only moderate success.

To be apprised promptly of the mine's operation, Perry required semiweekly telegraphed production reports, in addition to a Wednesday morning report containing the number of flasks of quicksilver recovered from each condenser. "This will give me early information," he wrote, "and a general line on the way the run will be for the week."[30] He also requested that another telegram be dispatched from the mine each Sunday night, giving the total week's production.[31]

Since the target of most of Perry's experimenting was the furnace, he ordered a report showing in detail the weekly furnace runs and the amounts of metal recovered from each condenser and retort. In his instructions he specified, "I want the wood burned to show each week, for amount of wood burned has not a little to do with furnace results." Perry described in infinite detail how the form was to be prepared. His twin objectives: reduced costs and increased profits. He wrote the management to "just have the printer

[28] Ibid.
[29] Hornaday, "Cinnabar Deposits," p. 1134.
[30] Perry to Chisos Mining Company, May 25, 1908, Chisos Mining Papers.
[31] Although the Chisos camp did not have telephone communications in 1908, it is assumed this traffic was carried by the Marfa and Mariposa telephone line to Marfa. A statement from the Marfa and Mariposa Mining Company to the Chisos Mining Company dated August 1, 1907, contains a $10.25 item for "The July Telephone Account" (Chisos Mining Papers).

set up the whole thing in type and type ruling instead of ruling with machine afterward, for it is much cheaper. . . . This allows them to set up and print the thing all at one time, without the after job of ruling."[32] His suggestions were apparently well conceived as this form, entitled the Monthly Furnace Record, was still in use in 1942 (see appendix D).

In developing this system of reporting, a method of coding the production messages evolved that bore Perry's indelible mark. Although secrecy became an integral part of his method, this cannot be attributed to him alone, because mining and secrecy have been synonymous throughout the history of the industry. One professional journal, noting this policy, reported: "The production of the year is problematical, as secrecy, which at time past raised a Chinese wall about the mines of many districts, is still in vogue."[33]

Whatever the source, personal or professional, the practice of secrecy invaded every phase of the Chisos operation. Explaining his transmission of information, Perry wrote Lafarelle, "Regarding the matter of average assays of ore at Tigner's. . . . We would better take a code word for 'Average of assays of ore from Monday morning to Friday morning of current week.' Please take the word 'Oxfly.' "[34] Key words containing vital information on the mine operation were entered in code books. One was kept at the mine and the other in Chicago or Portland. Fearing the code would be deciphered, Perry revised the code words periodically. As the system proved workable and satisfactory, its use continued as long as the mine was operative.

To ensure complete enforcement of his "gospel of secrecy," Perry not only forbade his employees to associate with employees of other mines, but whenever strangers visited the Chisos camp they were kept under constant surveillance. An associate of Perry's who once witnessed this technique in operation recalled, "The suspicion of every employee was aroused when a stranger appeared at the mine. An early attempt was always made to determine his business, but when this was established and his mission was a friendly one, he was usually left alone."[35] As Harry Fovargue, former mine superintendent, explained, "Perry was . . . a tight mouth when it came to the business of the mine. . . . He forbade his employees to

[32] Perry to Chisos Mining Company, May 25, 1908, Chisos Mining Papers.
[33] W. D. Burcham, "Quicksilver in Terlingua, Texas," *Engineering and Mining Journal* 103 (January 13, 1917): 97.
[34] Perry to Chisos Mining Company, May 25, 1908, Chisos Mining Papers.
[35] Interview with Harris Smith, Austin, Texas, July 7, 1965.

discuss any phase of the business. . . . Mr. Perry laid down the law and we hired hands had to abide by it."[36]

Fearing that other mine operators trading at the Chisos Store might gather information about some aspect of his operation, Perry issued strict orders at one time not to honor their patronage.[37] Robert Cartledge, while serving as Chisos Store manager, apparently failing to comprehend this edict, wrote Perry for clarification. "[I] did not know that you did not want to let him [E. A. Waldron] have any thing at all, that is some small amount of merchandise for hisself . . . but if you had rather not let him have any thing, we will not do it."[38] Financial sacrifice was the price Perry chose to pay for the concealing security of silence.

While Perry was successful in erecting the barrier of silence around the mining village, other uncontrollable factors gave him cause for grave concern. The fact that preyed most heavily on his mind was that quicksilver production declined after Phillips' departure. The 3,686 flasks reported by the district in 1907 represented a deficit of 1,075 flasks from the previous year.[39] While no one person could be held accountable for the drop, undoubtedly this fact impinged on Perry's confidence.

If this report caused a ripple on the surface of Perry's serenity, a tidal wave was destined to engulf his plans for immediate recovery. As autumn approached the stock market slumped and a financial panic spread throughout the country, sending business indicators to their lowest levels since the panic of 1893. Editorials in the financial dailies pointed an accusing finger at Theodore Roosevelt and his progressive legislation; economic historians, however, attribute the panic to speculation, inefficient business management, and banking rigidity so prevalent during the early years of the century. Whatever the cause, the nation's business came to a virtual halt, and with it, the demand for quicksilver. When the following year brought no relief, production dropped to the lowest level since 1901. The combined 1908 recovery of the Chisos and the Marfa and Mariposa mines, only 2,383 flasks, reflected the depressed economy.[40]

The spirit of optimism so long prevalent in the Perry corre-

[36] San Angelo *Standard-Times*, June 8, 1958.

[37] Smith interview, July 7, 1965.

[38] Cartledge to Perry, December 23, 1921, Chisos Mining Papers.

[39] J. Harlan Johnson, "A History of Mercury Mining in the Terlingua District of Texas," *The Mines Magazine* 36 (September 1946): 394.

[40] Ibid.

spondence disappeared completely and was replaced by foreboding of what was occurring in the quicksilver district. The following letter is a case in point:

> Your wire, indicating 21 flasks, came late last night and was indeed quite a shock. Confidentially between us three (Mr. Lafarell, Mr. Ferguson and myself) we are rapidly nearing a point where something has got to be done, or else submit to the inevitable. . . . We have certainly got to pull every string we have, to keep things going. I am aware that it is a nuisance to run that old retort, and the thought of the thing being run . . . almost makes me sick; but I know something else that thought of which makes me feel much sicker.[41]

Also, the 20-ton Scott furnace had been under construction over a year. This added expenditure, combined with declining recovery during a period of economic depression, was sufficient reason to fear "the inevitable." However, Perry's apprehension proved temporary and on June 23, 1908, he wrote the mine manager, "Your wire of yesterday received, and same brought good news indeed. It is fine. If we can hold up this high weekly average for the balance of the month we will have made 300 flasks. . . . Next shipment to St. Louis July 4th 70 flasks.[42]

By March, 1909, improved business conditions renewed Perry's former optimism, and the initial recovery from the new Scott furnace a few months later bore first evidence of his success as the actual leader of the Chisos. Only the Chisos and the Marfa and Mariposa mines remained operative throughout the year, and their combined production of 4,188 flasks of quicksilver nearly doubled that of the previous year.[43]

Perry was elated to learn in June, 1907, that the Marfa and Mariposa mine was operating at a loss and would eventually close.[44] Not only would this enhance his position in the local labor market, but also the Chisos camp was the logical site for relocating the Terlingua post office, which would bring increased trade to the Chisos Store.

Suggestions for renaming the post office were exchanged between Perry and the mine management. Far more interested, however, in acquiring the post office than in selecting a new name, Perry wrote Ferguson on December 8, 1908, "I think that it ought to be

41 Perry to Chisos Mining Company, May 25, 1908, Chisos Mining Papers.
42 Perry to Chisos Mining Company, June 23, 1908, Chisos Mining Papers.
43 Ibid.
44 Perry to Chisos Mining Company, June 8, 1907, Chisos Mining Papers.

easy for us to change the name of the Post Office somewhat later on. . . . The first thing to do is to get the Post Office."[45] In May, 1910, the Marfa and Mariposa mine finally suspended operations, and the post office was moved to the Chisos camp the following year. The mine management retained the name and henceforth Terlingua was used synonymously with the Chisos in the mining district.[46] C. A. Hawley, a former employee of the Marfa and Mariposa mine, joined the Chisos operation and became the first postmaster at the new location.

Shortly after he took office, the Postal Department notified Hawley that the Terlingua mail would be distributed from Alpine instead of from Marfa. Since this meant establishing a new mail contract, it was necessary to determine the exact distance from Alpine to Terlingua. Nearly three years would elapse before the first automobile made its appearance in Terlingua, and since Hawley had no ready-made device for measuring distances, he relied on his own ingenuity. "I procured a gadget," Hawley recalled, "that could be attached to the axle of a vehicle . . . and was so constructed that every revolution of the wheel was recorded."[47] With less than half of the distance to Alpine measured, the gadget broke and the remainder of the distance was measured by tying a handkerchief around the wheel of the company hack and counting each revolution. When Hawley arrived in Alpine, the number of revolutions were multiplied by the diameter of the wheel, and the resultant figure was just short of one hundred miles, approximately the present highway mileage.

When the Marfa and Mariposa mine ceased production in 1910, the Chisos became the oldest and most important mine in the district. As the only such permanent installation in the region, the thriving Chisos Store exercised a virtual monopoly on retail trade in the central Big Bend area. Although Perry enjoyed this position of eminence, it is doubtful whether he was aware at this early date that around him lay the dormant seeds of complete social and economic domination. But as his mine grew and his store prospered he discovered their potential and tended and nurtured their growth

[45] Perry to H. A. Ferguson, December 8, 1908, Chisos Mining Papers.

[46] Since the mines closed in 1945, the Terlingua Post Office has been moved several times, being located anywhere that someone could be found to act as postmaster. However, the name Terlingua still applies to the deserted village at the abandoned Chisos mine (Cartledge interview, January 22, 1966).

[47] C. A. Hawley, "Life Along the Border," *Sul Ross State College Bulletin* 44 (September 1964): 59.

until eventually he became the greatest controlling force in the region.

Another decade would intervene, however, before he realized the fruition of his strength, for in the immediate future there were counter forces at work that only the passage of time and an interruption of world peace could resolve. Although the Chisos No. 8 shaft penetrated the 550-foot level and exposed good ore bodies, production for 1911 declined to 2,326 flasks of quicksilver, indicative of the district's general decline. One year later, it dropped to its lowest point in the present century, only 1,990 flasks. Although the district experienced a general slump, these figures largely represent Chisos production, still sufficient to maintain a profitable operation. As Johnson reports, "Though there was considerable prospecting . . . and a few claims did change hands, practically all the other mines in the district showed little activity for a number of years."[48] Gloom pervaded the mining district, plans to build a railroad to Terlingua were abandoned, and the locally depressed economy produced the region's first and only violent labor troubles.

An independent mine owner working a claim on the Terlingua-Lajitas road approximately five miles west of Terlingua became the target for the first salary demands. Approached by a group of dissident Mexican employees demanding more money, he explained his profits were not sufficient to raise their wages, whereupon they attacked him. In the ensuing struggle he killed three of his assailants. Madison reports, "This was the first organized attempt to get a wage increase at the quicksilver mines and it brought disaster to the strikers."[49] Despite this initial failure the idea of higher wages was implanted in the minds of the Mexican employees, and this movement was certain to manifest itself again at some later date.

All remained quiet for nearly two years; then on January 21, 1912, the Chisos employees directed a letter to Perry, not seeking higher wages, but requesting one day off each week.

> Mr. Perry, appreciated Señor:
> It is for some time that all of the workmen have wanted you to allow us a favor. The favor is that we do not work on Sundays, that you tell us we are at liberty to rest, for working every day without rest wearies oneself. Then for to make up this work of Sunday, hire one seventh more men than there now are and it will be the same amount of work and the people go on Monday with more willingness to work. For they have rested. We wish distraction one day of each

48 Johnson, "Mercury Mining," 36 (October 1946): 447, 445.
49 Virginia Madison, *The Big Bend Country of Texas*, p. 191.

seven without an exception, Sunday, and another day would not be good. Study this problem and Saturday advise us what you decide.

The Workmen[50]

Whether this request stemmed from a desire for improved working conditions or from devout religious beliefs was never known, but the humble tone of the request was noticeably absent in Perry's reply. He directed a company employee to inform the workmen their request was rejected as it necessitated closing the furnace, and several days would be required to regenerate sufficient heat for metal extraction.

Nothing more was heard of the employees' grievances until later in the year when several dissident laborers banded together to prevent other employees from working. Robert Cartledge recalls that Perry was at the mine when the incident occurred and was alarmed at the prospect of being forced to treat with militant Mexican labor. To prevent possible destruction of property, Perry called Sheriff Allen Walton from Alpine and increased the night security staff. When Cartledge learned that most of the labor force was not in sympathy with the agitators, he and the sheriff entered the mine, identified the leaders, and with the threat of arrest, forced them to leave the camp.[51]

Although no more demonstrations occurred, the problem of employee-company relations remained foremost in Perry's calculations, not in the contemporary concept of improving the worker's social and economic position through better labor-management understanding, but simply in how the company could derive maximum productivity at the lowest possible cost. With a more-than-adequate labor supply moving unrestrained across the Rio Grande, Perry kept production costs at a minimum. Madison's research revealed that "at the wages Mr. Perry paid . . . an employe received $547.50, provided he worked 365 days, and for that wage he produced $4,381.31 worth of quicksilver."[52]

Fear of Perry and the ever-present possibility of being deprived of both employment and their living quarters kept the Mexican

[50] Ibid.

[51] Robert Cartledge rode horseback sixty miles to get Watty Burnham and Sam Nail to night watch during the labor disturbance. Andres Rocha, a loyal employee, told Cartledge that the leader of the agitation was Leonires Valenzuela, no relation of Feliz Valenzuela, a long-time employee of the Chisos mine, later killed aiding Cartledge in the apprehension of bootleggers (Cartledge interview, October 17, 1965).

[52] Madison, Big Bend Country, p. 191.

workers in restraint. Their apprehension was well founded. When the Chisos mine first opened, the men worked by the light of paraffin candles impaled on steel spikes driven into the shaft walls. The company furnished the candles, but when the Chisos changed to carbide lamps, Perry required the employees to purchase their own equipment. When the Mexicans resisted, Perry instructed the management to post an ultimatum on the doors of the mine office and the company store which stated that "any man who did not have his lamp when he came to the job next morning would be discharged and never be allowed to work there again and furthermore they would be driven out of their houses without delay and any man who wouldn't get out . . . would have it torn down about his ears." Perry recalled that though much excitement ensued, every man was there with his lamp, especially the Mexicans, who, he said, "are the greatest little bluffers in the world but rarely will stand pat."[53]

Perry's callous disregard for Mexican employees is best expressed by a former Chisos employee, C. A. Hawley, who professed admiration for both the Mexicans and his employer. According to Hawley, Perry "possessed very little of the milk of human kindness. He never learned a word of Spanish, and never manifested any interest in the Mexican people as fellow human creatures. With him it was simply a business relationship, sympathy and sentiment did not enter in."[54]

Any disregard for his fellow man that was manifest in Perry's relations with the Mexicans was not necessarily ameliorated when he dealt with his fellow Anglo. Perry's New York attorney, James B. Burke, stated succinctly, "I don't think he inspired affection, to say the least!"[55]

To inspire affection was not Perry's goal at Terlingua. Profit was his sole aim, and to achieve this, exploitation of human effort fitted logically into his concept of industrial enterprise. By offering the only permanent employment in a region of overabundant labor, Perry found he could dictate wages, hours of employment, and general working conditions, maneuvering each for the benefit of the Chisos Mining Company. Unionization was never manifest in Terlingua and by the middle of the second decade Perry had

[53] Perry to Cartledge, May 1, 1935.

[54] Hawley, "Life Along the Border," p. 59.

[55] James B. Burke, New York City, to Kenneth B. Ragsdale, September 28, 1965, in possession of author.

achieved almost total domination of the labor force, which he would keep firmly in his grasp until the advent of the National Recovery Administration (NRA) in 1934; then he would be forced partially to relinquish his control of one of the fundamental economic factors in the Terlingua Quicksilver District.

CHAPTER **5**

Facilities of
Civilization

AS optimistic production reports bore news of expanding Chisos recovery, they also contained information that living standards in Terlingua were improving in proportion to the ascending fortunes of the mine. When H. D. McCaskey, a member of the United States Geological Survey, visited Terlingua in 1912, he reported that "the Chisos Mine . . . has been supporting an important settlement of American operators and Mexican laborers for several years, . . . maintained law and order and has given the region all the facilities of civilization possible."[1]

By 1913 Terlingua citizens had access to a well-stocked commissary, an ice-making plant, public food and lodging facilities at the Chisos Hotel, several excellent dwellings, erratic telephone service, a dependable water supply, and United States Mail delivery three times a week. These predominantly urban conveniences, most of recent vintage, represented a noticeable improvement in Big Bend living standards. Drury M. Phillips, William B. Phillips' eldest son, who visited Terlingua first in 1904 and returned two years later as a Chisos employee, credited this transformation to the economic impact of the Chisos mine. Phillips recalled that in 1904 "Terlingua was a rough and rugged, almost no streets, very little water, not too good food. . . . In 1906, Chisos Mines had made a big difference. Instead of thatched huts and ragged tents, you saw adobe houses, some with flower plants in the front and clothes lines in back—even a few rocklined walks. There was a real nice "Company Store" . . . prices were well marked in dollars and pesos—you had no parking troubles."[2]

The urban appearance of the mining village was greatly enhanced in 1906 with the erection of the Perry mansion. When Perry

1 J. Harlan Johnson, "A History of Mercury Mining in the Terlingua District of Texas," *The Mines Magazine* 36 (October 1946): 445.
2 Drury M. Phillips, Huntsville, Texas, to Kenneth B. Ragsdale, January 17, 1966, in possession of author.

visited Spain to inspect the Almaden mine, the Moorish influence on the Spanish architecture so intrigued him that when he planned his Terlingua home, he attempted to incorporate these features into its design.[3] Instead of maintaining a pure Iberian theme, the structure developed into an incongruous mixing of rectangular simplicity and the curved delicacy of Moorish arches. The result was neither Moorish nor Spanish.

The primary significance of the Perry mansion is not its architectual origin but its revealing implications as an artifact of culture. Perry built the mansion to satisfy a desire to live lavishly. In erecting this structure he responded to the mores of an era that spawned him and his kind. He was one of nature's noblemen, a parvenu, the self-made man who, like his contemporaries, swarmed across the nation whenever abundance awaited the taking. The harvest was rich and the rewards affluent, and these fledgling entrepreneurs reaped sudden, unlimited, and unlabored wealth. To fulfill their commitment to a tolerant and approving society, they lived with spectacular abandon and became the darlings of the phrasemakers—"agents of progress," "captains of industry," and a little later, "business pirates" and "robber barons." And in their unrestrained pursuit of riches they granted new meaning to "popular consent," "conspicuous consumption," "watered stock," "gentlemen's agreement," and the "blind pool."

Their symbols of wealth followed a pattern of unimaginative uniformity: the yacht, the racing stable, the private railroad car, the "art" collection, and the institutional endowments. But the most consistent and pretentious evidence of the nouveau riche was the mansion. Whether it stood on Prairie Avenue, Nob Hill, Fifth Avenue, or a desolate Big Bend hillside, the symbolism was the same. Social historian Matthew Josephson succinctly explained this phenomenon: "Mr. W. H. Vanderbilt's palace and the adjoining one of his daughter on Fifth Avenue, extending the full block from Fifty-first to Fifty-second Street, like the mansions of the Astors was the visible trophy, the monument of a triumphant dynasty."[4]

[3] Interviews with F. L. Dahlstrom, Austin, Texas, November 16, 1964, and October 17, 1965. The original structure had only one story and no indoor sanitary facilities. However, after Harry Fovargue joined the Chisos mine in 1911, he added a second story and probably added a bathroom at that time. Dahlstrom described the two-story, nine-bedroom building as being about 120-feet long with a 90-foot front porch supported with nine 10-foot arches. When his staff occupied the house in 1943, the cellar was still stocked with Perry's fine imported liquors and "exquisite linens."

[4] Matthew Josephson, *The Robber Barons*, p. 330.

By comparison, Perry was neither a Vanderbilt nor an Astor, yet like them he performed his assigned role in the social and economic drama that was staged in the late nineteenth century. Perry learned his lines from the masters and continued the performance well into the twentieth century. Terlingua provided the setting for a successful run, albeit with a touch of the pre-Broadway try-out that never made it into town.

These men of wealth—and society as a whole—felt they were endowed with a basic obligation to live lavishly. However, since "money getting . . . is rarely met with in combination with the finer and more interesting traits of character," these symbols of wealth usually grew into monuments of cheap ugliness.[5] The mansion at Terlingua is Perry's "trophy." Though he rarely inspected his quicksilver properties, his mansion stood there symbolic of the power vested in the absentee sovereign. It still stands today in a state of gradual deterioration high on a hill overlooking the abandoned ruins of Terlingua and the Chisos mine, its angular and seemingly unplanned design relating a mute story of a gluttonous age that had scant regard either for the future or for taste.

Other structures in the village were less pretentious, as they served utilitarian functions. When Wayne Cartledge arrived in Terlingua in July, 1909, the Chisos Hotel consisted of a church tent stretched over a plank floor, and a cook named "Nigger" Charlie was the only employee.[6] Of necessity this facility offered a limited menu. Jeanette Dow, who taught school in Terlingua the same year that Wayne Cartledge joined the Chisos, "lived in a tent-house adjoining the white family that ran the *Mess Hall* [the hotel] and I mean it was a MESS! Canned corn and goat-meat was the main diet, and I couldn't even stand the thought of goat-meat, much less the smell. The engineer and his wife finally offered me room and board with them; and I have an idea that Mr. Hawley managed this move. He was a grand person." While food and housing at Terlingua fell short of Miss Dow's expectations, she nevertheless added that the morale in the mining camp remained high and practically everyone seemed to enjoy a rich and abundant life.[7]

Wayne Cartledge and his wife were equally unimpressed with the Chisos Hotel. They arrived in Terlingua just before noon and went to the hotel for lunch. Cartledge recalled they were seated at

[5] Edward C. Kirkland, *Business in the Gilded Age*, p. 13.
[6] Interview with Wayne Cartledge, Marfa, Texas, December 29, 1964.
[7] Jeanette Dow Stephens, Shreveport, Louisiana, to Kenneth B. Ragsdale, June 25, 1968, in possession of author.

a large table occupied by company employees—all men. "When asked what she [Mrs. Cartledge] wanted to drink," Cartledge recalled, "she said iced tea. Well, the idea of iced tea in Terlingua sent the men into fits of laughter and my wife broke into tears from embarrassment."[8]

While the Chisos Hotel had a modest beginning, its quality also improved with the company's growth. A two-storied frame building with corrugated iron roof replaced the tent soon after Cartledge arrived. As the only public hostelry in the entire quicksilver district, the Chisos Hotel in the ensuing three decades became the center of Anglo social activity in Terlingua, as well as the temporary residence of people visiting the mine. Long after the Chisos mine ceased to produce quicksilver, the Chisos Hotel continued to serve the community.

Residential dwelling also followed a similar pattern of growth. Jeanette Dow wrote that in 1909 "there were a very few good houses made of brick and adobe—one for Mr. Hawley, one for Dr. White, and one for a Mr. Quail [Quale], who was an engineer. The other white people lived in adobe huts or what was called 'tent-houses.' . . . We had a 10' × 12' for a kitchen and two 12' × 14' rooms for bedrooms. They were very comfortable."[9]

Wayne Cartledge's first home consisted of three rooms with 1" × 12" plank floors, framed walls, and a tent roof. Suspended above the roof was a chicken wire and sotol leaf ramada for protection from the intense summer heat. In 1912 the company replaced this structure with an adobe dwelling with tongue-and-groove flooring, plastered walls, beaver board ceiling, and a corrugated iron roof.[10] However, all company employees were not so commodiously provided for, as "Dr. Wilson and his wife lived in a double tent midway between our house [Hawley's] and the store."[11]

Visitors to Terlingua noted both the village's charm and the brisk activity that surrounded the Chisos mine. The journalist W. D. Hornaday, there in 1910, described the site as a "picturesque mining camp" with an estimated population of "about 700, all but a few of whom are Mexicans." The journalist also noted the company's extensive exploration program. Eight shafts, connected by six miles of underground workings, had penetrated the 500-foot level, and the

[8] Wayne Cartledge interview.
[9] Stephens to Ragsdale, June 25, 1968.
[10] Wayne Cartledge interview.
[11] C. A. Hawley, "Life Along the Border," *Sul Ross State College Bulletin* 44 (September 1964): 72–73.

company was surfacing "exceedingly rich cinnabar ore. . . . Also a new double-compartment shaft had just been finished and it will be equipped with an electric hoist. This will permit a large increase . . . in the ore output." Udden's confidence in Section 295 was well founded, as the 20-ton Scott furnace was reportedly yielding "upwards of $25,000 per month in quicksilver."[12]

Encouraged by good recovery and an active quicksilver market, Perry assigned 125 men to day and night shifts and began an around-the-clock operation. "Submitting to the inevitable" was a prospect long forgotten; the diminutive mine owner could now look forward to his second decade at Terlingua with renewed optimism.

Medical service for company employees was one problem of Terlingua life solved at the outset. The company employed a doctor and paid him a monthly salary. Every major quicksilver producer in the district also followed this policy.[13] Shortly after Hawley joined the Chisos in 1910, he engaged Dr. R. A. Wilson to replace a Dr. Smith. Hawley explained that the Chisos provided the "necessary drugs and supplies which were purchased from the San Antonio Drug Company." He added that normally the Chisos doctor had very little to do, and he "would now and then go out of camp to treat someone who was ill or had met with accident or injury, there being no other physician within one hundred miles."[14]

It is not known when the Chisos Mining Company began providing medical services for employees; however, in March, 1906, the company deducted a doctor's fee that amounted to approximately one day's pay per month. While there is no record at this early date of what services the employees received, they were, no doubt, limited. When Dr. Phillips' wife was injured in autumn of that year, Drury Phillips stated that the "nearest medical aid was 90 miles away in Alpine."[15]

The earliest known reference to a doctor in the Terlingua district appears in the April 5, 1905, issue of the Alpine *Times*.[16] That paper reported that Tomás Aguilar accidentally "shot and killed one of his little girls. . . . One of her legs was shattered, and Dr. Boyd

[12] W. D. Hornaday, "The Cinnabar Deposits of Terlingua, Texas," *The Mining World* (December 17, 1910), p. 1133.

[13] Hawley, "Life Along the Border," p. 71.

[14] Ibid., pp. 73, 72.

[15] Phillips to Ragsdale, January 17, 1966.

[16] Although the Alpine *Avalanche* first appeared in 1900 (earliest known issue, February 16, 1900), the Alpine *Times* provides the best coverage of the Terlingua Quicksilver District between June 22, 1904, and April 3, 1907.

amputated it but she had lost so much blood before his arrival he could not save her life." Later that year when Texas Ranger Tom J. Goff was shot near Study Butte while attempting to arrest a Mexican bandit, a Ranger captain and a doctor were called from Alpine. The Terlingua correspondent reported that Goff died the following day and "the body was placed in a hack and started on a 90 mile journey to Alpine, being met about half way by Capt. Rogers and Dr. Berkeley."[17]

Salivation is a special health hazard encountered by men working near a quicksilver furnace. Inhaled fumes emanating from the smelting ore stimulate the excretion of the salivary glands, and over a prolonged exposure period the teeth loosen and eventually fall out if the patient is not treated properly. Harris Smith, who operated a quicksilver furnace near Terlingua, was a salivation victim. "Every tooth in my head became loose, and I could no longer eat solid food. My total diet consisted of bean soup, crackers, coffee, and mouthwash."[18] Although Smith recovered from this rare malady, his future diet was contingent on dentures.

The recently established Terlingua jail, no doubt, inspired McCaskey's tribute to the Chisos Mining Company for maintaining law and order, two other "facilities of civilization" that came with the mine's prosperity. The recent urban amalgam of Anglo and Mexican, existing well beyond the pale of organized society, created a problem of law enforcement. Also, the lethal combination of liquor and guns in close proximity to a remote international boundary aggravated further the problem for the mine owners. Mexico provided both an abundant source of smuggled intoxicants as well as a haven for the lawbreakers. Hawley explained from firsthand experience that "liquor, together with the bad habit of carrying a gun, caused every violent death that occurred among the Mexican people that I knew in the mines about Terlingua. An armed man who has been drinking is a dangerous man in any country."[19]

Perry always accorded law and order a high priority, not because of any moral or philosophical posture, but because drunkenness and disorder were bad for business. He wrote the mine management in 1908: "I note that the Mexican Fourth of July passed off with proper rejoicing and not too much hilarity. I wish we only had something down there for recreation that had *no drink* to it—for

[17] Alpine *Times*, April 5, September 29, 1905.
[18] Interview with Harris Smith, Austin, Texas, July 7, 1965.
[19] Hawley, "Life Along the Border," p. 54.

men need recreation."[20] Perry apparently elected to leave the problem of recreation to the discretion of the Mexicans and then established a jail to restrict their excessive hilarity.

The following year "certain citizens" petitioned the Brewster County Commissioner's Court for a jail at Terlingua, and on November 8, 1909, that body authorized the judge "to bind the County to pay one half of the cost of erecting said jail provided the Chisos Mining Company, or the other citizens of Terlingua will pay the other one half."[21] Although the jail neither restrained the northward flow of sotol across the Rio Grande nor allayed the murder and violence that continued to erupt along the periphery of the district, it nevertheless became a static threat to the Mexican employee who looked to the Chisos mine for employment.

To make doubly sure that the law was enforced, a Chisos employee always served as the Precinct No. 4 justice of the peace. First Hawley, and later Robert Cartledge, became the symbol of law and justice in the quicksilver district. While justice may have been a fluid issue, the law was enforced. Elmo Johnson, who operated a store on the Rio Grande below Terlingua, explained that Cartledge's role as justice of the peace represented a positive good in that remote community. According to Johnson, Cartledge added a measure of social stability to an area of Brewster County that lay well beyond the perimeter served by organized law enforcement agencies.[22] Earl Anderau, a former Chisos Store employee, admitted there was much intoxication, fighting, and absenteeism among the Mexicans at Terlingua; however, on the whole he regarded them as a basically docile, law-abiding people. Anderau believed "a Mexican could be put in jail, told to stay there with the door unlocked, and he would be there when told he could leave." He added that he never knew of an Anglo being jailed in Terlingua.[23]

The symbol of progress enjoyed most by the cinnabar citizenry was the recently established Chisos-owned telephone line. The Marfa and Mariposa mine provided the quicksilver district's first telephone service with a line that the company built to Marfa. This facility, used by the entire district on a fee basis, was erratic at best, as a sixty-mile span consisted of the upper wire of a barbed wire

[20] Perry to Chisos Mining Company, September 25, 1908, Chisos Mining Company Papers, Archives of the Texas State Library, Austin.

[21] Minutes, Commissioners Court, Brewster County, Alpine, Texas, II: 190.

[22] Interview with Elmo Johnson, Sonora, Texas, June 3, 1966.

[23] Interview with Earl Anderau, Alpine, Texas, December 27, 1964.

fence.[24] When the Marfa and Mariposa mine closed in 1910, the Chisos management discussed acquiring that line. This action was delayed, however, when it appeared that Alpine might replace Marfa as the district's primary supply center. Sensing the commercial potential of this change, the Alpine *Avalanche* editorialized that "at the present time the freight wagons of this company [Chisos] pass through Alpine and go to Marfa for supplies. This should not be."[25]

Chisos superintendent Jim Lafarelle further stimulated this interest by announcing he was eager to get telephone communications with Alpine. If successful, the company's "intention [is] to do their trading here [Alpine]."[26] The Chisos management elected not to acquire the Mariposa's "barbed wire system" and subsequently built a telephone line from Terlingua to the 02 Ranch where the company joined a rancher's private line to Alpine. According to Wayne Cartledge, this service was never satisfactory: "between the company attempting to transact business and the ranchers' wives gossiping," tempers remained at a high pitch. The company discontinued this service in 1913 and built an extension from the 02 Ranch through Del Norte Gap to Marathon.[27] This provided uninterrupted telephone service to McIntyre's Store in Marathon, although it was limited to the hours the store remained open.

The mining company charged a small fee for personal calls. Telegrams could also be sent by telephoning the message to McIntyre's Store, where a store employee would, in turn, deliver it to the railroad telegrapher in Marathon for transmission. Earl Anderau, who went to Terlingua shortly after this line was established, explained that while the ranchers could no longer "listen in on the line," there was still little privacy. Since the only telephone at Terlingua was located in the Chisos Store just outside the general manager's office, "when someone made a personal call, everyone in the store listened."[28]

As the first vestiges of modern civilization appeared in Terlingua, a major political and social conflict erupted almost within sight of the mining village. Mexico had been racked by political strife almost constantly since gaining her independence from Spain in

[24] Hawley, "Life Along the Border," p. 21.
[25] Alpine *Avalanche*, March 31, 1910.
[26] Ibid.
[27] Wayne Cartledge interview.
[28] Anderau interview.

1821; however, in 1910 a new struggle for control of the government developed between President Porfirio Díaz, an ultraconservative, and Francisco I. Madero, an articulate liberal who advocated, among other measures, reforms in suffrage, land distribution, freedom of the press, and the single presidential term. (Díaz had become provisional president in 1876, and for the next three and one-half decades had dominated Mexican politics, most of the time as president.)

Madero opposed Díaz for the presidency in 1910 but was imprisoned by the opposition until Díaz's reelection was assured. Madero embarked immediately on a plot to overthrow the Díaz government. Agents of his insurgent party began spreading their doctrines throughout Central Mexico until on November 20, 1911, open hostilities began in Puebla and swept northward. Assassinations, imprisonments, vandalism, and banditry followed in the revolution's wake. Terlingua soon felt its stunning impact.

Mexican banditry had long been a fact of life in the Big Bend. While the early settlers had learned to cope with theft motivated by poverty and hunger, the revolution added a new dimension to this international larceny. In addition to food, the bandits now sought materials of war—horses, guns, and ammunition. Organized bands of raiders, numbering from twenty-five to as many as seventy-five, began attacking ranches and settlements along the United States side of the Rio Grande. W. D. Smithers, photographer, writer, and long-time resident of the Big Bend, witnessed this overflow from the Madero revolution and described its impact on life in that region:

> The bandits could cross the Rio Grande at many places where it was impossible for the Texas Rangers, mounted Customs officers, Immigration River Guards, and the cavalry troops to see them cross. Once they were on this side, they worked their way up the canyons and draws to the ranch they intended to raid. . . . Most ranches along the Border were raided at one time or another. . . . A Mexican named Chico Cano was probably the most notorious and worst-hated bandit in the upper part of the Big Bend during early days. He . . . operated with his men for at least seven years, from 1913 to 1920.[29]

In early 1907 the Alpine *Avalanche* reported a respite from these raids. A brief front page item noted that "all is quiet along the Rio Grande," as the Mexican outlaws "are conspicuously absent in these parts at this time." The newspaper cited Texas Rangers R. M.

29 W. D. Smithers, "Bandit Raids in the Big Bend Country," *Sul Ross State College Bulletin* 43 (September 1963): 75.

Hudson and Frank A. Hamer, who "have been 'keeping tab' on their movements on this side of the river."[30] Perry, who feared the bandits might pick the well-provisioned Chisos Store as their target, no doubt welcomed this good news. From New York City on July 8, 1908, he optimistically wrote the Chisos management, "I note that you say nothing about the uprising on the border, and I am inclined to think the trouble is dying out."[31]

Events of 1910 proved Perry wrong. As the insurgent campaigns swept northward through Chihuahua, many of its citizens fled across the Rio Grande and headed for Terlingua, the largest settlement in the Big Bend. By mid-January, 1911, about three hundred Mexican refugees were encamped at the Chisos mine. This sudden influx of Mexican nationals created many new problems for the Chisos management and Brewster County Sheriff Allen Walton responded to their request for assistance. Walton visited Terlingua and found the refugees "almost destitute . . . the people there are feeding them but . . . it will not last very long as a very serious problem will confront them."[32]

By March 20 the number of refugees seeking asylum in Terlingua totaled more than four hundred and Perry appealed to the military for aid and protection. In a characteristically long but well-organized letter, Perry cited for Brigadier General Joseph W. Duncan, Fort Sam Houston, Texas, conditions at Terlingua: the extensive mine installations, the well-stocked ($25,000) Chisos Store, the presence of Mexican refugees, and the remoteness of the location, and requested that "a small company of soldiers (mounted men) say (60) men . . . be located at some suitable camp within a few miles of our town."[33]

Under a heading entitled, "What Is Feared," Perry emphasized that "bad border men and bandits, both Mexican and white men will raid our store, or intercept our wagon trains with supplies." He also predicted a return of stable government in Mexico and reasoned that "these desperate characters will be driven across into Texas," which would mean that "our property which had taken hundreds of thousands of dollars to create would have to be aban-

[30] Alpine Times, February 27, 1907. Hamer later gained international notoriety for his role in the killing of the celebrated murderers, Bonnie Parker and Clyde Barrow.
[31] Perry to Chisos Mining Company, July 8, 1908, Chisos Mining Papers.
[32] Alpine Avalanche, January 19, 1911.
[33] Perry to Brigadier General Joseph W. Duncan, March 20, 1911, Old Military Records, Record Group 94, National Archives, Washington, D.C.

doned." While Perry feared attacks from Mexican bandits, revolutionary sympathizers in the mining camp gave him further cause for alarm. He explained to Duncan that "about 40% of the Mexicans in our camp were born between San Carlos and Ojinaga in Mexico (where there is fighting which we can often hear) and their sympathies are naturally with the Insurrectors."[34]

Under the paragraph marked "Please Hold as Confidential," Perry explained that "it might prove very unfortunate if it becomes known in our town or district or at any of the railroad towns . . . that we have written this letter or have desired any troops to be placed upon the border, for this would certainly reach our own workmen." Perry knew all too well that an order stemming from his request to move troops stationed at Alpine and Marfa to Terlingua would have multiple repercussions. First, it would be aggressively opposed by the merchants, who would not want to lose the economic benefits a cavalry troop afforded them. But what Perry apparently feared most was that the many Mexican nationals working for the Chisos would regard the arrival of United States troops at Terlingua as a hostile gesture. He continued:

> We regard our Mexicans generally highly and they do ourselves—
> We wish to retain this regard. It is equally for their benefit that we
> ask for troops but they, as a class, are a suspicious people and might
> not understand; for these reasons, we will greatly appreciate your
> kindly interest in the matter, to the end that we remain *as not having
> asked* for the stationing of troops in our district.[35]

Perry's apprehension was not shared by those at Terlingua. Contemporary Chisos employees recall that the Mexicans discussed the revolution openly, and while they appeared divided in their opinion of Madero, "most of them agreed that he was in the right and a real friend of the common people." But other than the interest the Mexican people exhibited in the issues, "the revolution in no way affected the course of life at the mines at Terlingua," according to Hawley. "Work went on as usual, and no apprehension was felt about our country being involved."[36]

Varying perspectives account for the differing views of Terlingua's security; that those nearest to the fighting felt the least concern was not without sound rationale. Company employees knew

[34] Ibid.
[35] Ibid.
[36] Hawley, "Life Along the Border," p. 74.

from firsthand experience that a community as large as Terlingua had never been attacked by Mexican bandits, whose previous targets had been unarmed and unprotected isolated ranches and stores. Perry, on the other hand, viewed the revolution in a different light. He saw the menace as the border backwash of a nation in turmoil. The enemy would appear, he believed, not as a few armed bandits, but as bands of organized revolutionaries terrorizing the border region, murdering people, looting farms, ranches, and villages, and disappearing unmolested to a secure sanctuary south of the Rio Grande. Perry, therefore, embarked on a long verbal battle with the army that eventually brought troops to Terlingua, and while that village was never attacked, his persistence was well founded. The organized bandit raids on the army-protected installation at Glenn Springs, not thirty miles from Terlingua, and the Villista-led Columbus, New Mexico, massacre that unleashed General John J. Pershing and 12,000 American troops into the heart of Mexico rendered a tragic verdict in Perry's behalf.

While there is no record of Duncan's reply to Perry's request, a series of categorical rejections thrust Perry into a new drama and a new role, which he played with great aplomb. He drew into his orbit of political maneuvering the nation's top military leaders: the United States Army chief of staff, the adjutant general, and a sampling of lesser generals, colonels, and majors. He also assigned roles to members of the nation's upper political echelon as well: the secretary of war, plus a host of senators and congressmen. For a man whose business philosophy was steeped in the theory of separation of government and business, this posture in one respect appears as an abrupt break with the past. In another respect, Perry was simply following in the footsteps of some of his more successful predecessors. As the nineteenth century drew to a close the strength of the United States government was flexed repeatedly both at home and abroad to protect private interests. And now, as Perry saw his property threatened on American soil by a foreign revolutionary force, he called for help.

The remainder of 1911 was ominously quiet at Terlingua, which reflected internal conditions in Mexico. The opposing factions agreed on terms of peace and on May 25 President Porfirio Díaz resigned his office. Madero's election to that post on November 15, instead of ushering in an era of peace, brought a resumption of hostilities. Díaz supporters embarked almost immediately on a counterrevolution and on January 31, 1912, the Juarez military garrison revolted, threatening the security of Americans on both sides

of the Rio Grande. The event had immediate international repercussions.

President William Howard Taft of the United States ordered
the secretary of war on February 4, 1912, "as a measure of precau-
tion, at once to increase the guard along the border, as quietly as
possible, to such strength as will amply insure the protection of
American citizens and their interests."[37] The adjutant general,
Henry P. McCain, in turn, sent telegrams to commanding officers in
Georgia, Arizona, Kansas, Arkansas, and Nebraska, alerting them to
"hold troops in your command, in readiness for immediate field
service, southern frontier. Strictly confidential."[38] Unknowingly the
cast of players was quietly assembling for the drama.

Conditions worsened along the Texas border. On February 9
two hundred revolutionists demanded the surrender of Presidio,
about fifty miles from Terlingua, and two weeks later a Colonel
Steever telegraphed the adjutant general from El Paso: "Exodus
foreigners from Mexico increasing. . . . Many Mexicans are ugly
toward Americans. Last year El Paso was unanimously pro Madero.
Now large portion Mexican population anti Madero and anti American, because apprehensive of intervention."[39]

Perry no longer represented a minority in his fears of border
attacks. Texas Governor O. B. Colquitt regarded the unsettled conditions in Mexico as a threat to border security and wrote Duncan
on February 29, requesting military protection from the "hostile
Mexican element" in the Big Bend area. Responding to the governor's request, Duncan telegraphed the adjutant general for authority "to grant requests for protection by sending United States
troops to Shafter and Terlingua in big bend country." Although the
southern department commander saw no immediate danger from
border incursions, he felt the presence of the military "would have
great moral and quietening effect on the excited citizens as was the
case in days of Indian troubles and such action would undoubtedly
popularize the Army."[40]

[37] William Howard Taft to the secretary of war, February 4, 1912, R. G.
94, National Archives.

[38] Telegram from Adjutant General McCain to commanding officers in
Georgia, Arizona, Kansas, Arkansas, and Nebraska, February 4, 1912, R. G.
94, National Archives.

[39] Telegram from Colonel Steever to the adjutant general, February 22,
1912, R. G. 94, National Archives.

[40] Telegram from Brigadier General Joseph W. Duncan to the adjutant
general, March 9, 1912, R. G. 94, National Archives.

While Duncan's decision did nothing to "popularize the army," neither did it satisfy Perry's demand for military protection. Perry regarded the eight-man cavalry detachment that arrived in Terlingua as what it was, a token gesture, and began mobilizing his forces to ensure that his demands would prevail. His first stop was the War Department in Washington. While there is no record of whom he saw, he personally delivered a letter "stating that eight cavalrymen are insufficient to accord protection" to Terlingua and that "an increased force ought to be sent there."[41]

This marks the beginning of Perry's war with the army and the entire military and political establishment in Washington and in Texas. And in the end Perry won his victory. But for all others involved in the debacle—the United States Army, the soldiers occupying the Terlingua post, the Chisos Mining Company officials, and especially the citizens of Terlingua for whom Perry demanded "protection"—it was a vacant triumph. But most of all, it was Duncan who misread the signs of the time; the operation calculated to "popularize the army" backfired.

Perry embarked on a two-pronged campaign: first, he wanted to secure the eight-man detachment at Terlingua; and second, he hoped to supplement this detachment with a full sixty-man troop of cavalry to assure the protection of his property. Since he undoubtedly realized that the lifting of the arms embargo weakened his position, he began a holding action, reserving his bigger demands for when conditions appeared more favorable. Perry's first ally was Texas Congressman William R. Smith, who began bombarding Secretary of War Lindley M. Garrison with correspondence pleading Perry's case. His appeal, dated September 17, 1914, was a replay of a familiar theme. Smith wrote: "Mr. Perry . . . states it is his opinion that 'to take away troops suddenly now will expose Big Bend country to danger' and that 'too early withdrawal of troops along entire border will make for further Mexican troubles in future.' "[42]

The troops remained and the military situation in the Big Bend was unchanged throughout 1914; however, early in 1915 the Terlingua area became the focal point of revolutionist activity. Perry sprang into action. On March 18, 1915, the Alpine *Avalanche* reported:

[41] Perry to the War Department, received by Major General W. W. Witherspoon, Chief Mobil Army Division, R. G. 94, National Archives.

[42] Congressman William R. Smith to Secretary of War Lindley M. Garrison, September 17, 1914, R. G. 94, National Archives.

Mexican revolutionists on this side of the river have been organizing for a raid into Mexico and as there are many desperate criminals included in their number. Mr. Perry, owner of the Chisos mines, became alarmed and sent a wire to Capt. Reamey of Troop F, stationed here, asking that he send a detachment down to disperse them. . . . Several of the raiders were arrested and were sent here under guard and will be sent to El Paso.[43]

The military recognized the deteriorating conditions along the Rio Grande and the new southern department commander, Major General Frederick Funston, assured Adjutant General McCain that "no troops have been ordered withdrawn from the Big Bend District. . . . On the contrary, orders have only recently been sent out to the various commanding officers along the border . . . to re-double their vigilance in an endeavor to prevent marauding bands from crossing the international boundary into the United States."[44]

Apparently Perry realized that conditions were favorable for initiating phase two of his plan to increase the military garrison at Terlingua. Ambitious plans demand people of comparable stature, and Perry began moving up the hierarchy to the top echelons of political and military power in Washington. First, he turned from the House of Representatives to the Senate and enlisted the support of Maine Senator Charles Fletcher Johnson. His ultimate target was the secretary of war, to whom Johnson wrote Perry a letter of introduction. The senator explained to Secretary Garrison that "I have been much interested in his [Perry's] statement in regard to conditions on the Texas border and the necessity of protecting and shall greatly appreciate your courtesy in giving him a hearing."[45]

While there is no record of Perry's audience with the secretary of war, it can be assumed that with the senator's legislative clout and Perry's persistence, Garrison received him. Three weeks later, however, Perry gained an appointment with the chief of staff, Major General Hugh F. Scott, to discuss conditions at Terlingua. Following that October 11 meeting, Perry wrote the general a letter confirming specific points covered in their discussions. This letter is a key document in the military experience in the Terlingua district,

---

43 Alpine *Avalanche*, March 18, 1915.

44 Major General Frederick Funston to the adjutant general, March 29, 1915, R. G. 94, National Archives.

45 Senator Charles F. Johnson to Secretary of War Lindley M. Garrison, September 17, 1915, R. G. 94, National Archives.

as it outlines Perry's argument that led to the transfer of a full troop of cavalry to protect the Chisos Mining Company. This letter is also noteworthy for two other reasons: first, it contains a new argument in Perry's plea for protection; and second, another point upon which Perry formerly based his appeal for aid is noticeably absent. Combined, they reveal something of the emotional and psychological temper of the time. Perry began: "I would like to write you somewhat fuller with regard to the further preservation of our business concern, the Chisos Mining Company . . . together with, and first of all, our white people—say 40 white men and women and little children."[46] Significantly, Perry's prime emphasis is "our white people" and noticeably absent is his former reference to "our Mexicans." Perry obviously had learned his lesson well. At the high tide of anti-Mexican sentiment in the United States, he carefully avoided any allusion to the "white man's burden," and especially to a professional military man, the person least likely to be moved by that argument.

Perry then moved in with the clincher. He explained to the chief of staff about a "new phase for the need for military protection." He adroitly pointed out, and correctly so, the small amount of quicksilver then available in the United States, as well as its critical need in time of war. Finally Perry emphasized that "these things surely indicate the propriety of calling attention of our Government to a mine like ours, producing easily one-fourth of all mercury made in this country, the destruction of which might prove a serious matter to the country. A moderate increase in number of cavalrymen at Terlingua will safeguard our property."[47]

While Perry attempted to clothe his argument in eloquent vernacular redolent of patriotism and humanitarianism, his motives were blatantly self-serving. Nevertheless Scott was moved by Perry's argument, as he submitted the matter to the secretary of war, who "believes on account of the importance and value to the United States of the interest at Terlingua, the garrison at that point should be increased by one troop."[48] Eight days later, on October 19, the commanding general at Fort Sam Houston assured the adjutant general that "additional troops have already been ordered into the

[46] Perry to Major General Hugh L. Scott, October 11, 1915, R. G. 94, National Archives.

[47] Ibid.

[48] Adjutant general to the commanding general, southern department, October 15, 1915, R. G. 94, National Archives.

Big Bend District and one troop will be designated for station at Terlingua."[49] Perry had won his war with the army.

Lesser skirmishes erupted immediately. Perry anticipated this and attempted to keep his maneuvers clandestine. However, the transfer orders directed to the Western Texas Cavalry Patrol District headquarters at Marfa on October 20 obviously "leaked." The following day Alpine Mayor Benjamin F. Berkeley telegraphed Perry's former ally, Congressman William R. Smith, that the War Department contemplated moving Troop A, 13th Cavalry, from Alpine to Terlingua. The mayor emphasized that the "business men's club and citizens urge you use influence to have troops remain here. . . . In the opinion of all our citizens Terlingua has less than one hundred population principally Mexicans. Greater necessity for troop to be located in Alpine."[50] While the mayor proved himself less a statistician than a politician, he too grasped the power of the racist argument.

That same day two other Alpine citizens, W. J. Yates and Jim P. Wilson, joined their mayor in trying to wield political pressure to hold the cavalry unit in Alpine. Their targets were congressmen Smith, John Nance Garner, and Morris Sheppard. Both Garner and Smith appealed to the secretary of war, who based his refusal on "the importance and value to the United States of the interests at Terlingua."[51] Unknowingly they had been playing a losing game as Perry out-maneuvered the local civic leaders as well as the cavalry commander at Alpine. Years later, Wigfall W. Van Sickle, Perry's Alpine attorney, recalled the dispatch with which the cavalry departed for Terlingua: "Early one morning I had a message from him [Perry] in Washington asking me had the two [one] calvary troops gone to Terlingua and I called the Major who was commanding officer and he told me he would be at my office in a few minutes . . . and inquired who this man Perry was and said that he, Mr. Perry, and the War Department had moved the troops without his knowledge."[52]

Perry's victory, however, was short-lived. Even before the troops left Alpine, the military commanders most familiar with con-

[49] Commanding general, southern department, to the adjutant general, October 19, 1915, R. G. 94, National Archives.

[50] Telegram from Benjamin F. Berkeley to Congressman William R. Smith, October 21, 1915, R. G. 94, National Archives.

[51] Telegram from W. J. Yates and Jim P. Wilson to Congressman John Nance Garner, October 21, 1915, R. G. 94, National Archives.

[52] Van Sickle to Cartledge, December 8, 1937.

ditions in the Big Bend vociferously opposed the transfer. They based their arguments on both the questionable need for military protection at Terlingua and the logistics involved in deploying a troop of cavalry to that remote region. Discussions began at the command level and proceeded upward.

Two days after receiving orders transferring Troop A from Alpine to Terlingua, the 13th Cavalry commander, Lieutenant Commander T. R. Rivers, dispatched a three-page memorandum to the commanding general of the southern department at Fort Sam Houston, questioning the advisability of the operation. The colonel emphasized that "the physical difficulties of supplying a troop at Terlingua are so great that I recommend . . . that the permanent station of Troop A, 13th Cavalry is not changed from Alpine to Terlingua but that a detachment of not to exceed twenty five (25) men be stationed at Terlingua. If the entire troop must go to Terlingua then I request steps be taken to supply extra transportation necessary to supply it."[53]

The colonel cited point-by-point the problems incident to that operation—transportation, shelter for men and animals, wood, water, and supplies—and implied that the need for additional protection at Terlingua was a illusion. The colonel explained:

> No untoward thing has happened at Terlingua and I do not believe there is any necessity for an entire troop at Terlingua. Mr. W. R. [Wayne] Cartledge . . . has stated to Captain Davidson that a detachment double the size of the detachment now there [approximately eight men] would be ample. There has been no change in conditions since he made the statement. I can supply twenty-five (25) men reasonably well with the transportation facilities at my command.[54]

General Funston, southern department commander, supported Rivers' position opposing the troop transfer. On October 25, Funston wrote Adjutant General McCain in Washington, explaining that while he "at once gave the order by telegraph for the movement to commence," he had also "taken the liberty of bringing this matter to the attention of the secretary of war because I do not believe that the conditions are thoroughly understood."[55] However, with the

[53] Commanding officer, Headquarters, Western Texas Cavalry Patrol District, Marfa, Texas, to the commanding general, southern department, Fort Sam Houston, Texas, October 22, 1915, R. G. 94, National Archives.
[54] Ibid.
[55] Major General Frederick Funston to the adjutant general, October 23, 1915, R. G. 94, National Archives.

troop already in transit the matter was purely academic and had progressed well beyond the limits of rational deliberation. Perry had maneuvered his Washington political forces into an impregnable position, and at this point there would be no turning back. A troop of cavalry would "protect" the Chisos Mining Company from a real or an imaginary menace that lurked along the Rio Grande, and the army would have to solve the problem of supply. On November 6 the commanding officer at the Galveston Port of Embarkation received the order to "send three motor trucks, now at Galveston Depot, to Alpine, Texas, for use supplying troops at Terlingua."[56] The vehicles were delivered according to instructions; the results were disastrous.

The military commander's reluctance to send Troop A to Terlingua was matched only by the troopers' disdain for their assignment. Theft, drunkenness, destruction of property, disturbing the peace, and general disorder became their pattern of conduct, and most people living in Terlingua soon felt that the presence of Mexican revolutionaries would be preferable to the soldiers of the United States Cavalry.[57] Since the troops were ordered to the border to protect the citizenry from the insurrectionists, every Mexican became the symbol of the soldiers' misfortune, and they soon vented their disgust and anger on any Latin who became the object of their attention. Whenever a river patrol encountered a "Mexican," regardless of his nationality or occupation, the pattern of apprehension followed an established routine: "rough'em up, 'confiscate' their money or tobacco, and tell'em to head for the river."[58]

The most flagrant episode occurred on December 1, 1915. Wayne Cartledge remembered the date vividly. It was the opening day of deer-hunting season, and he left the mine well before daylight for a day-long hunt. The incident that occurred in his absence had its beginning the previous day. November 30 was the soldiers' payday, which they celebrated with an all night bout of beer drinking and poker playing. Just after dawn of the following morning the soldiers observed several Mexican children driving a herd of goats across the crest of a mesa behind the Cartledge house. The animals silhouetted against the gray morning sky presented a real challenge to their marksmanship. Still in a state of drunken revelry,

[56] Acting quartermaster general to the adjutant general, November 3, 1915, R. G. 94, National Archives.
[57] Anderau interview.
[58] Wayne Cartledge interview.

the soldiers grabbed their carbines and opened fire on the goats. Mrs. Cartledge, hearing the report of the rifles and the sound of ricocheting bullets, fled the area and sought protection in another part of the village. The firing continued as long as an animal remained alive on the mesa.

Cartledge returned to Terlingua late that night, and when informed of the incident, called the commanding officer in Marfa and demanded that disciplinary action be taken. When the officer sent to settle the claims arrived at Cartledge's office and was apprised of the facts of the shooting and confronted with the carcasses of fifteen dead goats, he appeared unmoved by the entire episode. However, when told the Mexican owner wanted $2.50 per head for his animals, the officer lost all composure expected of a person of his command and stormed out of Cartledge's office. His parting words were to "send the goddamned savages back to Mexico." No disciplinary measures were imposed on the soldiers nor was the Mexican compensated for his losses.[59] If the United States Cavalry gave Perry a feeling of security as he gazed across the blue expanses of Casco Bay 2,000 miles away, his serenity was not shared by those whom the cavalry was ordered to protect.

Robert Cartledge was justice of the peace during the military occupation and it was his duty to try to enforce the law. The primary source of trouble was the perennial conflicts between the soldiers and the Mexicans. Reports of racial violence became routine in the Alpine *Avalanche*: "A Mexican was killed at Terlingua Tuesday night and a soldier [has] been placed under arrest. . . . the soldier was stabbed twice before the Mexican was killed, and he is now being brought to Alpine on a stretcher to receive medical treatment."[60]

The one episode that stands out most vividly in Robert Cartledge's recollections involves rescuing a Mexican woman from an attempted assault by a young drunken soldier. Cartledge recalled being interrupted in his office one day by a Mexican girl, who told him excitedly that a woman was screaming for help. Grabbing his pistol, he ran to the woman's house. Finding the door locked, he forced his way into the adobe shack and confronted the young drunk violently struggling with a feeble, old Mexican woman. Cartledge, expecting a fight, forcibly grabbed the soldier but was sur-

[59] Ibid.
[60] Alpine *Avalanche*, April 1, 1915.

prised when he offered no resistance. He then led him to the street, where without warning the soldier wrestled from his grip, struck him, and started to run. Angered by this personal assault, Cartledge drew his pistol and took aim, but realizing the difficulties that would arise from shooting a United States soldier, he holstered his gun instead and gave chase. Catching him, Cartledge knocked the young drunk to the ground, and in the middle of a dusty Terlingua street he jumped astride his sweating body and began a methodical "whippin'" calculated to divert his attention from the more casual pursuits of his Terlingua assignment.[61]

The episodes Cartledge described were pathetically similar—stories of undisciplined young men stationed on an international border waiting for action that never occurred and seeking recreation with a people for whom they shared a mutual dislike and distrust. Throughout military history such situations have always precipitated trouble. Nearly three-quarters of a century before the cavalry came to Terlingua, Frederick Law Olmsted observed the same chaotic mixing of races and cultures in San Antonio. His philosophic conclusion was that "where borderers and idle soldiers are hanging about drinking places and where different races mingle on unequal terms, assassinations must be expected."[62] Those were prophetic words.

After two months of grappling with the frustrating problems of supplying a field combat unit challenged by nothing more than a dedicated justice of the peace trying to uphold the law, Troop A received orders to return to Alpine. Thus Perry greeted the new year with a familiar theme. On January 1, 1916, he telegraphed the chief of staff from the Congress Hotel in Chicago "Have wire from Chisos Mine that Terlingua troop with the exception of twenty men have been recalled to Alpine. I hope you will kindly consider having them . . . returned to Terlingua at least before trouble threatens." Panic was still the "name of the game" for Perry. He continued: "If Villa's utterances reported are true then the awful massacre of sixteen of our citizens near Chihuahua is but the beginning and which will extend to our border and across our border to Terlingua."[63] But this time Scott was not moved. Two weeks later Funston telegraphed the adjutant general that Troop A had been

---

[61] Interview with Robert L. Cartledge, Austin, Texas, January 15, 1966.

[62] Frederick Law Olmsted, *Journey Through Texas*, p. 84.

[63] Telegram from Howard E. Perry to Major General Hugh L. Scott, January 1, 1916, R. G. 94, National Archives.

reassigned to Alpine and that one officer and twenty men remained in Terlingua.[64]

Although Perry's persistence continued, it yielded nothing but rebuffs, as the hostilities he had built up among members of the military establishment began to surface. In response to a telegram from Perry, Funston wrote Adjutant General McCain that he had "never been of the opinion that there was any especial necessity for the stationing of troops at Terlingua," as there had been no raids in that vicinity. Funston reserved his strongest words for the matter of supply:

> The supply of the troop of cavalry that has been stationed at Terlingua has been one of the greatest embarrassments in the way of administration that we have had to deal with in this department, it being exceedingly expensive owing to the great distance from the supply point. . . . The twenty [men] left there are more than ample to meet any contingency likely to arise, and it is more than probable that an experienced and capable officer like Lt. Col. Rivers, thoroughly familiar with local conditions, is in a better position to be a judge of the matter than is the writer of the enclosed telegram, who seems to be in Chicago.[65]

Perry's answer came not from Funston, but from the adjutant general, who reiterated the military opinion that "there has never been any special necessity" for stationing troops at Terlingua. The adjutant general, quoting Funston, closed his letter in a manner calculated to terminate the entire issue: "The department commander [Funston] further states that there are scores of mines and hundreds of ranches all along the border that have no guards whatever, and suggests that the mining company at Terlingua take some steps for its own protection, if they deem it advisable."[66] Thus ended Perry's war with the army.

The abrasive impact that the military occupation had on the people living in the Big Bend is best revealed in the following incident. Soon after the troops were withdrawn from the Terlingua area, Wayne Cartledge established a ranch and a general store at Castolon. When Cartledge asked his Mexican employees to suggest a name for the ranch, one observed, "Señor Cartledge, with the

[64] Telegram from Major General Frederick Funston to the adjutant general, January 14, 1916, R. G. 94, National Archives.

[65] Major General Frederick Funston to the adjutant general, January 19, 1916, R. G. 94, National Archives.

[66] Adjutant general to Howard E. Perry, January 25, 1916, R. G. 94, National Archives.

soldiers gone, all will be peaceful along the Rio Grande. Let's call it La Harmonia Ranch." Cartledge agreed and the name was adopted.[67]

In the second decade of the twentieth century military operations became an almost universal experience. At Terlingua, as in Mexico, Europe, and Asia, the imminence of armed conflict dominated the life of much of the world's population. Social and political upheavals spawned an era of prolonged unrest that effected changes in all facets of human endeavor. As old political orders passed into oblivion, new ones took their places, and attitudes, values, goals, and symbols shifted with the power structures. This was a time of opportunity and expectancy for some and an era of hardship and privation for others caught in the wake of armed havoc. Consequently, as men of vision looked to the future in 1914, they could foresee both chaos and progress. On August 3 Germany declared war on France and invaded Belgium. Two days later newspaper accounts described the opening of the Panama Canal as the beginning of a new era of worldwide economic growth. Although both incidents seemed remote and unrelated to Terlingua, they did not go unnoticed by Perry. Since world struggle and international commerce were the lifeblood of the quicksilver industry, he understood the vast ramifications of these events.

This was also a period of timely occurrences. Late in 1914 the Chisos began to "stope a pipe-shaped ore body which proved to be the largest and richest in its history"—the famous Chisos "ore chimney" had been discovered.[68] By the time Germany began its attempted blockade of England on February 18, 1915, the Chisos Mining Company had begun installing a new 30-ton rotary furnace to meet the increasing wartime demand for quicksilver. Labor was cheap and plentiful and, with unprecedented ore reserves blocked out, the company experienced its most successful year. Of the annual total of 4,177 flasks of quicksilver, nearly 1,200 were produced in November and December alone. One year later ordnance

---

[67] Wayne Cartledge interview.

[68] Clyde P. Ross, "Preliminary Report on the Terlingua Quicksilver District, Brewster County, Texas," in *The Geology of Texas*, vol. 2, *Structural and Economic Geology*, by E. H. Sellards and C. L. Baker, *University of Texas Bulletin No. 3401* (January 1, 1934), p. 561. The magnitude of this discovery was contained in a candid statement Robert L. Cartledge made as he and E. P. Schoch, professor of chemical engineering at the University of Texas, listened to a Texas House of Representatives hearing in the Capitol in Austin: "You could put this whole damned Texas House of Representatives in the Chisos 'ore chimney' excavation" (Robert L. Cartledge interview, January 13, 1966).

demands boosted the price of quicksilver from a low of $40.25 per flask to a quoted high of over $300.00, but mounting inventories later stabilized the price at $125.00. During the year the Chisos workers penetrated the 700-foot level and still found good ore. Of the 6,306 flasks reported for the district in 1916, the Chisos mine recorded the major portion and became the second largest producer in the nation, exceeded only by the New Idria mine in California.[69]

The favorable quicksilver market stimulated renewed activity in the district, and by 1916 many plants, some idle for almost a decade, began production. The Marfa and Mariposa mine, the Colquitt-Tigner mine, and the Big Bend mine under contract to the Study Butte Mining Company reported substantial recovery by the end of the year.[70] From late 1916 until the end of the conflict, the Alpine *Avalanche*—while announcing "Alpine To Have A Winter Chautauqua" and "Your Uncle Samuel Has Got Your Number"— trumpeted the financial glories of Perry's mining enterprise:

> A solid car of the metal, valued at more than $25,000, was shipped from here last week by the Chisos Mining Co.

> The second car of quicksilver valued in the neighborhood of $25,000, was shipped from here this week by the Chisos Mining Co. . . . This makes approximately $50,000 worth of the liquid metal that has been shipped from Alpine by the Chisos within the past ten days.

> Another car of quicksilver was shipped by the Chisos Mining Company. . . . This is the second car since the first of January. This car carried 480 flasks or about $35,000 worth of the metal, making a total of over $80,000 worth of mercury shipped from Alpine within fifteen days.

> Five big loads of quicksilver came in from the mines Saturday, amounting to about $40,000. H. E. Perry of Chicago, owner of the Chisos Mining Co., was in town Saturday on his way to the Chisos Mines.[71]

The Chisos shipping orders tell an even more graphic story of the almost monotonous regularity with which the mine produced its precious metal.

July 19, 1918, 342 fl.	Jan. 6, 1919, 405 fl.
August 5, 1918, 342 fl.	Feb. 1, 1919, 391 fl.

[69] Johnson, "Mercury Mining," 36 (October 1946): 446.
[70] Ibid.
[71] Alpine *Avalanche*, September 21, 28, 1916, January 25, February 15, 1917.

August 26, 1918, 361 fl.
Sept. 16, 1918, 345 fl.
Oct. 4, 1918, 346 fl.
Oct. 31, 1918, 357 fl.
Dec. 2, 1918, 450 fl.

March 22, 1919, 400 fl.
April 19, 1919, 401 fl.
May 26, 1919, 444 fl.
Aug. 8, 1919, 370 fl.[72]

Total production of the Chisos mine, like that of the entire district, was problematical, as secrecy remained the watchword of the quicksilver industry. Records were kept with extreme vigilance, few reports were filed, and even members of the United States Geologic Survey were forbidden entrance to some producing shafts. It was reported that when the geologist F. L. Ransome, representative of the Critical Metals Resources Board, went to Terlingua to estimate the quicksilver reserves, Perry refused him admission to the mine. At that time only one geologist not permanently employed by Perry had been permitted under ground, and that was Udden. Following Perry's refusal, Ransome asked if he might use the telephone. In response to Perry's question as to whom he wished to call, Ransome answered, "I am going to call Washington, and if necessary, get a detachment of infantry to see that I get into your damn mine!" Ransome said he entered the mine; Robert Cartledge, who remembered the incident, said he did not.[73]

The survey's statement that Brewster County produced 10,759 flasks of quicksilver valued at $1,141,745 in 1917 is indicative of the great wealth accruing to the mine owners. Perry's claim that the Chisos mine "furnished our country more than twice as much quicksilver as any other mine in America" was probably boastful exaggeration.[74] While the total value of the wartime production can never be determined, statements of coeval company employees indicate the vast wealth that accrued to the Chisos' owner. An assistant furnace operator reported that during the early war years "we bottled 30 flasks daily, at $60 per flask. During the last year

---

[72] Chisos Mining Papers.

[73] Interview with Dr. Ross A. Maxwell, Austin, Texas, October 1, 1964.

[74] Perry to R. V. Rinehart, June 25, 1941, Chisos Mining Papers. Total United States quicksilver production for 1917 was 36,159 flasks; while California led five other states with 23,983 flasks, Texas remained in second position. During 1917 the United States imported 5,207 flasks of quicksilver and exported 10,770. This was the last year that exports exceeded imports, a factor destined to affect the industry adversely during the coming decades. A major portion of the year's domestic production went into the production of blasting caps for high explosives; additional amounts were used in the manufacture of drugs and chemicals.

[of his employment, 1916] 101 at $300.00 per [flask]. It was certainly a going and paying mine at that time."[75] Hawley claimed he knew "Perry's mine averaged profits of $2,000 a day for a while, yielding more than $1,250,000 in quicksilver in its best year."[76]

Although the war continued through most of 1918, there was a noticeable decline in the district's production: 8,451 flasks compared to 10,791 flasks in 1917.[77] Preoccupied with production during the early war years, many operators had neglected prospecting and development and found themselves without adequate reserves. Wartime labor shortages were another obstacle to production, and several mines were on the verge of closing during the greatest boom the quicksilver industry ever experienced. The Chisos mine, more fortunate, maintained production throughout the year. However, reduced recovery necessitated closing the recently installed rotary furnace and most metal came from the 20-ton Scott furnace.

Despite the brisk demand, continuing reports of the great Allied offensive in Europe made many aware that Terlingua would soon have to readjust to a peacetime economy. But their attention was suddenly diverted from thoughts of industrial endeavor to survival. As preparations were made to sign the Armistice, a domestic catastrophe—an influenza epidemic—struck the nation with stunning impact. Death, which during the war had seemed remote, now became a common occurrence throughout the country. Terlingua was not excepted.

The epidemic struck first in the urban settlements in Texas, and by mid-October Fort Stockton, a ranching village approximately 150 miles northeast of the Terlingua district, had succumbed to the disease. Local physicians, unable to attend all cases, appealed to Alpine for assistance and "fortunately Dr. Barrett of Amarillo was in our city and he volunteered his services."[78] The October 31 issue of the Alpine *Avalanche* contained Robert Cartledge's statement that no cases of influenza had been reported in the Terlingua area.

When by late October the epidemic had spread to Alpine, Mayor Berkeley issued the following appeal: "The physicians of this city report . . . families and individuals are badly in need of nurses and even tho such helpers may be unskilled a useful and humanitarian purpose is served in bringing sick ones over a critical period.

[75] J. E. Colcord, Portland, Maine, to Kenneth B. Ragsdale, September 15, 1965, in possession of author.

[76] Hawley, "Life Along the Border," p. 87.

[77] Johnson, "Mercury Mining," 36 (October 1946): 447.

[78] Alpine *Avalanche*, October 17, 1918.

... Your mayor deems it his duty to call upon a number of good women to list their names as volunteers for this noble work."[79]

The "good women" responding to the mayor's appeal were organized in groups, each contributing some specific service: cars, broth, and day and night nursing. The local Red Cross unit made influenza masks that were supplied to all those needing them. Despite this unified effort, the disease took a steady toll and the number of "Cards of Thanks" in the Alpine *Avalanche* grew steadily.[80]

At Terlingua the Chisos employees, reading with cautious interest the devastation that the epidemic wrought in other areas of the state, wondered whether their village would be spared. Their answer was not long coming. Robert L. Arthur, the rotund cook at the Chisos Hotel, had been "under the weather" several days but his incapacity attracted little attention. "Guess Arthur's been hittin' the bottle pretty heavy of late" was a local nonprofessional diagnosis. However, his condition worsened and late one afternoon he entered a coma and died a few hours later. "Died pretty quick after he got down," was Robert Cartledge's recollection of Arthur's illness.[81]

When awakened at his home and told of Arthur's death, Wayne Cartledge gave instructions to wake the driver of the mail truck, Theo Savell, and to prepare to leave immediately for Alpine. He thought that if Arthur had died of influenza, the health of the other company employees would be best served by removing the body immediately. Since caskets were not available, Cartledge went to the hotel where he wrapped the enormous body of Arthur in blankets and tightly bound them with rope. With the aid of Mrs. Arthur and Savell, he lifted the body to the bed of the truck, where he securely lashed it for the ninety-mile trip to Alpine.

Savell had already started the engine of the mail truck when he remembered he had promised the Catholic priest a seat in the truck for a trip to Alpine. Cartledge firmly opposed any further delay, as

[79] Ibid., October 31, 1918.

[80] Ibid., October 24, 1918. The 1918 influenza was a respiratory infection, probably due to streptococci. Most deaths, however, resulted from a bacteria type of secondary invader that accompanied the influenza virus and was also probably streptococcal in nature (pneumonia). Other factors probably contributing to the high death rate were poor diet, lack of sanitation, and lack of interior environmental control. These factors were present to an even greater extent prior to 1918; however, the wartime mobility of the national population accentuated them and an exceptionally high mortality rate resulted (interview with Dr. Donald E. Pohl, Austin, Texas, January 8, 1966).

[81] Robert L. Cartledge interview, April 9, 1966.

by then it was almost daylight and the day shift would soon be reporting to the mine for work. However, Savell remained adamant and the priest was awakened. When he arrived at the hotel, Savell and Mrs. Arthur were already seated in the mail truck and the only space remaining for the priest was the truck bed, on which lay the tightly wrapped body of Arthur. Seeing there was not room for him in the seat, he tossed his suitcase in the truck, climbed on the truck bed, and seating himself comfortably on the corpse, shouted above the noise of the motor, "I'd just as soon ride on this bundle as in the seat, if it won't hurt anything." Cartledge hastened to answer, "It won't hurt a thing." And with that, the three people and the body of Arthur began the day-long journey to Alpine.[82] The date was October 30, 1918, and Terlingua had just registered its first influenza death.[83] Within two weeks the entire quicksilver district would be caught in the throes of the dreaded epidemic.

Among the first to contract the disease was Wayne Cartledge. On October 14, 1918, the Alpine *Avalanche* reported his inability to take the oath of office as county commissioner, and the following day Perry wrote Robert Cartledge, expressing his concern for Wayne and others who were ill. "I am wondering this morning how Wayne is. . . . Your telegram of yesterday reassured me somewhat. . . . but still I am very anxious about him, as I am about all who are very ill with the thing." The letter continued in typical Perry fashion, offering detailed instructions for the procurement and use of whiskey in the treatment of influenza. "The use of whiskey for patients recovering from influenza and pneumonia has been found in the East extremely useful." He added that he had telegraphed Judge Van Sickle in Alpine to get a prescription from Dr. J. R. Middlebrook, and would then wire some nearby point for a case of whiskey that would be consigned to Robert in Terlingua. Perry emphasized two points: one, the whiskey would be charged to the company; and two, it was to be consumed only by the infirm. Perry's generosity achieved even greater proportions; he concluded the letter by offering the use of his house to those in Terlingua caring for the sick. He wrote Robert Cartledge: "I got to thinking yesterday about how crowded they must be at Wayne's house, and wired you that perhaps your Mother, and possible also the doctor, would be more comfortable in my house and if so, they would be

---

[82] Wayne Cartledge interview.

[83] Ibid., death certificate, County Clerk's Office, Brewster County, Alpine, Texas.

welcome."[84] Perry's offer of both the house and the whiskey were enthusiastically accepted. These were all firsts for Perry!

As conditions worsened in Terlingua, there was evidence of the epidemic's mitigation in Alpine. The *Avalanche* announced that "the influenza situation has improved so rapidly the last few days that the County Health Officer R. E. Taylor has lifted the quarantine. . . . The school will open Friday morning. . . . Sunday school and preaching will be held at all churches Sunday." The health officer's optimism proved ill-founded and by Thanksgiving the quarantine had been restored. However, by Christmas the epidemic had spent its force, and though many remained ill, new cases declined steadily. The tragic toll of the epidemic is dramatized by a macabre announcement by the Livingston Undertaking Company, which "is expecting soon a carload shipment of caskets to be distributed to their five stores, at Alpine, Marfa, Fort Davis, Marathon and Sanderson."[85]

In assessing the epidemic's impact in the Terlingua area, Robert Cartledge stated there were very few graves in the Terlingua Cemetery prior to 1918, but in November and December "they sure filled that damned graveyard up." This was also accounted for by the dead in the district's other mining villages who were buried at Terlingua. Regarding the Chisos fatalities, Cartledge added, "I lost two good men, possibly three, I just don't recall. This didn't count Mexicans. Don't know how many of them died." The disease exacted its greatest toll from the latter, and although Perry exhibited little regard for the Mexicans as fellow human beings, this epidemic, striking at the very heart of his mining empire, moved him deeply. Cartledge recalled that Perry returned to the mine early in 1919 and asked to be driven to the cemetery. Cartledge remembered it was a cold day and as Perry stepped from the automobile he pulled on his old gray, oversized overcoat with the pockets, as usual, bulging with papers. He then walked alone through the cemetery, viewing the mounds of fresh brown earth and pausing from time to

[84] Perry to Robert L. Cartledge, November 15, 1918, Chisos Mining Papers. The doctor referred to in the letter was Dr. N. O. Brenizer of Austin. When Wayne Cartledge first became ill, his father, Eugene Cartledge, sent Dr. Brenizer to Terlingua, where he stayed about two weeks and was paid $1,000 (Wayne Cartledge interview).

[85] Alpine *Avalanche*, November 14, 28, 1918, January 1, 1919. The Mexicans who died in Terlingua were interred without the services of an undertaker. Caskets were fabricated in the Chisos carpenter shop from materials purchased in the Chisos Store: pine planks, size 1″ × 12″, blue sateen, and big head brass tacks (Robert L. Cartledge interview, October 17, 1965).

time to read the inscriptions—some in Spanish and some in English —on the plain wooden crosses. He finally returned to the waiting automobile, got in, said nothing, and Cartledge drove him back to the mine office.[86] The epidemic had taken its toll.

Following the Armistice, the Terlingua Quicksilver District entered a postwar lull. Although the Chisos mine remained operative throughout the year and the directors voted to pay a $100,000 dividend, the combined production of five major mines was only 5,019 flasks of quicksilver, a 40 percent drop from the previous year.[87] The slight business slump early in 1919 proved temporary and by 1920 the economic indicators pointed toward continued business growth; however, this vigor failed to carry over to the quicksilver industry as it reached its lowest point in history.

Of the 6,339 flasks produced in the United States in 1921, 3,172 were credited to the Chisos mine, now the district's only producer. Thus in the depths of the industrywide slump, the Chisos mine achieved the dubious distinction as the largest producer of quicksilver in the United States. A number of factors contributed to this break in the industry's economic health: "(1) a tremendous drop in domestic commercial demand, (2) greatly increased production in Spain and Italy, (3) a very unsettled world market, and (4) considerable stocks in the hands of domestic producers and users."[88]

Being the leader of a faltering industry was not sufficient encouragement for the Chisos management to look to the future with confidence. The lucrative "ore chimney . . . which sustained a high production in the war years continued to yield, but not at the former high rate."[89] The company now furnaced a poor grade of ore that not only yielded less metal, but required additional personnel for hand sorting. With renewed emphasis on prospecting, the Chisos management hoped to maintain operations until richer ore was located or market conditions improved. This was the challenge of the coming decade. The district's dwindling population, now centered largely at Terlingua, looked with growing apprehension to the Chisos Mining Company for their economic security. Their interests centered on its progress in the future. None was more interested than Howard E. Perry.

[86] Robert L. Cartledge interview, January 4, 1966.

[87] *Cause No. 688*, Chisos bankruptcy proceeding, Federal Records Center, Fort Worth, Texas; Johnson, "Mercury Mining," 36 (October 1946): 447.

[88] Johnson, "Mercury Mining," 37 (March 1947): 28.

[89] Ibid.

CHAPTER 6

The Lengthening Shadow

BY 1921 the Chisos mine had passed its apogee of glory. Production records established during the preceding decade were never repeated, and within the immediate future the company relinquished its honored position as the nation's leading quicksilver producer. As the Chisos mine entered its third decade and passed the midway point in its brief life span, the usually optimistic Perry found little encouragement in an uncertain future. Both quicksilver production and prices had declined steadily during the preceding four years, and following the buildup of World War I reserves, little demand remained for the Chisos product. From the 1917 record yield of 10,791 flasks of quicksilver, district recovery declined to 3,436 flasks in 1920, mostly Chisos production. The price trend was equally dismal; after peaking at $125.89 per flask in 1916, the quicksilver market declined to $105.32 the following year, recovered to $117.50 in 1918, but dropped to $79.66 by 1920.[1]

Perry understood well the message these statistics conveyed. He could, however, enjoy the fruits of a temporarily affluent present while savoring the rewards of the recent past. In addition to previous stock dividends and apparently unlimited marginal benefits— no records of which exist—Perry voted himself an annual salary commensurate with the mine's productivity: 1917, $40,000; 1920, $30,000; 1921, $10,000; and similar amounts in 1924, 1925, and 1926. He said he received his last annual salary, also $10,000, in 1930.[2] Four years later two Texas geologists, E. H. Sellards and C. L. Baker, cited the three-decade Chisos recovery at $12 million.[3] Although a definitive appraisal of Chisos profits will remain forever

---

[1] U.S. Department of Interior, *Mineral Resources of the United States, 1917, Part I: Metals*, p. 367.

[2] *Cause No. 688*, Chisos bankruptcy proceeding, Federal Records Center, Fort Worth, Texas.

[3] E. H. Sellards, "Economic Geology of Texas," in *The Geology of Texas*, vol. 2, *Structural and Economic Geology*, by E. H. Sellards and C. L. Baker, *University of Texas Bulletin No. 3401* (January 1, 1934), pp. 529–530.

a matter of conjecture, the mine was a lucrative enterprise for many years, and by whatever criterion company production is evaluated—government reports, professional papers, legal testimony, newspaper accounts, or employee recollections—Perry was richly rewarded for his efforts.

By the time the 1920s began to roar many changes were in the offing. Although the Chisos produced steadily into the 1930s, it was no longer a bonanza and announcements of record productions were only recollections of the recent past. After basking in industrial glory and financial success, Perry, no doubt, found it difficult to accept a lesser position. His struggle to increase Chisos production and maintain a profitable operation is an underlying theme of the ensuing two decades.

While Perry failed in these efforts, he was, nevertheless, successful in other areas. These activities, though not altogether financially remunerative, exercised profound influence on life in the Big Bend. As the lengthening shadow of Howard E. Perry slowly and inevitably covered the village like a blanket of distrust, suspicion, and fear, his influence touched every facet of Terlingua life—the economy, social life, law and order, schools, politics, and even the intimacies of domestic life. While no citizen of the village escaped the pervasive mystique of the absentee landlord, it did not detract from their zest for pursuing the good life. Most could look back on the Terlingua experience with pleasant recollections, and some would later ponder inquiries about the hardships and personal restrictions imposed in the mining villages: "Well, there *was* no electricity, it *was* hot in the summer, and the glare *did* hurt your eyes. I *guess* those were hardships. But *we had a good life*. We *made* our fun."[4]

The quicksilver district enjoyed indigenous fun making early in its history. In 1907 the "Terlingua Record" column in the Alpine *Times* reported the antics of a young schoolmistress, Miss Luda Jones, who, disguised as a "coal black 'cullud lady,'" seated herself in the dining room of the company hotel. The report noted that the other guests "immediately shouted a vigorous protest declaring that they would not stand for any Booker Washington fete in Texas." When the ruse was revealed, "the joke was hugely appreciated and as Miss J. was pronounced by all a very fine specimen of feminine

---

[4] Interview with Mrs. Hattie Grace Elliot, Alpine, Texas, December 29, 1964.

Ethiopian beauty the boys prevailed upon her to sit for kodac photographs before discarding her disguise."[5]

Community fun was not limited to racist pranks. Sports—horseshoes, tennis, basketball, and baseball—were important recreational activities for Chisos employees. Baseball, however, attracted the most interest. Several mine employees organized a baseball team, and since the company provided no funds for recreational activities, they solicited donations from supervisory personnel to purchase equipment. All departments were represented on the team roster; for example, mine superintendent Marcus Hulings, bookkeeper Arthur Ekdahl, and store clerk Earl Anderau alternated pitching duties.[6] The "Miners" played most of their games at Terlingua, but when the Study Butte employees formed a team, games were alternated between the two villages. An intersectional game with Alpine, however, led to a rout of the home team. Offensive play apparently outshone defensive techniques, as this game ended with a score of 22-44 in favor of the visitors.[7]

Although the team roster contained both Anglo and Mexican names, most social activities in Terlingua followed strict ethnic divisions. Racial segregation was philosophical, as well as geographic, and while both communities enjoyed a rich social life they seldom shared their pleasures. The company store, where both communities met on equal terms, also marked the division between their two separate worlds: everything west of the store was Anglo and everything north and east of it was Mexican. Industrial installations filled the area south of the store.

Since Mexican social life focused mainly on domestic activities unrelated to company operation, it remained relatively free from Perry's interference. Music and family visitation, frequently interrelated, were the prime social functions in the Mexican sector of the village. In March, 1907, the Alpine *Times* reported that "a musicale was given in the social realms of the Mexican quarters Monday night last. Guitar music and singing, supplemented with sotol refreshments, made the affair hilarious and seemingly enjoyable to those present."[8]

H. C. Hernandez recalled much visiting in the evenings in Terlingua. "Nights were pleasant most of the year," he explained,

5 Alpine *Times*, March 6, 1907.
6 Interview with Earl Anderau, Alpine, Texas, December 22, 1964.
7 Alpine *Avalanche*, March 2, 1934.
8 Alpine *Times*, March 13, 1907.

"and after supper the old folks would sit around outside and talk while the kids played." One special recollection of those evenings in Terlingua was of "an old boy who showed up with one of those old-timey phonographs that had a big old horn hanging on the thing. It was an old Edison phonograph and the records were round like quarts of oil. *That was something special.*" Pets were an important part of Terlingua home life, according to Hernandez. "Everyone had pets," he explained. "Kids especially. About a million dogs and a thousand burros. Cats, and some even had coyotes. They would catch them when they were little fellers and tame them, but they wouldn't stay around the house very long. They'd take off. I remember one old boy that had a pet deer. Raised him on a bottle."[9]

Mack Waters, who also spent his youth in the mining village, emphasized the work ethic of Terlingua recreation: "Most of us had to work when we were out of school. We had some cattle and horses. No fences. Sometimes we would have to ride all day to find a cow. All Mexican boys had work to do, haul wood, water. After a kid got up to six or seven years old he had to work. *Everybody in Terlingua worked, women, children, and all.*" Waters noted, however, one exception to his domestic chores. In the late 1930s after the company established a motion picture theater adjacent to the Chisos Store, he attended the movies on weekends, "mostly on Saturday nights. They showed movies in both English and Spanish."[10]

As elsewhere, Christmas was the highpoint of the Terlingua social season. On Christmas morning before going to the Chisos Hotel for dinner, all supervisory personnel stopped by each Anglo home in the village to spread the Christmas cheer. There were so many homes between their dwelling and the hotel that, according to Mrs. Ed Babb, "by the time we arrived there for dinner I didn't care whether we had turkey or not."[11] Recalling an earlier Christmas at Terlingua, Hawley wrote: "Christmas Day, 1911 is a day I shall never forget. . . . On that occasion every white man at the Chisos Mine but myself was pretty well tanked up and the entire camp was in an uproar. Even . . . the steady and effecient . . . foreman at the smelter, was drunk as a Lord."[12]

[9] Interview with H. C. Hernandez, Alpine, Texas, October 23, 1972.

[10] Interview with Mack Waters, Panther Junction, Big Bend National Park, October 24, 1972.

[11] Interview with Mrs. G. E. Babb, Sanderson, Texas, December 26, 1964.

[12] C. A. Hawley, "Life Along the Border," *Sul Ross State College Bulletin* 44 (September 1964): 76.

But Christmas at Terlingua was for the children and the Chisos employees began a Christmas tradition in the 1920s that brought joy and goodwill to all segments of the mining colony. Since so few Anglo children lived in the village, everything was planned for the Mexican children. On the night before Christmas the Babbs would go to the Chisos Store and fill small paper bags with candy and nuts. Then early Christmas morning they would take these gifts to the Chisos Hotel, where the Mexican children would be waiting in line. As the children passed through the line, Mrs. Babb handed each one a sack, while her husband gave them an apple or an orange. She recalled how appreciative the children were and added that sometimes the very old grandparents—never the parents—would join the children in line. While they did not understand the meaning of Christmas as celebrated by Anglos, they nevertheless wanted to share the pleasures with the children. As Mrs. Babb explained: "The spirit of Christmas was for everyone, young and old alike, Mexican and Anglo, and we served all who stood in line. Yes, Christmas *was something* in Terlingua."[13]

While most facets of social life remained relatively free from Perry's control, he wanted to know what functions were held, who attended, and who visited whom. Domestic relations also received a particularly high priority because this was a prime factor in employee stability and efficiency; therefore, when the ship of romance encountered troubled waters, the Terlingua-Portland correspondence assumed the character of a metropolitan tabloid gossip column. Some examples follow:

I am at present standing a good chance of losing Fred. Not that Fred is not satisfied, but on account of his wife. . . . [Her] brother. . . . married a Mexican that has a bad reputation and it is embarrassing to [her]. She is continually complaining to Fred that she wants to leave here.

Waters had a little sad thing to happen to Hazel, she ran off and got married and it seems that Waters took Hazel away from her husband and took her with him to San Antonio.

I wrote several months ago that Henry was married. Alice came down here not long ago. At first she slept at Mrs. ———. However, it now seems that she is sleeping up with Henry.[14]

---

13 Babb interview.
14 Cartledge to Perry, undated, Chisos Mining Company Papers. Archives of the Texas State Library, Austin.

All was not tranquil along the Rio Grande, and Cartledge's stoic reports on marriage, divorce, or domestic instability usually aroused Perry's paternal instinct. Frequently he dispatched a wordy letter from Portland. In every case the objective was to calm an unsettled household that threatened the operation of the Chisos Mining Company.

Dancing was probably the most universally enjoyed form of social recreation in Terlingua, but, as with most other activities in the village, ethnic origin partitioned villagewide participation. The Mexican dances, however, were sometimes the exception to this rule, and the Anglos, even if they did not participate, frequently came to view the color and excitement of the Mexican *bailes* in Terlingua.

Dances, like almost everything in Terlingua, were scheduled to mesh with the industrial needs of the mining company. The Mexican employees, who made up a major portion of the Chisos staff, worked in three eight-hour shifts, which they alternated every two weeks. Since these bimonthly shift changes occurred on weekends, Hawley observed that "no work was done on Saturday night, and that was an occasion for dancing and merrymaking."[15]

"Nothing was more colorful and exciting than a Saturday night *baile* in Terlingua," Petra Benavídez, a former resident of Terlingua, recalled.[16] Excitement could be felt throughout the Mexican section of the village all day Saturday, but the first visible signs of the celebration appeared late in the afternoon, between sundown and dark, with the procession of the señoritas through the village. According to rigid Spanish social custom, young women were escorted to all social functions by older women in the community, who kept them under prying scrutiny throughout the evening. Mexican dances, even in this remote mining village, were severely proper affairs and were conducted with extreme formality and decorum.

Mack Waters has vivid recollections of these events. Waters, however, occupied a tenuous social position in Terlingua; his father was Anglo and his mother Mexican. And while he identified well with both communities, he was never completely accepted by either. But socially, in Terlingua Waters was a Mexican. He explained the restraints observed at the dances. "Girls couldn't come to the dances alone. There had to be two or three older women with them. You would see that little light going from house to house.

[15] Hawley, "Life Along the Border," p. 24.
[16] Interview with Petra Benavídez, Study Butte, Texas, October 25, 1972.

They would be picking up the girls to take them to the dance."[17] The escorting adults carried an oil lantern as they led the procession of the señoritas to the dance hall. Mrs. Ed Babb recalls that the procession was a thing of rare beauty. Each young lady dressed in her best dress reflected every color of the spectrum—blues, yellows, greens, reds, oranges, and white—as she made her way through the winding streets.[18] Paz Valenzuela, who once visited in Terlingua when a dance was being held, recalls the startling incongruity of a lovely young Mexican girl in her mid-teens wearing a pink party dress, emerging from a cave her family occupied as a residence and joining the procession to the *baile*.[19]

The dances were held in a large red corrugated tin building with a concrete slab floor, located on the old Alpine road east of the Chisos Store. The mining company erected the building specifically for this purpose, and while no fee was charged for its use, Robert Cartledge's permission was necessary to schedule a dance. The dances were also free; however, as a rule members of the Mexican community would get together and chip in to pay the musicians.[20]

The processions began in different sections of the village and arrived at the hall at about the same time. As the señoritas entered the hall, the lines dispersed, and they seated themselves on benches lining the perimeter of the dance floor. The young men then entered the hall; each proceeded to the person with whom he wished to dance first and stood there until he was accepted or rejected by the señorita. Those rejected proceeded around the room until everyone had a partner. Then the *baile* was ready to begin.

A local Mexican orchestra provided the music. While the instrumentation varied with the talent available, these groups usually consisted of several violins, at least two guitars, and sometimes a saxophone or trumpet. The music they played, like the dances the young people performed, were traditionally Mexican. "Their dancing was all round dances, such as the waltz, the polka and schottische. The square dance they knew nothing about," Hawley explains.[21] Earl Anderau, an Anglo who sometimes attended the Mexican *bailes* as a youth, recalls that these events were very important social events for the young people in Terlingua. While he

[17] Waters interview.
[18] Babb interview.
[19] Interview with Mrs. Paz Valenzuela, Alpine, Texas, October 27, 1972.
[20] Hernandez interview.
[21] Hawley, "Life Along the Border," p. 24.

enjoyed the dances, he noted even in his youth that frequently the quality of the music was noticeably lacking. "The Mexican tunes were ok," Anderau explains, "but when they attempted to play a request of some popular American tune, it usually turned out pretty bad."[22]

The same couples danced together throughout the first set, which usually consisted of three or four selections, says Anderau. At the end of each set the orchestra leader would shout, "*Dá le los gracias, todos*" ("Everybody thank them"), whereupon the young men thanked the young ladies for the dance and selected a new partner—always at the lady's discretion—for the next set. To request permission to dance again with the same partner was considered a severe breach of etiquette; however, the young man could return to his original partner for the third set.[23]

The Mexican dances attracted wide attention in the village from both Anglos and Mexicans. While these were essentially Mexican functions, Anglos did attend, both as observers and participants. Little interracial socializing transpired, even at these dances, as segregation, if not a consistently enforced policy, was involuntarily observed. "We'd all dance together. Maybe Anglos would dance with Anglo girls. We really didn't mix up too much but we'd all be together," Waters, explaining this tenuous relationship among the youth of Terlingua, states.[24]

While Waters saw racial concord in just being together, the Mexican dances also contained elements of strife and disunity. That much of the crime in Terlingua stemmed from these Saturday night dances was not a coincidence. Though racism was one factor, alcohol in its various exotic forms—sotol and tequila—was another. With the mine operating with a reduced work force during these bimonthly shift changes, the entire village was free to celebrate, and since the employees had access to the free flow of illicit liquors across the Rio Grande, these weekend holidays frequently erupted into fiestas of violence and tragedy.

Expressing the pacifist viewpoint, Waters explains: "None of the fellows wanted any trouble because if there was any kind of disturbance the ladies would take all the girls home. And that would end the dance."[25] H. C. Hernandez was more candid in his view of crime in the Mexican village. He recalls there was frequent trouble

[22] Anderau interview.
[23] Ibid.
[24] Waters interview.
[25] Ibid.

at the dances, "fist fights and sometimes they would cut one another, and sometimes they would shoot each other." He remembers two brothers specifically, Winslow and Jack Coffman, who frequented the Mexican dances "always looking for girls and trouble and always trying to pick a fight with somebody, especially the Mexicans."[26] They were successful on all counts. Soon after Winslow was severely stabbed at a Terlingua dance, he and his brother, in the company of two Mexican girls, met an early death in front of a Mexican's Winchester during a gunfight at Study Butte.[27]

While most Anglos recognized the periodic violence that occurred east of the Chisos Store, they regarded Terlingua as a wholesome community. As Anderau explains, the town had "no organized vice or crime. But there was always plenty of liquor, even during prohibition. Smuggled across the border. But it was a good town. A good place to live."[28] Anderau's praise for the quality of life at Terlingua was an involuntary tribute to Howard E. Perry. The master of the Chisos professed an abhorrence for violence and moral laxity, not as a matter of principle but as a deterrent to company efficiency. For nearly a quarter of a century his demands for the Chisos version of law and order fell on the already overburdened shoulders of his general manager, Robert Cartledge. And while Cartledge became a controversial figure in the Big Bend, he almost singlehandedly and in the name of Perry and the Chisos Mining Company gave Terlingua and the surrounding area its only viable law enforcement. By whatever standard Cartledge and his Terlingua justice is evaluated, his unceasing efforts improved the quality of life in Terlingua for both Anglos and Mexicans.

Reports of theft, violence, and disorder parallel the early history of the Chisos Mining Company. In 1904 the Alpine *Times* reports that while I. A. Dewees was camped near the Chisos, "his Mexican stole a suit of clothes, a pair of shoes, his sixshooter and about $40 in checks." A later report noted that in a "fight at the Chisos Co. mining camp last Tuesday a Mexican by the name of Louis Sosa was dangerously shot." The item adds that Sosa's assailant escaped. As the village grew, the newspaper reported deaths by gunshot—legal and illegal—with frightening regularity. An item headed "Killing at Chisos Mine" states that Brewster County Deputy Sheriff Guillermo Rodríguez "shot and killed Máximo Carbojal, who was

26 Hernandez interview.
27 Interview with Robert L. Cartledge, Austin, Texas, October 2, 1965.
28 Anderau interview.

resisting arrest." The story explains further that "the dead man has not borne the best of reputation."[29]

Early in the twentieth century the Terlingua country was conducive to crime and disorder. Well beyond the pale of organized law enforcement and in close proximity to an unguarded international boundary, that region existed in a strange twilight zone between two national cultures. While most violators were Mexican, Hawley, who served as justice of the peace at Mariposa before joining the Chisos mine in 1910, sought a philosophical justification for their misdeeds. He maintained that "race, color, or nationality . . . have little to do with the essential qualities of human beings." Instead, he cites "a common saying among the white people at Terlingua that the only really bad Mexican was the one who had lived for some years on this side of the river. Such a one had acquired all our American vices but none of our virtues." "Now and then one Mexican would kill another," he admits, "but that was invariably the result of gun toting and liquor." Hawley, however, sustains Anderau's belief that Terlingua was a wholesome community and explains that his docket consisted mainly of cases involving disorderly conduct, though "that too was caused by too much liquor."[30]

While Hawley may have minimized the evils of liquor and disorderly conduct, Perry on the other hand came to recognize these as obstructions to progress. If the Mexicans elected to murder each other at the Saturday night dances, this was, for the most part, remote and unrelated to the profit and loss columns in the Chisos ledgers. But bootleggers, sotol, absenteeism, tardiness, and employee inefficiency were clear-cut matters of economic reality with which Perry and Robert Cartledge could grapple personally.

Newspaper accounts and employee recollections provide the only records of Terlingua law enforcement during the early years of the century. Hawley states that he accepted an appointment as justice of the peace while employed by the Marfa and Mariposa Mining Company (he joined that firm in 1907), which he apparently brought with him to the Chisos mine in 1910.[31] Robert Cartledge joined that firm a year later, and when Hawley left the

---

[29] Alpine *Times*, October 12, 1904, January 16, 1907; Alpine *Avalanche*, April 27, 1911.

[30] Hawley, "Life Along the Border," pp. 33–34, 62.

[31] Ibid. Hawley's only reference to other law enforcement agencies in the mining district included the Internal Revenue and the Texas Rangers, for whom he expressed low regard.

Chisos mine in 1912, Cartledge replaced him as justice of the peace.[32]

Precinct No. 4, the Terlingua area, was created in 1916 and two years later Cartledge's name first appears on the Brewster County voting records. He held that position unopposed until 1940, when Vina Brown defeated him, twenty-two votes to fifteen.[33] An ironic symbolism is contained in these statistics. After twenty-nine years of dedicated service to the Chisos mine, Cartledge's dismissal by Perry came almost simultaneously with his defeat at the election polls. Both events, obviously related, reflect a cultural and economic change that was occurring in the Big Bend. One era was about to end, while another had begun many years before; Cartledge had played a leading role in this changing drama.

"Demon Rum" in its many forms lay at the root of most crime in Terlingua. Perry recognized this early in his tenure in the Big Bend when in 1908 he wrote the mine management lamenting the absence of some acceptable form of nonalcoholic recreation.[34] From his vantage point along the Rio Grande, Cartledge realized that to much of the Chisos labor force, liquor and recreation were virtually synonymous. He also knew that if he was to maintain an efficient operation, he would have to separate the men and their drink. His target: the bootleggers.

As expressed in the Eighteenth Amendment, the prohibition of alcohol stemmed from a puritanical impulse to reweave the frayed moral fiber of the American character by denying its citizens free access to intoxicants. Prohibition, Terlingua style, while struggling for the same elusive goal, held no such lofty purposes. Perry wanted maximum yield with "not too much hilarity" from a labor force which probably could not comprehend the meaning of the National Prohibition Act and whose cultural traditions tolerated and condoned the very custom that Americans were attempting to eradicate through legislation. But whether one sought moral restoration or economic efficiency seemed not to matter, for in Terlingua, as in Trenton, Tallahassee, Tucumcari, and Tarrytown, prohibition failed. And while many well-meaning lawmen attempted to enforce the statutes to the letter of the law, this was destined not to be. This law ran counter to the will of the majority and as the twenties length-

[32] Cartledge interview, October 2, 1965.

[33] Election Records, Office of the County Clerk, Brewster County, Alpine, Texas.

[34] Perry to the mine management, September 25, 1908, Chisos Mining Papers.

ened toward the thirties, Cartledge must have realized that he was fighting a losing battle. But instead of accepting defeat, he called for help.

Others than Cartledge realized the futility of his position. In the spring of 1928 a Terlingua citizens group filed a petition with the Brewster County Commissioners Court requesting police protection from the lawless element in that immediate region. They specified that the court forward their request to the adjutant general in Austin and request that two Texas Rangers be stationed at Terlingua. At that time only two rangers were assigned to the entire lower Big Bend. One was stationed at Glenn Springs and the other at Presidio; neither was within a responsive distance to Terlingua.[35]

When the citizens' appeal failed, Cartledge embarked on a one-man campaign to get help. He first wrote to his father in Austin, suggesting that he use his political influence to have a Texas Ranger stationed at the mining camp. This appeal came in the wake of an armed holdup of Sublett's store a few miles south of the village on Terlingua Creek. Cartledge wrote: "Three men, evidently from Mexico, robbed Sublett last Saturday night. They took all the dry-goods he had. . . . Sublett himself was in town. His wife and daughter being the only ones at home." As Cartledge reasoned, "there is not but one way to stop it [lawlessness] and that is the service of someone that will lay out at night on trails and catch . . . the mexicans that ride at night."[36]

Bootleggers accounted for most of this nocturnal traffic, and with an eager Terlingua market awaiting their contraband liquors, Cartledge must have realized that he alone could do little to stem the northward flow of sotol. He observed to his father that "these robberies and stealing are increasing and consequently more of my time is taken up." In desperation he implored the elder Cartledge to "make one more last and final effort to see if we can get [T. C. (Creed)] Taylor appointed Ranger Sargeant . . . and stationed down here." He suggested that his father "talk to General [Henry] Hutchin[g]s [adjutant general] once more and see if he cannot get the Governor [Miriam A. Ferguson] to appoint Taylor."[37] The growing pressure from Perry's demand for law enforcement at Terlingua

[35] Alpine *Avalanche*, May 18, 1928.

[36] Cartledge to Eugene Cartledge, May 3, 1933, Chisos Mining Papers.

[37] T. C. Taylor was the son of Creed Taylor, a veteran of the Texas Revolution, who joined the Texas Rangers in 1841, under John Coffee ("Jack") Hays. During the Mexican War, again with Hays, he fought in the battles of Palo Alto, Resaca de la Palma, Monterrey, and Buena Vista.

was evidently taking its toll, for Cartledge confessed to his father, "I . . . have been a nervous wreck the past two weeks, but I hope I can pull myself together again soon."[38]

If Cartledge appears melodramatic in emphasizing his demands, just reason existed for his trauma. As he attempted to prevent "too much hilarity" among the Mexicans, he placed himself squarely between the bootleggers, armed and eager to sell their wares and disappear across the border, and their anxiously awaiting customers, ready to gamble their rights to Chisos employment in exchange for a bottle of cheap sotol. The situation was volatile and tempers were growing short. Cartledge went armed, seeking the violators, always ready to exercise his mandate from Perry. In almost childlike devotion he wrote Perry, "I was out all last Saturday night to hold down drunks and in this connection I like to have had to kill two Mexican bootleggers. In order to avoid killing them I took a terrible chance of them killing me."[39]

Whether Cartledge realized it, he and the bootleggers were moving along a collision course, without compromise and without solution. Without additional law enforcement officers, which were not forthcoming, the illicit traffic would continue; and while Cartledge probably knew he could not stem the flow, he would continue "to do more work on Saturday nights in trying to prevent men from getting drunk."[40] Cartledge's latest brush with the Mexican bootleggers was prophetic. They would meet again, like helpless pawns in an unequal contest of strength, and the issue would be resolved temporarily with gunfire.

While neither the governor nor the adjutant general granted Cartledge the assistance he sought, other lawmen frequented the mining village. Itinerant Texas Rangers appeared in the quicksilver district on special assignment and the United States customs officers, who shared Cartledge's unofficial vigilance along the Rio Grande, made periodic investigations in Terlingua, though their interest was contraband merchandise, not bootleggers and drunks. Reappraising his predicament, Cartledge shared his frustrations with Ranger Taylor:

> Bebee [a customs officer] tells me that you are expecting to come down here this month. If you all can search a house, I will sign complaint, try to get down early some Saturday morning.

38 Ibid.
39 Cartledge to Perry, January 22, 1935.
40 Cartledge to Perry, June 18, 1937.

I have a place here that I know sells sotol, but I have layed off searching it to try and catch a time that I was sure they had it. I have told Bebee of a couple of times that I felt sure that there was some there, but he apparently shows no interest about it. It seems that our laws are now such that all that a customs man can do is draw a salary. . . . I heard about third hand, that Bebee said that he was not supposed to bother with Booze, that all he was to look out for was merchandise.[41]

When the showdown came, however, Cartledge was not alone. A rotting wooden cross in the abandoned Terlingua cemetery bears the following terse inscription: "Feliz R. Valenzuela, Feb. 22, 1882; June 19, 1938."

Valenzuela had joined the Chisos mine as a freighter before World War I and over the years he and Cartledge became close friends. Through Cartledge's patronage Valenzuela had his choice of preferential job assignments with the company, and in the 1930s the two men entered into a joint ranching venture in the Terlingua area. A former Chisos employee, describing their friendship, explained: "Feliz was Bob's closest friend. In fact, he may have been Bob's *only* friend."[42] At Cartledge's suggestion Valenzuela entered the race for constable of Precinct No. 2 in 1924 and won the election unopposed, receiving only one vote.[43] With the two men constituting the precinct's only permanent law enforcement agency, each began to assume the responsibility for the other's safety. Elmo Johnson, who operated a ranch near Terlingua and knew both men well, explained that "Cartledge was the justice of the peace but had no authority to make arrests. His deputy [actually a constable], Feliz Valenzuela, had that authority." Johnson, pausing as if to phrase his thoughts carefully, added, "Feliz idolized Bob. Would have died for him. In fact, he did."[44]

On Saturday night, June 19, 1938, two dances were in progress in the quicksilver district, one at Terlingua, the other at Study Butte. Cartledge and Valenzuela patrolled both sites during the evening in Cartledge's Buick sedan but encountered no trouble. Returning to Terlingua around midnight they met Octaviano López, who told them that his two brothers and Marcario Hinojos, Jr., had just departed for the dance at Study Butte with a load of sotol.

41 Cartledge to T. C. Taylor, August 1, 1936, Chisos Mining Papers.
42 Interview with Elmo Johnson, Sonora, Texas, June 3, 1966.
43 Election Records, Brewster County, Alpine, Texas.
44 Johnson interview.

Cartledge and Valenzuela left immediately, hoping to intercept the bootleggers with their cargo of contraband liquor.[45]

Before leaving, however, Valenzuela reminded Cartledge to re-check the manner in which he was carrying his gun. The western style gunbelt was never part of Cartledge's attire; when on duty as a peace officer, he carried a .45-caliber Frontier Model Colt pistol in his right front pocket with the handle and trigger guard exposed. Valenzuela had previously suggested to Cartledge that, if the owner is right-handed, he should carry his pistol on his left side in a "cross draw" position with the handle reversed. While Cart-ledge never fully understood his constable's reasoning, he agreed on this particular night to reverse his usual custom and carry his gun in his left front pocket. Although Valenzuela did not realize it at the time, this advice, calculated to protect his friend's life, re-sulted in his own death later that night.

Their trip to Study Butte failed to produce Hinojos and the López brothers. Since the dance there was unusually quiet, Cart-ledge and Valenzuela returned to Terlingua. By that time it was approximately one o'clock in the morning, and with the Terlingua dance over, Cartledge suggested that they go to bed. Valenzuela, however, insisted that he should go back to Study Butte alone just to make certain that they had not missed Hinojos. Cartledge refused to allow his constable to go alone. He had good reason for his concern.

When at a previous dance at Terlingua Hinojos had become in-volved in a drunken altercation, Valenzuela, attempting to restore order, had slapped him. Hinojos had then attacked Valenzuela and had had to be restrained by some of his friends. Cartledge, standing nearby, had told the elder Hinojos to take his son home and put him to bed, but had added that he wanted to see him the following day in his office.

When the young bootlegger arrived at Cartledge's office the next day, Cartledge reprimanded him for his misconduct and warned him that if any more such disturbances occurred, he would personally lock him in the Terlingua jail. When asked to explain his threats against him and Valenzuela, Hinojos, a little embarrassed, shrugged his shoulders and mumbled in broken English, "Aw, meester Bov, I'z jees a leetle drunk." This short visit, though seeming to end cordially, did little to cool Hinojos' burning resentment for

[45] This account of the death of Feliz Valenzuela, unless otherwise cited, is based on a series of interviews with Robert Cartledge, dating from December, 1964, through February, 1972.

the two men who had embarrassed him publicly in the presence of his friends.

Shortly after this meeting two immigration officers shot a bootlegger friend of Hinojos'. Cartledge stated many years later that he believed that this event was symbolic to Hinojos. To some border Mexicans bootlegging and all forms of international larceny were a way of life, and occasional conflict with the law that sometimes erupted in gunfire was simply an occupational hazard. "The law had killed his bootlegger friend," Cartledge rationalized, "and now Marcario began to brag that he was out to get them—meaning me and Feliz." Cartledge's awareness of this resentment was the basis for his decision to accompany Valenzuela to Study Butte.

On the second trip they found their suspect. The dance had ended and Hinojos, having taken a guitar from a member of the orchestra, was attempting unsuccessful to play it. Valenzuela in turn took the instrument from Hinojos and returned it to the owner. This encounter produced no exchange of words between the two men; however, when relieved of the purloined guitar, Hinojos got in the automobile with his friend and drove west toward Terlingua. Cartledge and Valenzuela followed. As the two automobiles approached the Lajitas-Terlingua intersection, instead of turning right toward his home—which Cartledge anticipated—Hinojos speeded up as he turned left toward Lajitas. Cartledge explained later that he thought that this border village was Hinojos' source of sotol "and if he was carrying some in his car he hadn't been able to dispose of that night, he was returning it to his supplier. This I intended to stop." Cartledge accelerated his car, passed the suspected bootleggers, and forced their automobile into the ditch. When the two automobiles slid to a stop, Cartledge leaped from his Buick and ordered the occupants of the other vehicle out and began an immediate search for the suspected sotol.

The vehicle in which Hinojos was riding was a two-door sedan, and to avoid the obstruction of the steering wheel, Cartledge opened the right front door, titlted the seat forward, and leaned inside to determine if any liquor was concealed on the rear seat. In Cartledge's haste to search the vehicle, he forgot momentarily to keep the suspected bootleggers under surveillance, and in that one brief unguarded moment pandemonium erupted beside the Lajitas road. As Cartledge extended his body into Hinojos' vehicle, his external view was temporarily obstructed, and his left side was both exposed and unprotected.

Cartledge recalled later that as he felt Hinojos whip the gun from his exposed left pocket, he first sensed the impending tragedy. His shout to Valenzuela, "Look out Feliz! Marcario's got my gun," was drowned out by two quick pistol shots, and almost simultaneously the López brothers attacked Cartledge and threw him to the ground beside his automobile. The *Avalanche* reported the tragedy in four terse sentences the following Friday:

> When the two officers stopped the car, Hinojos jumped out, it was said, grabbed Mr. Cartledge's pistol from his belt and shot Valenzuela twice, once through the right lung. While Mr. Cartledge was struggling with two of Hinojos' companions in the darkness nearby, Hinojos took Mr. Cartledge's car and made his escape to the river where confederates were apparently awaiting him with horses, and made his get-away across the river.
>
> Two Lopez brothers of Terlingua, in the car with Hinojos and believed to be his accomplices in the illegal liquor running, also escaped and are still at large. . . . Valenzuela is survived by a large family.[46]

Disarmed and afoot, alone and defeated, Cartledge walked back to the mine office and called the sheriff in Alpine, but for all practical purposes the case was closed. Cartledge found his automobile the following day abandoned at Lajitas. No one had seen Hinojos, nor was he ever seen again in Terlingua. The López brothers, whose only charge was assault, were never brought to trial. Cartledge had enforced Perry's law, but though he had achieved success, his was a vacant victory. Valenzuela was dead, Cartledge had lost his best friend, Hinojos was exiled for life, and bootleggers continued to sell sotol in Terlingua as long as buyers remained.

On the next afternoon Valenzuela was buried in the Terlingua cemetery and a few weeks later the family placed a concrete cross at the site of the tragedy. But the death of Feliz Valenzuela lingered long in the minds of those living in Terlingua as a symbolic event. While the ensuing debate followed a traditional rationale—a tragic death, a senseless killing, a life sacrificed for economic expediency —it also drew the role of Cartledge the lawman into sharper focus. For whatever Cartledge was to Terlingua—and he was a pivotal figure—his tenure as justice of the peace affected life greatly in the quicksilver district. Different people expressed differing views of the man; most had opinions, none were neutral. Yet the one area of consistency was the inconsistency of motivations that they believed drove Cartledge relentlessly to give his adopted community law

[46] Alpine *Avalanche*, June 24, 1938.

and order. Ambition, insecurity, belligerence, desire for power, fear, and admiration for Perry—all surfaced as reasons for the dominant role Cartledge played in life at Terlingua. And when all the data is analyzed and evaluated, it can be concluded that because of Cartledge's—and Perry's—demand for law and order, for whatever reasons, they made Terlingua a better place to live.

Pablo Sandate, who came to Terlingua from Sierra Mojada, Mexico, in 1912 to work at the Chisos mine, pronounced crime in Terlingua "no worse than any place else," because "we were watched over by the law [Cartledge]." Sandate then added philosophically, "He who would look for a fight would find it, but he would stay at home, nothing would happen to him [my translation]."[47] Mack Waters, who represented a later generation of Chisos workers, saw Cartledge in a somewhat different light. He cited the Chisos general manager's bellicose manner as the key to his attitude toward law enforcement. As Waters explained, "He tried to be tough." Then rationalizing his views, Waters adds, "Any law enforcement officer has a few enemies and I think Bob had a few. While I never saw Bob hit a man I have heard him talk pretty rough."[48]

Racism and authoritarianism were two dimensions to Cartledge's personality noted by Mrs. Elizabeth Bledsoe, a former Big Bend school teacher. She believes that "Bob liked the authority of the justice of the peace. He would be overbearing with the Mexicans but I don't believe he could get by with it with white people."[49] Rubye Burcham, who also went to the quicksilver district as a school teacher, sustains her former colleague's interpretation. She explains that while Cartledge "probably didn't manhandle them [the Mexicans], they were afraid to do anything they thought he didn't want them to do."[50]

Apparently Perry's Terlingua law was based on a racist concept. His was a law for Mexicans. Anglo names seldom appear on Cartledge's court docket. That the village population was approximately 98 percent Mexican provides only a partial explanation. A more fundamental reason stems from regional mores, images, and concepts for which neither Cartledge nor Perry could be held accountable and which they did not comprehend. Hawley, quoting an early day Del Rio attorney, remarks, "In this state [Texas] we have one

---

[47] Interview with Pablo Sandate, Alpine, Texas, October 24, 1972.

[48] Waters interview.

[49] Interview with Mrs. Elizabeth Bledsoe, Alpine, Texas, October 25, 1972.

[50] Interview with Mrs. William D. (Rubye) Burcham, Alpine, Texas, October 24, 1972.

set of laws for white people and one for Mexicans, all in the same words and in the same book."[51] And while Cartledge may or may not have read Hawley's philosophy, he was part of the regional syndrome that accepted, and probably applied, this interpretation of the law.

On the other hand, Cartledge frequently tempered justice with mercy, and while observing the spirit and intent of the law, he could "forget the letter" if conditions seemed to warrant it. Cartledge may have also viewed his legal obligations through the eyes of a politician and public servant who at times must please all segments of his constituency. Such was the case of a young couple who confessed their devotion for each other and begged Cartledge to marry them immediately. If he should refuse, they threatened to elope. The prospective bride's parents also approached Cartledge and pleaded with him to use his influence to prevent the elopement. Cartledge reportedly pondered the case and arrived at a decision calculated to satisfy both parties: after apprehending the eloping couple, he locked them together overnight in the one-room Terlingua jail![52]

While Cartledge, as well as Perry, regarded his legal position and the Terlingua jail as the keystones of local law and order, traditional penological procedures were frequently abridged in favor of convenience and economy. As Waters explains:

> I remember one thing they were doing that my daddy didn't agree with them on. They would get them boys and put them in jail and keep them there for four or five days. But at noon and at suppertime they would turn them loose and let'em go home to eat. I had a cousin there named George Scott who got in some kind of trouble— fightin'—and they had him in jail. One day at noon they sent him home to eat. My daddy didn't like it and he told Bob, "You got this man a prisoner and you're gonna feed'im." Of course, he didn't keep him long after that.

Waters added that he never recalled an Anglo being jailed in Terlingua.[53]

Mrs. Paz Valenzuela, a distant relative of Feliz's, reports visiting Terlingua and seeing "many Mexican men tied to iron stakes outside of the Terlingua jail. I believe they were mean to them

---

[51] Hawley, "Life Along the Border," p. 9.

[52] This is an often-repeated story of Terlingua life. One version was described in the Waters interview.

[53] Ibid.

down there."[54] H. C. Hernandez, grandson of the Terlingua jailer, substantiates Mrs. Valenzuela's account and explains that he helped feed the prisoners chained in front of the jail. He describes this restraint as "an iron stake set in concrete with a large iron ring at the top of the stake . . . a long chain would be run through this ring and, in turn, would be wrapped around the ankle of each prisoner and locked in place."[55]

Hernandez is candid in his appraisal of the Terlingua justice of the peace and the varied responsibilities he discharged in the village. He explains that while some people feared Cartledge, others looked to him with confidence. He counseled the villagers, both Anglo and Mexican, on domestic matters, financial problems, "he jailed'em, married'em, and I remember one old Mexican couple who even wanted Bob to divorce'em. Bob was out and I told'em that a divorce would cost them quite a bit. Well, they talked it over and decided, since they were getting old, they would just go on living together." Hernandez, who gained his sobriquet, "Little Judge," while serving as Cartledge's first court interpreter, explained that he felt that Cartledge was just in his relationship with the Mexicans. Responding to the question of whether Cartledge was "hard on the Mexicans," he explains:

> In a way he was and in a way he wasn't. I look at it this way, since they didn't have no law down there, he figured that by keeping those bootleggers out of those camps he would be doing those people some good. Which he was. Some of those people would get drunk and stay drunk for a couple of weeks and wouldn't work. So I think Bob decided to go ahead and do what he thought was right and clear all of those bad elements out of that part of the world.[56]

W. D. Smithers, a photographer who lived at the Elmo Johnson ranch in the lower Big Bend, believes that Cartledge's Terlingua justice was oriented toward the Chisos Mining Company and its economic well-being. He recalls that when a young man named Jones died at the ranch, Smithers went to Terlingua to report the death to Cartledge. Smithers explains:

> It was about two o'clock in the morning when I got to the mine. I woke Bob up and told him that a boy had died down at the ranch. Bob was furious. He said, "What the hell do you have to wake me up at this time of night to tell me that?" I said, "Don't you have to hold

---

[54] Valenzuela interview.
[55] Hernandez interview.
[56] Ibid.

an inquest?" And Bob said, "Hell no! Bury him!" And with that he slammed the door in my face.

Smithers added that when some time later another death occurred at the ranch, "I sure as hell didn't bother Bob Cartledge this time."[57]

Smithers' friend and landlord, Elmo Johnson, lived in the Big Bend for more than a quarter of a century and knew the people, Anglo and Mexican alike, from firsthand experience. He and the Mexicans along both sides of the Rio Grande enjoyed a mutual admiration and respect. Johnson, explaining this, says, "I went down there to live among'em, and I lived among'em." In appraising Cartledge's contribution to the community welfare, Johnson could probably speak with greater candor and objectivity than many of his neighbors. He believes that Cartledge's role as Terlingua's controversial justice of the peace represented a positive good for the community. Choosing his words carefully, Johnson has stated, "Bob was the dominant figure in the Big Bend. He *was* the law west of the Pecos. The Mexicans feared him and I guess it was a good thing, that far from the law. I don't believe the sheriff was at my house in twenty years."[58]

Rubye Burcham, while sustaining both Johnson's and Hernandez's views on Cartledge's contribution to community stability, saw Perry as the motivating force. "Just *anything* that Mr. Perry wanted him to do," she explained, "he [Cartledge] would do it to the best of his ability. And Mr. Perry wanted law and order."[59]

That Perry wanted law and order—even if his motivation was economic—resulted in a social climate that fostered a full, rich lifestyle for those who came to Terlingua to serve the Chisos mine. While isolation was an ever-pervasive factor, the company employees, their families, and friends created a self-contained social environment that, according to former Chisos employees, transcended limitations imposed by remoteness. And for whatever restrictions Perry may have pronounced on the day-to-day life in Terlingua, they were negligible. As Hattie Grace Peters, apparently speaking for the majority, remarked: "*We had a good life. We made our fun.*"[60]

---

[57] Interview with W. D. Smithers, Alpine, Texas, June 7, 1966.
[58] Johnson interview.
[59] Burcham interview.
[60] Elliot interview.

CHAPTER 7

School Bells and Ballot Boxes

IF Perry fostered law and order for the indirect benefit of the Chisos Mining Company, his interest in public school education was even more closely aligned with the profit incentive. Although public school instruction came early to the Big Bend in the wake of the quicksilver development, school politics did not become a public issue until midway in Terlingua's life span. Not until a steadily increasing transient population gave schools, politics, and profit a singular identity in the mid-1920s did public education attract the interest of the mine owners. None was more interested than Howard E. Perry.

Newspaper accounts and the recollections of former school teachers and mine employees provide the earliest record of organized education in the quicksilver district. When William B. Phillips went to Terlingua in 1906 as Chisos superintendent, he found no schools and his wife taught their youngest son, Laurance, in their home.[1] The following year, however, the Alpine *Times* reported that "Miss Lydia Jones, a very accomplished and popular young lady . . . has proven herself a successful teacher in conducting the Terlingua public school during the present term." Schools were apparently in operation in other areas of the Big Bend as the same item noted that "the charming little school mistress, Miss Cora Smith, who has charge of the school at Lajitas, is giving good satisfaction and is greatly beloved by both her pupils and the people of that place."[2] A later report reiterated the multiple benefits—social, as well as educational—that public education brought to the region. The Alpine *Times* reported that "the school at Big Bend [Study Butte] is in a flourishing condition under the tutorship of a Miss Kerns. This charming young lady is very popular with everybody in the Bend—especially with the young men."[3]

[1] Interview with Laurance V. Phillips, Austin, Texas, December 14, 1964.
[2] Alpine *Times*, February 27, 1907.
[3] Ibid., April 3, 1907.

Despite these optimistic reports, frontier life was harsh for these young schoolmistresses, conditions were primitive, and the financial rewards were small. Jeanette Dow Stephens taught school at Terlingua during the 1909–1910 school year, received $65 per month, and held classes in "a large old 'tent-house' with long bench-like seats. I can't remember a blackboard, but we probably had one of some kind. Everything was rough and primitive."[4]

The 1911–1912 Brewster County Annual Statement of School Funds, which provides the earliest statistical data on Big Bend education, substantiates Miss Dow's minimal salary claims. While this report mentions neither Terlingua nor the Chisos Mining Company, the cost entries for Common School District No. 4, which included Terlingua, list the teacher's base salary as $65.00. The Chisos Mining Company, however, appears first in the 1912–1913 report, when it received $120.00 for rent and $24.25 for supplies. While the company apparently received no rent the following year, income for supplies increased to $192.03.[5] Although these relatively small transactions mark the beginning of the company's intrusion into local school matters, change was in the offing, as Perry had heard the school bells ringing in Terlingua, especially those attached to the cash register.

The modest budgets and low teaching salaries at Terlingua were reflected in equally lax professional standards. Miss Dow continued:

Mr. Hawley . . . was our everything in Terlingua. . . . [He] got us some easy primers with one line of Spanish and the same line in English. I am afraid that I learned more Spanish than the Mexicans learned English. I was so inexperienced in such teaching that I am not so sure that I helped the little Mexicans very much. They probably learned to write their names, and to count and make the figures and to sing little songs in English. I had five or six white children and ten or twelve Mexicans, maybe more. One Mexican boy spoke fairly good English and helped me quite a bit.

Miss Dow remained in Terlingua only one year. Her reasons for leaving were that she "wanted to see a train! Secondly, I was good and homesick. Thirdly, I was anxious to get back to civilization and continue my education."[6] She was probably in a minority, as many

[4] Jeanette Dow Stephens, Shreveport, Louisiana, to Kenneth B. Ragsdale, June 25, 1968, in possession of author.
[5] Annual Statement of School Funds, 1911–12, 1912–13 and 1913–14, Brewster County, Records Division, Texas State Library, Austin, Texas.
[6] Stephens to Ragsdale, June 25, 1968.

young women who went to Terlingua to teach found the experience rewarding, married men employed by the mines, and became permanent citizens of the region, if not of Terlingua. Eleanor Schley (later Mrs. Robert L. Cartledge), Rubye Richardson (later Mrs. William D. Burcham), Katheryn Turney (later Mrs. Earl Anderau), Hattie Grace Peters (later Mrs. Dave Elliot of Alpine), Mrs. Elizabeth Smith Weaver (later Mrs. Eligah Bledsoe), and Annie Lee Adams (later Mrs. Ed Babb) are representative of that group of immigrant educators who, through both formal and informal education, contributed greatly to the social and cultural life, as well as the general welfare, of the region. Unknowingly and without plan, Howard Perry bestowed education as one of the permanent benefits on the region he occupied temporarily.

Hattie Grace Peters began her teaching career in Terlingua in 1923, when she taught fifty-three students—fifty-two Mexicans and one Anglo—from pre-primer through the fourth grade in one adobe classroom. She was paid $100 per month, from which the Chisos Mining Company deducted $30 for room and board.[7] Despite the low pay and inconvenience of frontier life, she pronounced the experience "a marvelous year." Like Jeanette Dow, she questioned her professional qualifications for the assignment, as language was a barrier to learning. Her knowledge of Spanish—which the school board forbade her to use—was limited to basic high school and college requirements and few of her students spoke or understood English.

If language retarded the students' academic progress, nothing restrained their mischievousness. She recalled a young Mexican boy who raised his hand and exclaimed aloud, "Teechair, Teechair, I be 'scused?" After receiving permission to leave the room, he departed and quickly returned with a live, wiggly, menacing—but totally harmless—vinegaroon, well calculated to frighten the young school mistress. His success was rewarded by his being made to stand in the corner and hold two sticks of stovewood above his head until he could hold them no longer. Miss Peters recalled later that she tried the punishment herself, decided it was cruel, and never used it again.[8]

She later turned to a more orthodox punishment when Santiago Jacques insisted on reading his lesson aloud during class, "virtually screaming at the top of his lungs," she recalls. When young Jacques

[7] Interview with Hattie Grace Elliot, Alpine, Texas, December 29, 1964.
[8] Ibid.

failed to respond to her admonishment, she led him out of the class-room with one hand and a leather belt in the other. Realizing that he was to be punished for his misconduct, he began jumping up and down and crying, whereupon she instructed him to "be still so I can whip you where you are supposed to be whipped." As she swung, Jacques, failing to heed her well-meaning advice, jumped, and the leather belt missed its intended target and struck his bare leg instead. The impact ruptured an infected boil and "blood simply gushed down his leg. When he saw the blood he thought I had killed him. Well, he broke loose and ran home and I didn't see him until after lunch when he returned with every relative he had in camp." She was confronted by "*mama y papa, aguello y aguella, y tio y tia, y muchos primas.*" Everyone talked at once in Spanish and waved their hands violently in Hattie Grace's direction, and then, their wrath vented on the surprised and speechless teacher, departed in a group as suddenly as they had appeared. She heard nothing more of the incident and did not see Jacques again for eight weeks, but when he finally returned to school "he was a model student. It was a lesson to us both. He didn't read aloud and I never whipped anyone again."[9]

Miss Peters termed the Terlingua experience as "one of the highlights of my life." She shared a cottage with another teacher and their furniture consisted of a wooden bedstead and a dressing table, which they later supplemented with apple boxes in lieu of chairs. She recalls laughing:

> We covered them with bright colored prints from the Chisos Store. And we bathed in a number two washtub with water heated on a three-burner Perfection kerosene range. The first bar of Cashmere Bouquet soap I ever owned I bought at the Chisos Store for twenty-five cents. . . . My, that *was* luxury! When I got married a year later I made my wedding dress from white satin I bought there. *It was a good life!* If there were hardships, we didn't know it.[10]

The Terlingua schools accorded the Mexican children their first indoctrination to the life and culture of the United States, and while they gained some knowledge and information from those who taught school in Terlingua, the young Mexican immigrants acquired their attitudes, ideals, and concepts as well.

The experience of Annie Lee Adams dramatizes the impact of acculturation. She went to the Big Bend in 1925 and her first teach-

9 Ibid.
10 Ibid.

ing assignment was at Molinar, a farming community south of the mining village on Terlingua Creek. She occupied two rooms in the Alberto Molinar residence. One was her living quarters and the other was the classroom. The following year she transferred to Terlingua, where she taught a nine-month term for $100 per month. She points out that she made the dresses she wore to school, which in design and materials were unlike any the young Mexican girls had ever seen. With this new concept of feminine charm provided by their teacher, they began to think of the calicos and domestics they formerly purchased at the Chisos Store as no longer acceptable. Sensing this change in public taste, the buyer at the company store solicited Miss Adams' advice in selecting new lines of piece goods. The rewards were multiple. With her aid the company began stocking more current merchandise, which was soon reflected in changing dress styles among the Mexican girls. In the process romance blossomed: the young school mistress married Ed Babb, the buyer for the company store who solicited her help.[11]

By the 1927–1928 school year the Terlingua school had acquired a staff of four teachers who each taught a nine-month term for $100 per month.[12] Although the financial statement for that year omits the number of students enrolled in the Terlingua district, it reveals the growth of public education in Terlingua and the increased financial benefits accruing to the Chisos mine. The 1927–1928 term was the most lucrative since the establishment of the Terlingua school, as the school account reached $786.39. The combined "rental and supply" items totaled $511.87, and a single $274.52 entry, dated November 24, 1927, indicated "supplies, desks, etc." This steadily increasing item on the company ledger, no doubt, attracted Perry's attention. Riches could be recovered from above the ground, as well as below.

The expanding population in the quicksilver district dramatized for Perry the economic potential of establishing a consolidated school at Terlingua, and he began promoting that idea with the Brewster County School Board, the agency that administered the common school districts. The board scheduled a meeting at Terlingua to discuss the proposed consolidation, and on April 25, 1930, Cartledge reported to Perry the success of that meeting: "I am very muchly pleased with the County School Boards' attitude in the school matter. They arrived here last Monday evening and spent

[11] Interview with Mrs. Ed Babb, Sanderson, Texas, December 27, 1964.
[12] School Funds, 1927–28.

the evening up at my house drinking a little beer and discussing school matters. Col. [John] Perkins [Alpine attorney] took the lead and expressed the idea from the start that the only way to have good schools was to centralize them here at Terlingua."[13]

Cartledge pointed out to the group the economic fallacy of a decentralized school system and emphasized the difficulty of acquiring competent teachers if they had to "either batch or live with some Mexican." He carefully avoided mentioning the school at Study Butte and the potential obstacles to consolidation that this adjacent mining village might hold. Cartledge noted that "Col. Perkins wanted to know what I thought [William D.] Burcham [operator of the quicksilver mines at Study Butte] would think about sending children of Big Bend [Study Butte] over here and I told him that I felt Burcham would be against it."[14]

Cartledge, well aware of the tension between the two communities, did not accompany the school board to Study Butte. The group, however, "had allready planned how they were going to talk to Mr. Burcham, and elected Col. Perkins . . . speaksman." Although both Burcham and his wife were opposed to consolidation, "Col. Perkins won them over and they agreed that the advantages of a centralized school overcame the disadvantages of having to come over here to school." With consolidation virtually assured, Cartledge concluded that "if we do centralize the school here this will mean . . . remodeling or building a new school house."[15]

The school bells now rang loud and clear for Perry. A larger school building would yield larger rentals, and with his propensity for lucrative investments he quickly reevaluated the profit potential of the emerging consolidation. Perry's move, however, was ill-timed and expressed in a manner inimical to the attitudes of Brewster County's political decision makers. Although he built his new school and assessed the county increased rentals, the Terlingua consolidated school was never forthcoming. A fragmented system of public instruction remained in the quicksilver district, and accusing fingers began pointing toward Terlingua and the Chisos Mining Company.

What occurred in the months following the Terlingua School Board meeting is not entirely clear, as no record remains of the ensuing transactions. But those who recall the episode believe Perry

[13] Cartledge to Perry, April 25, 1930, Chisos Mining Company Papers, Archives of the Texas State Library, Austin.
[14] Ibid.
[15] Ibid.

acted in bad faith to take unfair advantage of the county school administration. Harris Smith, a Big Bend rancher, quicksilver mine owner, and member of the Brewster County School Board, recalls that the board recommended consolidation on two primary assumptions: one, that the county construct a permanent bridge across Terlingua Creek to assure access to the Terlingua school; and two, that Perry continue renting the Terlingua school building at the previous rate. Members of the Commissioners Court, according to Smith, gave verbal approval for the bridge project, which they summarily withdrew when Perry announced that the rental for the new facility would be doubled.[16] Mrs. W. D. (Rubye) Burcham, who taught school at Study Butte, also recalls the incident and sustains Smith's explanation of the consolidation debacle. She apparently expressed the local consensus of Perry's economic motivation that retarded education in the lower Big Bend—"he would do anything for a dime."[17]

In 1930 the Chisos Mining Company erected a new school facility in Terlingua appropriately named the Perry School. This attractive four-room stucco building was probably opened the following school year (1931–1932) and accommodated 141 students during the nine-month term.[18] That year the Chisos Mining Company's school account jumped to $1,069.70 (previous year, $452.20), of which $368.00 was identified as rent and the remainder designated for new equipment and supplies. The following year the company's school income dropped to $701.58; however, rentals increased to $600.00.[19]

The affairs of District No. 4 were administered by a board of trustees whose key function was budget making, a role that determined the amount that would be spent annually for education, specified the expenditures, and designated the recipients of political favoritism. And Perry, not one given to leaving anything to chance, always succeeded in having a Chisos majority appointed to the board. Mrs. Elizabeth Bledsoe, who taught school at Study Butte and Terlingua (she preferred Study Butte), explains that the board members were always from Terlingua, "always company people and they ran the schools to suit themselves. Bob Cartledge had always

[16] Interview with Harris Smith, Austin, Texas, February 2, 1966.

[17] Interview with Mrs. William D. (Rubye) Burcham, Alpine, Texas, October 24, 1972.

[18] Superintendent's Annual Report to the State Superintendent of Public Instruction, 1931–32, Records Division, Texas State Library, Austin, Texas.

[19] School Funds, 1931–32 and 1932–33.

been the one who told them what to do. [He] told the trustees when they would have a meeting." Mrs. Bledsoe added that "the only thing that made him quit was when he and his wife got in a ruckus over those kids being segregated." She explained that the problem arose when Mrs. Cartledge told the teachers that she did not want her children going to school with Mexican children. "Mexicans and Anglos were going to school together," she added, "but nobody thought anything about it except Mrs. Cartledge. I told her one day [that no one but her objected to integrated schools]—and I made her mad when I did." Mrs. Bledsoe explained that, "I didn't know it until we had a bridge party and I won the prize and she couldn't find the prize. Afterwards some women told me that's the way she does. She gets mad at you and she just won't find that prize." Then she concluded philosophically, "It was probably something I didn't want anyway so I didn't worry about it."[20]

By the early 1930s the Terlingua school budget had grown to a five-figure item, which the Chisos management, in the name of the school board, supervised with paternal guardianship. The budget for the 1931–1932 school year totaled $10,566, which was increased to $11,029 the following year.[21] Frank Lewis, Chisos mining superintendent, was president of the Terlingua School Board in 1932 and felt the dictatorial pressure from the absentee mine owner. Cartledge, after dispensing Perry's orders at the mine, replied, "I asked Mr. Lewis to raise his estimated budget and he said that he would. However, he did not like it."[22]

Cartledge sensed the growing resentment of the company's control of school affairs, especially in the matter of rentals that Perry insisted be allowed some of his local friends. He reported to Perry that "there seems to be quite a bit of dis-satisfaction from the County School Board about the $100.00 rent allowed Mr. Skaggs [for the Lajitas school house]." Cartledge's concern stemmed from Chisos income property, rather than school economy, which he explained to Perry. "I note in your letter of the 12th also that you want Mr. Lewis to include in next years budget $100.00 for Mr. Skaggs. In this connection, I feel that we should not try to do anything in trying to see that *others than ourselves get a good rental* [emphasis

[20] Interview with Mrs. Elizabeth Bledsoe, Alpine, Texas, October 25, 1972.
[21] James M. Day, "The Chisos Quicksilver Bonanza in the Big Bend of Texas," *The Southwestern Historical Quarterly* 64 (April 1961): 442.
[22] Cartledge to Perry, June 17, 1932, Chisos Mining Papers.

added]." He added that "the County School Board has been making inquiry at Lajitas to see as to whether or not Pinedo could not lease them a building suitable for the school for less money."[23]

Although Skaggs received his $100 rental according to Perry's instructions, and Pinedo's name is noticeably absent from the school records, a tide of opposition arose outside of Terlingua to Perry's control of school affairs. Mrs. Elizabeth Bledsoe, who mobilized the liberal forces at Study Butte, recalled:

> I came to town here [Alpine] with another man from Study Butte and met down there at the county courthouse with the county superintendent and we just said that we would like to have a trustee over there [Study Butte]. We never had one that we could go to when we really needed one. . . . So he says, "Well, in the next election you pick out the one you want to be the trustee and we'll see if we can't get'em elected." Well, we picked out Rubye [Burcham] and we all voted on it and got her elected.[24]

Mrs. Burcham's election was a vacant triumph, for Chisos employees continued to dominate the local school board.

The lower Big Bend was another center of dissent. In the early 1930s the county school census enumerated thirty-five students living in the vicinity of Elmo Johnson's ranch, about twenty miles south of Terlingua. Johnson, concerned that the state appropriated money for their education yet no instruction was provided, appealed to the Chisos-dominated District No. 4 school board for a school, and receiving no satisfaction, carried his campaign to County Judge C. D. Wood in Alpine. Although Judge Wood failed to give Johnson the help he sought, Johnson nevertheless gathered information at the Brewster County courthouse that was to have vast repercussions when he made his final appeal to State Superintendent of Schools L. A. Woods in Austin. While Johnson sat in Woods' office in the state capitol, the state superintendent dictated a letter to Judge Wood in Alpine, stating that all state school funds were being withheld from Brewster County until instruction was provided for all students living within District No. 4. Johnson spoke with an air of smugness as he recalled, "We got the school. The teacher lived with us and I furnished the schoolhouse. Back then Brewster County was run about like Duval County is run today."[25]

23 Ibid.
24 Bledsoe interview.
25 Interview with Elmo Johnson, Sonora, Texas, June 3, 1966.

By the mid-1930s the friendship between Skaggs and Perry had apparently waned and the Sage of Lajitas wrote State Representative E. E. Townsend, long-time political leader of Brewster County, citing the sordid conditions of public school administration in the quicksilver district. Skaggs speaks out as a concerned citizen; however, like Perry, some of his statements stem from economic self-interest.

> The main obstacle to overcome down there in school matters is that Mr. Perry admittedly, frankly, and certainly controls the School Trustees and two are on his payroll—and the mine runs its school to suit Mr. Perry. One result of building the school house at Terlingua, it was hoped, would be that there would be only one school in that area, and therefore the Mexicans would move to the Chisos and work in the mine, as labor continuously becomes more scarce and less tractable. The other idea, and which was perhaps more important, was that a better school would keep in the camp the children of the white employees, so the wives would not move to town and thus disrupt the family life of the American workmen. Another factor affecting the trustees, is that Mr. Perry charges such a rental for the Terlingua School House they are afraid there is not enough money to have other schools more than four and a half months a year or to pay adequate rentals for other school buildings. Mr. Perry demands a thousand dollars a year but takes or accepts about $800 a year rental, or did the last I heard. Although he threatened each year to raise it.[26]

While Skaggs erred in some of his data—Perry never received more than $600 annual rental for the Terlingua school building—he nevertheless reflected the public consensus in his opposition to Perry's control of the board of trustees. Chisos self-interest, however, blocked the channels of democratic action, the minority voices of dissent went virtually unheard, and the company continued to operate the schools according to detailed instructions from Portland. A case in point is Hannah Weinkaupf, a Terlingua school teacher in whose professional career Perry exhibited an inordinate interest. When it appeared Hannah's temporary teaching certificate would expire because of her failure to acquire the necessary six semester hours of college courses during the 1932 summer semester, Perry interceded in her behalf. His simultaneous appeals to Sul Ross Normal College President H. W. Morelock, Alpine attorney Wigfall

[26] Virginia Madison, *The Big Bend Country of Texas*, pp. 191–92.

W. Van Sickle, and Austin attorney Eugene Cartledge[27] apparently yielded results, as Hannah Weinkaupf's name remained on the District No. 4 payroll.

Perry accorded Hannah's tenure a high priority and passed this responsibility to Cartledge. Prior to the opening of school in 1935 Perry wrote his mine superintendent to "please see to it that there is no failure to reelect Hannah."[28] Cartledge responded two weeks later that the trustees "hired Hannah and I understand *paying her what you told them to* [my emphasis], which I understand is $95.00 per month. In this connection, I have told [Ed] Babb [store employee and school board member] they would just have to raise the salary of the other teachers to this amount, as they would sure find out what Hannah is getting." Hannah's position was apparently secured the following year with Cartledge's return to the school board. Again, Chisos interest lay at the crux of the matter. He reported to Perry that "all teachers were elected as per your instructions. I did not expect to qualify as a Trustee but in order to get some money owed us, I accepted."[29]

In the depths of the Depression the Terlingua School Board continued to administer substantial public funds, as the 1936–1937 school budget totaled $13,444.59. The Chisos school account increased despite a concurrent decline in students. Rent and supplies netted $734.59 in 1935–1936, $765.75 in 1936–1937, and $818.62 for the following year.[30] The census reports for the same period, however, tell a different story. The population of the mining district, which had apparently remained fairly stable for almost two decades, had begun to decline. The 427 students enumerated in Common School District No. 4 in 1935–1936, dropped to 366 the following year, increased to 421 in 1937–1938, but reached the lowest point in recent years in 1938–1939, when only 291 students were reported.[31] While it is doubtful whether Perry had access to this data, he was well aware of the district's changing economy. School enrollment told a singular story of great social and economic changes in the Big Bend. Early in January, 1938, he inquired of

[27] Perry to Eugene Cartledge, August 25, 1932.

[28] Perry to Robert Cartledge, July 17, 1935.

[29] Robert Cartledge to Perry, July 23, 1935, August 4, 1936.

[30] School Records, Office of the County Superintendent, Brewster County Courthouse, Alpine, Texas.

[31] Superintendent's Annual Report, 1935–36, 1936–37, 1937–38, and 1938–39.

Cartledge, "has there ever been any comment . . . in the matter of so many school teachers and so few people. . . ? If there has been any talk about the matter, I shall be glad to know about it."[32] While Cartledge's obvious answer is unrecorded, by 1938 the matter was purely academic. Time was fast running out for Perry and his Chisos Mining Company, and the decline of the Terlingua schools, to which Perry had looked so long as a source of income and political exercise, simply dramatized openly what was soberingly apparent on the company ledger. There had been talk, but as the decade drew to a close, scarcely anyone cared to listen. The Terlingua school bells no longer sounded the same to Perry.

If company control restricted democratic administration of Terlingua schools and opened new avenues of personal profiteering, four decades of industrialization and urbanization had nevertheless brought to a largely primitive populace new opportunities for social and educational growth. As the academic offerings increased from basic elementary subjects at the turn of the century to a more comprehensive high school experience by the early 1930s, peripheral benefits of a more fundamental nature also accrued to the Mexican people. For instance, while Mrs. Bledsoe realized the importance of teaching the Mexican children the language of their newly adopted country, she felt equally committed to acquaint them with health care and personal hygiene. Soon after she went to the mining district she, together with an itinerant school nurse, obtained vaccine and the services of Dr. W. N. Kelley, a Chisos company doctor, and inoculated the Mexican children for smallpox and other communicable diseases. They established a clinic in the Study Butte school and "in all that vicinity, we got 'em to come in there and we got 'em vaccinated. Nobody before had ever been interested in those people like that. They just didn't care. I was just so glad that I was able to do that and they had enough trust in me that they brought their babies."[33]

Mrs. Bledsoe also taught Mexican mothers how to sterilize bottles for their babies, and when there was an outbreak of impetigo "we just turned one of the rooms there in the school building into a bathroom, and believe me, I made 'em take a bath." When she later discovered an infestation of head lice she, on orders from Dr. Joel Wright in Alpine, mixed a solution of salad oil and coal oil and applied it to the children's hair with a small paint brush. She later explained the problem: "Well, they were just scared to death of coal

[32] Perry to Cartledge, January 1, 1938.
[33] Bledsoe interview.

oil. They think it will kill'em. But we got that on'em and we got the lice cleaned out. And you know, the grown people began to do that."[34]

Mrs. Bledsoe also taught the Mexican children a practical application of some of their academic training. With the passage of the Immigration Law of 1923, parents began coming to school with their children for help in filling out forms to establish United States citizenship.

Elmo Johnson also shared Mrs. Bledsoe's high regard for the Mexican children's intellectual potential and recognized the practical aspects of Anglo education in the lower Big Bend. He recalled that his wife taught the Mexican girls to cook, sew, can fruits and vegetables, read, and write. Johnson explained, "the cookbook and the Sears catalogue were their primers. All they needed was an education."[35]

In later years, however, Mrs. Bledsoe began to question the value of superimposing American culture on a foreign people who crossed the turgid Rio Grande for the economic benefits of industrialization and remained to grapple with the massive problems of Anglo acculturation. As she explained:

> The Mexican people had lived there along the Rio Grande always. They didn't know any better and they were happy. I said one time that I thought we were doing an awful thing to educate'em. I have seen those little Mexican children pick wild flowers and bring them to me like they were bringing me an orchid. And they appreciated nature. Now we are teaching them to want things that they were just as happy without. And I don't know that I exactly approve of that.[36]

Unknowingly, and without prejudice, Mrs. Bledsoe, the champion of education for all, had succumbed to the same philosophic malady that afflicted Howard E. Perry. Theirs was a racist society nurtured in an Anglo-dominated environment that recognized the priority of a single race. The Mexican was still a transplanted primitive, a foreigner, a temporary resident north of the Rio Grande not entitled to all the benefits society affords. But since social progress had outrun the regional racist credo, many Mexicans remained and prospered. And while few of those who went to school at Terlingua, Study Butte, Lajitas, Molinar, or Johnson's Farm can even recall

[34] Ibid.
[35] Johnson interview.
[36] Bledsoe interview.

such names as Perry, Cartledge, Bledsoe, or Burcham, United States citizenship became a permanent symbol of achievement.

Of all areas of local involvement—excepting, of course, economic—nothing gave Perry greater concern than politics, both county and state. The taxing authority vested in the Brewster County Commissioners Court and the state legislature held the key to profit and loss, second only to productive ore. And as the assay reports declined in the 1930s and the margin of profit grew thin, Perry became increasingly sensitive to news from Alpine and Austin.

While Perry remained a stranger in the Big Bend, he was by no means a stranger to the world of politics, having acquired his skills in the rough-and-tumble arena of late nineteenth-century Chicago municipal elections. From experienced professionals he witnessed the rewards of personal protection and financial gain that accrue from a carefully guided electorate. During the decades of Republican ascendency he also remained close to the party hierarchy and frequently turned to political allies to enhance his economic posture. Perry had already demonstrated this in his quest for military protection during the Mexican border unrest.

But now Perry was engaged in a different drama, on a different stage, and with a different cast of characters. To function effectively politically one needs friends, and the lack of them was Perry's greatest weakness in the Big Bend. He never progressed beyond being a distrusted stranger in an unfamiliar environment, and this inherent weakness increased the magnitude of everything he undertook in the quicksilver district. But Perry had two prime political assets: money and a large staff of submissive employees who looked to the company for economic security. With these two compensating factors Perry marshaled his forces in the quicksilver district and played the game of politics with great dexterity. Motivated by a fear of declining profits and of being exploited by a group of people with whom he shared a mutual dislike and disrespect, Perry pursued a course well calculated to remove the element of chance from the process of money making.

Perry's first encounter with the Big Bend political establishment occurred during a period of affluence and expansion. The Brewster County tax records chart a collision course toward his confrontation with the Commissioners' Court. Prior to the peak production years during World War I, Perry's tax payments were relatively modest. The ten-year average, 1905–1915, was only $354.42; however, the 1916 evaluation, $106,048.00, was almost double that of the preceding year. Tax payment increases were equally dramatic, jumping from $698.82 in 1915 to $1,102.91 in 1916.[37] While quicksilver production almost doubled the following year—from 6,306 flasks in 1916 to 10,791 in 1917—the price average declined from $125.89 to $105.32, so that Perry's taxes remained virtually unchanged.[38] The blow-up occurred the following year when the company was assessed $5,954.22 in taxes, based on a $522,300.00 evaluation (1917 evaluation, $106,822.00).[39] This more-than-quadruple increase spurred the diminutive mine owner into action.

Perry began in a manner least likely to succeed. He instructed Robert Cartledge to appeal to Brewster County Judge A. M. Turney to lower the Chisos taxes. Cartledge, who did not relish the assignment, long remembered Turney's harsh verdict, "Hell no! We might even raise'em another $250,000."[40] But defeat was only temporary, as Perry must have realized that an outsider challenging a closed political institution could expect little more than Turney's blunt rejection. Chicago had taught Perry a lesson in political strategy, and as the war in Europe wound down and production consequently declined, he studied the Big Bend political climate with growing interest. His decision to challenge the local establishment, while earning him the relief he sought, touched off a series of political skirmishes, court battles, and internecine struggles that were remembered long after Perry's departure from the Big Bend. Again he turned to a Cartledge—this time, Wayne.

Wayne Cartledge, the Chisos general manager and Perry's handpicked candidate for Precinct No. 4 county commissioner in 1918, assumed the office unopposed.[41] Perry moved quickly into action, and the Alpine *Avalanche* revealed the incensed mine owner's strategy. A February 27, 1919, item stated:

> Chisos Mining Company has filed suit in Equity Court, El Paso (Judge W. R. Smith), against Judge A. M. Turney, Commissioners Hord, Fletcher, Cartledge, and Crawford, County Attorney Martin and Tax Collector E. E. Townsend. The suit is presumed to be for

[37] Tax Records, Office of the Tax Assessor and Collector, Brewster County, Alpine, Texas. Hereafter cited as Brewster County Tax Records.

[38] U.S. Department of the Interior, *Mineral Resources of the United States, 1917, Part I: Metals*, p. 371.

[39] Brewster County Tax Records. Assessments of other Big Bend mining companies in 1918 are as follows: Big Bend, $392.74; Colquitt-Tigner, $17.10; and Marfa-Mariposa, $269.04.

[40] Cartledge interview, May 7, 1965.

[41] Election Records, Brewster County, Alpine, Texas.

the purpose of restraining the collection of state and county taxes on the basis of the increased assessment of the company's property as made by the board of equalization last May.[42]

The victory was only marginal; while the 1919 evaluation dropped to $376,670.00, the company still paid $4,859.05 in taxes. Whether because of Cartledge's reelection in 1920 or the general depressed condition of the quicksilver industry, the company assessment dropped to $2,110.15,[43] as the entire district's production declined to an eight-year low of 3,436 flasks.[44] In that year the Chisos management dispatched only 712 flasks of quicksilver[45] at an average price of $76.66,[46] yielding an annual income of approximately $57,717.92. The following year the Chisos mine fared even better when the company paid only $988.89 in taxes,[47] less than half the preceding year's assessment, while recovering 3,172 flasks, over half of the nation's production.[48]

When Cartledge did not stand for reelection in 1922 the Precinct No. 4 commissioner's post passed to William D. Burcham, a mining engineer and mine operator at Study Butte. While politics were not neglected by the Chisos management that year, other matters demanded immediate attention. Much water was encountered below the 800-foot level in the company's rich No. 8 shaft, and all attempts to seal off the flow failed. Through graduated pumping operations, the water was lifted first to the 750-foot level, where a second series of pumps boosted it to the surface. Although the pumps operated fourteen hours daily and surfaced approximately 94,000 gallons of water, the power plant was inadequate to activate the pumps at the lower levels, so that the workings below 800 feet had to be abandoned. This pumping operation, which necessitated closing the mine for two and one-half months during the summer, reduced production. Of the 2,834 flasks of quicksilver produced that year, 745 were recovered during cleanup operations while the mine was closed. There was, however, encouraging news for the quicksilver industry during the year. Tariff legislation placed an $18.00 per flask tax on all foreign quicksilver imports, which

[42] Alpine *Avalanche*, February 27, 1919.

[43] Brewster County Tax Records.

[44] J. Harlan Johnson, "A History of Mercury Mining in the Terlingua District of Texas," *The Mines Magazine* 36 (October 1946): 447.

[45] Furnace Records, 1920, Chisos Mining Company, Chisos Mining Papers.

[46] Johnson, "Mercury Mining," 36 (October 1946): 447.

[47] Brewster County Tax Records.

[48] Johnson, "Mercury Mining," 37 (March 1947): 28.

advanced the 1922 price to $59.74, a $13.67 per flask increase over the previous year.[49]

Following Burcham's election in November, 1922, Robert Cartledge assessed the political climate of the Big Bend for Perry and concluded that the company's abstinence from the recent election was wise. He reasoned that "if we had run and won, we would loose more in the long run as we would stir up that ill feeling that has existed against us, but which seems to be dying out." He added that he did not foresee any immediate tax increases; however, "in the meantime we can be watching the sentiment and if necessary make the race next election."[50] Perry no doubt felt that Burcham represented the mine owner's interests on the Commissioners Court, as for the next decade the Chisos management remained politically inactive and Burcham held his office unopposed. During this period the Chisos produced consistently, if not spectacularly, while the quicksilver market moved steadily higher, reaching $123.51 in 1928, when the company reported 1,679 flasks.[51] The Brewster County tax records for that period substantiate the accuracy of Cartledge's political analysis; during the decade of Burcham's tenure in office, Chisos tax payments remained fairly stable, never fluctuating lower than $980.02 in 1922 or above the $1,234.38 maximum reached in 1929.[52]

The political equipoise Perry was apparently enjoying was soon abruptly jolted. While the company had not been actively supporting political candidates during this period of calm, both Cartledge and Perry remained vigilant for local political announcements. In July, 1934, Perry alerted Cartledge, "[I] understand no one [is] running against Billy [Burcham] and that is good, but if anyone does finally run, we must go the limit for Billy or we may get dreadfully hurt."[53] If Perry was attempting to anticipate the political thinking of the county fathers, he appears clairvoyant, for less than one month later Burcham informed Perry that an eight-man committee had been appointed to study the tax valuations of Brewster County. The committee consisted of J. J. Roberts, Clay Holland, T. J. Cartwright, Homer Wilson, H. T. Fletcher, Herbert Pruitt, Sam Nail, and Guy Combs. Perry wrote Cartledge, "I am sending you this for it is something to think about as we are liable to run up against

[49] Ibid.
[50] Cartledge to Perry, November 14, 1922, Chisos Mining Papers.
[51] Johnson, "Mercury Mining," 37 (March 1947): 30.
[52] Brewster County Tax Records.
[53] Cartledge to Perry, July 25, 1934.

these men and we should try to cultivate and steer them somewhat in the right way if opportunity arises." On the basis of the committee's makeup, Perry rationalized that they will probably try "to get the valuation down on cattle and lands in the south part of the county and raise the values near the railroad." But Perry also knew from painful experience that "these men may also try to take a crack at the mine."[54]

As the 1934 primary election approached, all appeared well for the Chisos management, as Burcham had no opponent. However, in the interim between the primary and the November general election a silent opposition gathered momentum. An anti-Chisos coalition selected Elmo Johnson, Robert Cartledge's recent adversary in the school administration conflict, as their write-in candidate to oppose Burcham for the Precinct No. 4 commissioner. Johnson later explained that while he personally liked Burcham, his supporters were fighting the quicksilver establishment controlled by Perry and the Chisos mine. He added that if you did not work for the company, you were virtually disenfranchised in the Big Bend. "That end of the country," Johnson added, "was run for the benefit of the Chisos Mining Company. Bob Cartledge was afraid if I was elected I would spend money on road improvement in the precinct which might raise the company's taxes. I tried to talk to Bob one time about the poor road conditions and he stormed out at me, 'You've been gettin' over'em, haven't you.' "[55]

The success of a write-in campaign depends on secrecy and the element of surprise—especially one planned to defeat Perry in his own bailiwick—hence only oral records of Perry's close call with defeat remain. Johnson, smiling, later dramatized his near victory with the traditional two-finger symbol of political success. "They beat me by two votes."[56] If this represented a moral victory for the anti-Chisos, anti-Perry coalition, it also sent a warning to the company. Much soul searching ensued within the upper Chisos managerial echelon as its members assayed the near disaster. Perry wrote from his home in Portland that "if Johnson were elected and soaked us on taxes, as we got soaked like we did many years ago . . . it might easily be the straw that would cause me to close the mine down."[57]

[54] Perry to Cartledge, August 16, 1934.

[55] Johnson interview.

[56] The Brewster County Election Records show Burcham a three-vote winner, 38 to 41.

[57] Perry to Cartledge, January 4, 1935.

Perry had no thoughts of closing the mine for the Chisos continued to produce steadily. In 1934 1,178 flasks of quicksilver were recovered and the following year the number increased to 3,243 after exceptionally rich ore was encountered on the leased Mariposa property.[58] Nor could he complain about the work of the tax study board as the company's assessment remained virtually unchanged. Perry's scare tactics had the desired results; Cartledge busily analyzed the company's political posture while preparing for the next election, two years hence. He assured Perry that he would see that "everyone pays his poll tax when and if we have another election between Johnson and Burcham."[59]

But as Cartledge began shoring up the faltering Chisos political machine, he encountered a surprising dissident undercurrent within the company management. It stemmed from a belief in the free functioning of the democratic process. Cartledge reported to his superior that Frank Lewis, the mine superintendent, was having trouble indoctrinating the men under his supervision with the company's political philosophy. According to Cartledge, the source of the trouble was Lewis himself, who "thinks that a man should vote as he wants to."[60] This postelection analysis also revealed that some of Johnson's strength came from the other mine operators as well as from within the Chisos staff. Cartledge, who prior to this election had been able to produce votes almost at Perry's bidding, then dispatched the following letter, which he probably did not relish. On April 13, 1935, he wrote Perry: "I have learned that Starr [Rainbow Mines] voted for Johnson though he let me believe that he was going to support Burcham. There are about four votes that Johnson got that I can't account for. It looks as tho . . . two of our people have voted for Johnson." Nor did Cartledge ignore the privacy of a husband and wife relationship in scouring the village for the politically disloyal. A Chisos employee named Dagget assured Cartledge that he voted for Burcham; however, Cartledge later explained to Perry that he "does not know how his wife voted."[61]

As the Chisos management evaluated the company's political future in the Big Bend, news from the 44th Texas Legislature suddenly diverted its attention from Alpine to Austin. The workings of the Brewster County Commissioners Court suddenly appeared small and inconsequential in comparison to the penalties which

[58] Johnson, "Mercury Mining," 37 (July 1947): 21–22.
[59] Cartledge to Perry, January 8, 1935.
[60] Ibid.
[61] Cartledge to Perry, April 13, March 29, 1935.

House Bill No. 338 portended. On January 30, 1935, Representative Lonnie Smith of Fort Worth introduced a bill levying "an occupation tax of $5 per flask on all quicksilver produced within the State of Texas; providing for the reports and records imposing forfeitures and penalties for failure to keep records; providing for penalties for failure to pay tax."[62]

Perry's response was predictable: defeat the bill at all cost. He forthwith outlined to his Alpine attorney, Wigfall W. Van Sickle, and Cartledge, a two-fold plan of action: one, entice a House investigating committee to Terlingua for an on-the-spot view of company operations; and two, apply all possible political pressure on Smith to force him to withdraw the bill. From his Alpine office Van Sickle coordinated plans and began preparing a "guest list" for the trip to the Big Bend. The key person in the plan was Representative Stanford Payne of Del Rio, who promised to bring his legislative colleague, Smith, on the fact-finding sojourn. To make the journey more attractive to Payne, Cartledge "gave Judge [Van Sickle] $25.00 to send Payne for gasoline and hotel expenses for the trip out here."[63] Van Sickle, in turn, outlined his plans for Cartledge in Terlingua:

> I have told Mr. Payne to include as our guest Mr. [Coke] Stevenson [Speaker of the House of Representatives and later Governor of Texas] . . . and the boy [name unknown] who told us that he would make Lonnie Smith withdraw the quicksilver bill, so rest assured as everything is being done.
>
> Likely I will be unable to go down to Terlingua, but you want to give them a party . . . and show them desolation . . . deserted shafts, and everything that will indicate a hard time. . . . show them everything that is necessary, but we will not show the production unless we are compelled to, but we can show the losses.[64]

Cartledge, realizing he was involved in a volatile matter, concludes his typewritten response to Van Sickle with the following postscript, "As Mr. Perry says, burn this up. Commands me."[65]

Phase one of Perry's plan stalled when Lonnie Smith refused to be lured to the Big Bend. Van Sickle and Cartledge then moved ahead immediately with an alternate plan, as Cartledge reported to

---

[62] *Texas House Journal*, Regular Session, 44th Legislature (1935), p. 229.
[63] Cartledge to Perry, February 26, 1935.
[64] Van Sickle to Cartledge, February 27, 1935, Chisos Mining Papers.
[65] Cartledge to Van Sickle, March 1, 1935, Chisos Mining Papers.

Perry that he and Van Sickle were returning to Austin via Fort Worth "so as to get a line on Lonnie Smith, also to see if we can not find some close friends of his that we can get to try and induce him to withdraw his bill."[66]

Before leaving Alpine, however, Van Sickle and Cartledge met in Van Sickle's office to interview a former assistant attorney general of Texas, who offered to lobby for the defeat of the quicksilver bill, claiming to know the members of the legislature intimately. The lawyer assured them that with unlimited expenses to purchase "plenty of good whiskey, thick steaks, and pretty women, he could pass or defeat any bill ever introduced in the Texas legislature."[67] Cartledge and Van Sickle, however, refused the offer, as by then they believed they were gaining sufficient support to defeat the bill.

Accompanied by Alpine Mayor M. L. Hopson and County Judge R. B. Slight, Cartledge and Van Sickle departed for Austin to lobby for the defeat of House Bill No. 338 and attend the hearing before the Committee on Revenue and Taxation.[68] Cartledge's Austin assignment was Stanford Payne, and when he discovered that Payne was an exception to the lawyer's thesis of corruptibility, Cartledge confined his favors to food, taking the Del Rio representative to lunch daily at Leslie's Chicken Shack on the old Dallas highway in north Austin. Luncheon talk focused on the forthcoming committee hearing at which Payne agreed to speak for the quicksilver producers.[69]

Cartledge's high hopes for Payne's support faded when the committee members began questioning the Del Rio lawmaker. "He made a poor showing before the committee," Cartledge recalled. "Never have been so surprised in my life. The man who saved our lives was a representative named [Walter Elmer] Polk from Corpus Christi." Cartledge explained that while Polk represented the coastal area, he also owned some property south of Marathon and had a personal interest in that region. "He made the strongest case against the bill of anyone that was questioned," Cartledge stated. "As I recall he said that it was not in the best interest of the people making a living off the quicksilver mines to pass such a bill in the middle of the Depression."[70] Again the Terlingua-Portland correspondence

[66] Cartledge to Perry, March 8, 1935.

[67] Cartledge interview, January 13, 1966.

[68] House Bill No. 338 was referred to the Committee on Revenue and Taxation on Wednesday, January 30, 1935 (*Texas House Journal*, p. 229).

[69] Cartledge interview, January 13, 1966.

[70] Ibid.

conveyed the good news that Perry had long become accustomed to receive: on Tuesday, March 19, 1935, the committee filed an adverse report on House Bill No. 338.[71]

Returning to the Big Bend, Cartledge and Van Sickle knew they could not rest on their laurels. They needed to win other elections. With the near defeat of the write-in campaign still fresh on their minds, they knew that 1936 would bring another political challenge. By mid-October, however, Burcham still did not have an announced opponent for commissioner and Perry, realizing the pitfalls of a single-candidate race, warned Cartledge that "the danger lies in voters going to sleep and not going to the polls for they naturally . . . think their man is safe enough for reelection. This is false security." Perry added that he had "heard rumblings" that the anti-Chisos coalition might "get some cowpuncher to write his name on the ticket." Thinking back to the recent write-in campaign, Perry concluded with threatening emphasis, "If that damned skunk Johnson would ever get in he would try to put us out of business and there are others who would prove to be about as bad."[72] Perry's fears, although well founded, were all for naught. The silent opposition failed to mount an offensive, and Cartledge reported with just pride that "Burcham got 59 or 60 of the 61 votes cast here."[73]

As long as the Chisos mine maintained full employment, a company-endorsed candidate enjoyed a numerical superiority and the democratic process remained a philosophic ideal in the Big Bend. By 1938, however, the region was in transition, the population declining, and many of those who had formerly given Perry carte blanche endorsement at the polls no longer worked for the Chisos. Burcham, the quicksilver district's perennial candidate for commissioner, also had forsaken residence at Study Butte for life in Alpine. Again Perry emphasized to Cartledge that he would have "to use some extraordinary means to get Burcham down there to talk to voters. . . . This is the most important thing we have at present."[74] But as Perry must have suspected, the balance of power had shifted, and instead of the virtual "walk-in campaigns" Burcham had enjoyed since 1922, he now had two opponents, and of the eighty-one

[71] *Texas House Journal*, p. 229.
[72] Perry to Cartledge, October 19, 1936.
[73] Cartledge to Van Sickle, November 5, 1936.
[74] Perry to Cartledge, July 18, 1938.

votes cast, he polled eighteen.[75] Only four votes came from Ter-lingua.[76]

With Charles G. Burnham, a Big Bend rancher, and Eugene Cartledge, Robert Cartledge's nephew, campaigning for a post that was becoming increasingly important to the Chisos Mining Company's survival, the runoff campaign posed some delicate problems. For Robert Cartledge, his was both a struggle of family against family and family against the company for which he had given, and given up, so much. But it was Perry, not Cartledge, who resolved the issue. With finality he wrote his general manager, "Eugene, I do not want." He reasoned with the young candidate's uncle that Eugene, like his father, possessed an abrasive personality that would limit his usefulness. Perry reminded Robert Cartledge of "what Wayne did many years ago when we elected him. . . . He antagonized the entire Commissioner's Court and was not able to do anything." That mistake would not be repeated. Instead Charles G. Burnham would be the company candidate. Perry forthwith instructed Robert Cartledge to "reduce any probable votes for Eugene by discharging men who have their living at Chisos. . . . And, yet when it comes to election, vote in a way which displeases us." He encouraged his general manager to seek votes elsewhere than in Terlingua, "even sending some Mexican cars and big Jumbo [a company truck] if needed to bring every voter in the entire district to the polls provided you have good grounds for believing they will vote for Charlie [Burnham]."[77] Cartledge apparently selected the voters well, for when the ballots were counted a Cartledge had defeated a Cartledge and for another two years company policy would no doubt prevail.

Early in January, 1940, Cartledge, looking forward to another campaign, wrote Perry, "[I] am busy getting out a bunch of application[s] for Poll Taxes. . . . This year is a voting year and I want to try to be in a position to elect anyone you want for commissioner."[78] Cartledge's devotion, however, went unfulfilled and unrewarded, for on the following August 24 Perry dismissed his ever-faithful general manager, and one year later the company was bankrupt.

By whatever criterion Perry's Big Bend politics are evaluated,

[75] Cartledge to Perry, July 26, 1938.
[76] Perry to Cartledge, August 15, 1938.
[77] Perry to Cartledge, August 2, August 17, 1938.
[78] Cartledge to Perry, January 26, 1940.

the result is a sordid commentary on the democratic process. Through devotion, fear, submission, and ignorance, Perry manipulated the ballot for the economic benefit of the Chisos Mining Company, while nullifying all other interests in the region. Only the dollar mark established political priority in Perry's world, and the power vested in that symbol stifled any minority voices that struggled to be heard. Perry brought to the Big Bend an eighteenth-century Federalist political concept articulated in a late nineteenth-century Republican vernacular, which believed that "as the wealth of the commercial and manufacturing classes increases, in the same degree ought their political power to increase."[79] The Chisos payroll was the basis of the Terlingua economy, and to benefit from company employment one had partially to surrender his rights of citizenship. Frank Lewis' belief "that a man should vote as he wants to" was totally foreign to Perry's political precepts. And while this Federalist power-property logic may have lost currency in more enlightened environments, the seeds of political and economic serfdom flourished along the banks of Terlingua Creek. Elmo Johnson had walked there; he experienced the futility of challenging the "aristocracy of property."

The manipulation and restraint of personal freedom is an often-repeated chapter in the history of United States economic expansion in the post-Civil War years. An abundant and eager—and in many cases untrained and ill-informed—labor force solicited industry for employment and accepted a commitment that frequently compromised some basic human principles. But social and economic change was in the offing. Two years before Jack Dawson reported the first flask of quicksilver in the Terlingua district, Samuel Gompers as the first president of the American Federation of Labor embarked on a crusade to preserve for the individual worker those basic rights that most Chisos employees forfeited to enter Perry's service.

As the new century grew older the urban laborer for the most part assumed a new role in the scheme of things, but conditions remained virtually unchanged in the fledgling urban cells of the Big Bend. Here the average worker was illiterate, unenlightened, a despised foreigner willing to accept voluntary bondage for the opportunity to grovel in the earth's bowels for the pittance Perry elected to pay. His heritage was steeped in a tradition of mass submission that continued north of the Mexico-Texas international boundary. If Perry is condemned for exploitation, the

<hr/>

[79] G. J. Clark, ed., *Memoir of Jeremiah Mason*, p. 350.

exploited came begging for the opportunity. Willingly and without coercion, both parties entered into a contract for their mutual benefit.

Even the victims could be candid. As H. C. Hernandez, whose father was forced to surrender his freighting business to the company, explains the workings of the democratic process in Terlingua: "In the old days they used to slip a marked ballot to a guy [a Mexican] and all they would do would just run'im through a votin' line, and that was that." Asked if the voter realized what he was doing, Hernandez replied, "I guess he did. That's how they played the game."[80] Mrs. W. D. Burcham, whose husband received Chisos political support while rejecting most of Perry's social policies, adds that while only a small portion of Mexicans voted, those "who did vote knew who and what they had to vote for. They [the Chisos] bought votes every election. Everybody knew that the mine employees—especially the Mexicans—were told how to vote." Then pausing, she concludes, "They really had no choice if they wanted to work."[81] Economic restraint was the single key to Perry's political successes, and if this control should weaken, one fail-safe mechanism always remained, as Van Sickle candidly explained to his client: "We will ride the winning horses even if we have to buy the leader in action."[82]

As usual Mrs. Bledsoe, the vociferous and articulate Study Butte school teacher, represents the spirit of independence in the quicksilver district: "The only time I ever voted down there was one summer and Bill Burcham was running for county commissioner. They told me I couldn't vote and I said 'I'm going to.' That was over at Terlingua and there would have been one wild hullabaloo if they didn't let me vote."[83]

With that action Mrs. Bledsoe withdrew from the mainstream of Terlingua political life. Without obligation—social, political, and especially economic—she was indebted to no one and could vote for whomever she chose. When her rights were challenged, she threatened to raise "one wild hullabaloo." In this one instance Howard E. Perry and Chisos management submitted. But she was the exception. Usually the voter, not the private company, provided the submissive party.

---

[80] Interview with H. C. Hernandez, Alpine, Texas, October 23, 1972.
[81] Burcham interview.
[82] Perry to Cartledge (quoting Van Sickle), January 4, 1935.
[83] Bledsoe interview.

CHAPTER 8

The Biggest Store between Del Rio and El Paso

THE company-owned mining town in the American West followed a virtually unvarying pattern. Whether one visited Tombstone, Tubac, Tonopah, or Terlingua, the same urban image emerged. Describing the consistent yet unplanned design, Western historian James B. Allen writes:

> If a person suddenly found himself in the middle of a company-owned town, he would have little difficulty identifying it as such, for certain features stood out. . . . In a prominent location . . . would stand a larger, more imposing structure: the home of the company manager or superintendent [or owner]. . . . [But] the town seemed to center around a focal point where a store, a community hall, school, and other public buildings were located. The company store usually dominated the group.[1]

Many of these villages, although stereotypes, frequently possessed unique individuality. Their location (mountain or desert terrain), the product that created and sustained them (gold, silver, zinc, copper, lead, or quicksilver), their ethnic structure (Cornish, Welsh, Irish, Indian, Chinese, Negro, or Mexican), as well as the varying personalities of the management and ownership, frequently gave these mining villages identity.

The recovery of the rare liquid metal gave Terlingua identity, as did the abrasive personality of its owner, Howard E. Perry, who single-handedly controlled the village from Chicago, Portland, or Florida, or Canada, or aboard his yacht, or somewhere in transit. When Allen's mining town stereotype is superimposed on the urban organization of Terlingua, however, a striking ambivalence emerges: the Chisos-owned store, while providing traditional services to company employees, also served another population living well beyond the village perimeter. This quaint and unusual institution attracted trade from a radius of more than one hundred miles on both sides

---

[1] James B. Allen, *The Company Town in the American West*, pp. 79–80.

of the Texas-Mexico border, while the ever-increasing demands for its extensive merchandise gave currency to a sobriquet in which Perry took just pride: the Chisos Store was "the biggest store between Del Rio and El Paso,"[2] 425 miles apart.

The success of this company affiliate also gave rise to another regional saying in which the little mine owner probably took even greater pride: "Perry's mine was a silver mine, but his store was a gold mine."[3] While probably composed in jest, this statement was founded on a sound economic basis. The Chisos company store not only enjoyed a longer life span than did the mine, but it consistently yielded profits long after quicksilver recovery became a deficit operation. Robert Cartledge, who began his Chisos career in 1911 as a store clerk, reports the company store always showed a profit of at least $12,000 annually, and in one year following World War I profits exceeded $21,000.[4]

Fragmentary store records substantiate Cartledge's claims. The Chisos Mining Company monthly statement for September, 1924, shows an operational deficit of $1,574.04 on quicksilver sales of $6,080.00, while the store, with a minimal inventory of $20,419.31, yielded a $1,679.82 profit on $8,390.76 sales.[5] In 1929 with the economy booming, production up, and quicksilver averaging $122.15 per flask, the company posted a $24,570.19 annual profit and the store operation yielded 20.27 percent profit on a total of $89,317.03.

[2] Interview with Dr. Ross A. Maxwell, Austin, Texas, October 1, 1965. The Chisos Store, in addition to serving company employees, was also the main link in a chain of Big Bend border stores located on the north bank of the Rio Grande whose trade came largely from Mexico. The two largest stores, whose gross volume was second only to the Chisos Store, included the one at Castolon operated by Wayne Cartledge in partnership with Howard E. Perry, together with Elmo Johnson's store, located a few miles downriver from Castolon. (Cartledge estimated his largest annual gross at $80,000; Johnson pointed out that most transactions were barter, with the store accepting furs, stock, and produce in lieu of cash.) Other Big Bend border merchants and their locations were: Mrs. Maria (John) Daniels, Boquillas; J. T. Sublett, at the confluence of Terlingua Creek and the Rio Grande; Cipriano Hernandez, "below the bluff" at Castolon; and Thomas V. Skaggs and others, Lajitas. No record exists of stores operating below Boquillas. Another chain of stores and trading posts began at Fort Leaton and extended upriver to the Candelaria area. The Marfa and Mariposa mine, the mines at Study Butte, and later the Waldron mine on Section 248 offered periodic commissary services patronized mainly by their own employees.

[3] Maxwell interview.

[4] Interview with Robert L. Cartledge, Austin, Texas, October 10, 1965.

[5] Monthly Statement, September, 1924, Chisos Mining Company Papers, Archives of the Texas State Library, Austin.

This annual sales figure, added to the end-of-the-year inventory, revealed cash flow of $112,841.04 for 1929.[6] If not the proverbial gold mine, the Chisos Store was indeed a productive enterprise.

Like similar emporiums in other western mining towns, the Chisos Store was born of necessity. Alluding to the universal problem of supplying the mining districts, Allen explains, "A problem of particular concern to the mining industry is raised by the relative remoteness verging on actual inaccessibility of many mining regions."[7] Inaccessibility, with its concomitant social and economic complexities, was not confined to mining districts, as other industries during the formative years of America's economic development also faced problems of remoteness. In the eighteenth century with water the cheapest power source, New England mills and plants had to be located in remote regions, near river rapids, and the company owners were forced to provide employee services normally found in urban areas.[8] Consequently, when Perry invaded the nearly inaccessible Big Bend of Texas to establish his quicksilver mining empire, he carried with him almost two centuries of entrepreneurial tradition of operating normally urban enterprises in areas of isolation. When faced with the problem of providing all basic needs for his employees living almost one hundred miles from the nearest supply center, Perry established his Chisos Store.

Chisos Mining Company records offer no clue to when the company store was established. The 1905 Brewster County tax records, however, show the company rendered a $700 item identified as "Goods, Wares, and Merchandise of Every Description." This statistic, which increases to $1,500 in 1907,[9] reveals the expanding vitality of this fledgling enterprise, described that year by Chisos employee Drury M. Phillips as a very nice company store.[10] A large tent apparently housed the facility Phillips patronized and served as the company store for several years until a permanent building was erected.[11] The earliest documentable reference to this structure is contained in a letter from Perry to the Chisos management, dated

[6] Profit and Loss Statement, 1929, Chisos Mining Papers.

[7] Allen, *Company Town*, p. 6.

[8] Thomas R. Navin, *The Whitin Machine Works Since 1831*, p. 73.

[9] Tax Records, Office of the Assessor and Collector, Brewster County, Alpine, Texas. Entry for 1906 is omitted.

[10] Drury M. Phillips, Huntsville, Texas, to Kenneth B. Ragsdale, January 17, 1966, in possession of author.

[11] C. A. Hawley, "Life Along the Border," *Sul Ross State College Bulletin* 44 (September 1964): 62.

June 23, 1908. He writes, "I suppose you have already painted the roof of the store and porch. If not please do it at once; do not delay; for once the rust gets under the thin paint now on there it will be impossible to save the roof."[12]

This new structure, one of the few company buildings that survived the quicksilver mining era, reflected the expanding population drawn to Terlingua to work at the Chisos mine, as well as the growing economic importance of the store operation. In rendering its 1909 taxes, the Chisos management claim a $2,000 stock (no doubt a modest evaluation), which increases to $5,000 the following year, and $8,000 in 1911, the figure which remains through 1913, when this entry is discontinued in the tax records.[13]

By its physical location and multiple services to the community, the Chisos Store closely adheres to Allen's Western company town thesis: the "company store [was] . . . usually located near the center of population and . . . served many functions."[14] Located near the geographic center of Terlingua, the Chisos Store became the focal point of many community activities. Situated on the dividing line between the Anglo and Mexican sections of the village, it was easily accessible to all citizens. The store building also provided office space for the Chisos general manager, the company purchasing agent, the post office, as well as the Precinct No. 4 justice of the peace. And while Terlingua lacked a "community hall and public buildings," the covered porch that shaded the broad store front evolved as the unofficial meeting place where current issues were discussed and friends could visit on topics of masculine appeal. Women—especially Mexican women—seldom entered the store. Although Terlingua was a segregated village, the company store was the one location where the company's economic interests temporarily removed racial barriers.

The store building was also the dominant structure in the village, even overshadowing the larger quicksilver smelters partially secluded in an arroyo to the south, and the smaller Perry mansion which overlooked the mining village on the hillside to the west. Through both its location and its several functions, the Chisos Store not only became the nerve center of the huge Chisos quicksilver complex, but exercised vast control over the company employees who traded there.

12 Perry to Chisos management, June 23, 1908, Chisos Mining Papers.
13 Brewster County Tax Records.
14 Allen. Company Town, p. 128.

The store occupied a strategic location in relationship to the mine office where the men, emerging from the underground workings, were paid at the end of each shift, and in turn, walked directly to the company store and purchased their daily household needs. This payroll method, the employee identification system, and the purchasing habits of the Mexican miners provide a perceptive insight into the industrial philosophy of the company, as well as the cultural and domestic habits of the employees.

Before reporting to his job assignment each employee went first to the mine office and removed from the workboard a small, round, tag-like brass token, a *ficha*, which bore the employee's identification number. The corresponding number, printed on the board beneath each *ficha*, became visible when the employee removed his *ficha*. Former Chisos employee Mack Waters explains this operation, "Now each employee had his number. He would go by there in the morning and take his tag, put it in his pocket, and when they were all gone the timekeeper would look at the board to see who was working." Stating that the workboard was the key to the payroll system, Waters continues, "That's how they got paid. When they came out of the mine they turned in their tag, held out their hands, and the payroll clerk would dump the money in their hands from a small envelope. They always kept the same envelope in the mine office. I don't know why." When asked what recourse an employee had if he lost or forgot to remove his *ficha* from the workboard, Waters explains, "If he forgot his tag in the morning he just didn't get paid. Company policy, they said."[15]

The five o'clock trek to the company store was one of the more dramatic and exciting events in daily life at Terlingua. As the workers emerged from the mine and received their wages, they formed a steady stream of humanity between the mine office and the store. This was a daily necessity as the Mexican home economy, based on a day-to-day concept, coincided with the company payroll policy. Robert Cartledge describes the five o'clock scene at the Chisos Store:

> It was just like some kind of a religious thing with them. Some would even come in after work—and some who weren't working— even if they didn't have any money to spend. It was kind of a herd-like thing. Just wanted to do what everyone else was doing. The Mexicans were great talkers. It was one mess when they all hit there

[15] Interview with Mack Waters, Panther Junction, Big Bend National Park, October 24, 1972.

at five o'clock, talkin', pushin' to get to the counter, just like they thought the store was fixin' to close.[16]

The company employed four or five full-time clerks who spent most of each working day preparing for this daily onslaught. Former Chisos Store clerk Earl Anderau describes the packaging method as follows:

> Of course we knew just about what the Mexicans were going to buy so we were ready for them with their daily staples wrapped in small packages: rice, 10¢; beans, 15¢ or 25¢; dried whole corn for tortillas, 25¢; dried chili for seasoning, 5¢; for coffee they had a choice of either a package of Arbuckles for 25¢, or for 15¢ they could buy a package of raw coffee beans which they would roast and grind themselves; and always last on the list was *dulces* [candy]. This was a daily must. The Mexicans were not too choosy about quality, but bought what they could get the most of for their money.[17]

The "biggest store" verbal accolade was apparently well deserved, as the company "sold everything from a spool of thread to a windmill. Everything but fresh meat," according to Anderau. Barbecued *cabrito* (goat) is traditionally the Mexican's favorite meat staple, much of which they raised and butchered at home. Cheese was their daily substitute for meat and "the Mexican loved his cheese." The company purchased bulk cheese in circular blocks weighing from ten to twenty pounds and dispensed from a wooden cutting block after being sliced with a fixed blade cleaver-like cutter. Anderau remembers that "for ten cents you almost missed the block of cheese. Hardly cut off enough cheese to bait a mousetrap."[18]

Only two noticeable breaks in the daily store routine occurred: one was Saturdays when customers from the other mining camps and farms and ranches on both sides of the Rio Grande came to the company store to purchase their weekly supplies; the other was Lent. Since most of the company employees were Catholic, the Lenten period on the church calendar created abrupt changes in the Mexican's purchasing habits. "Ed Babb, the store manager, was really on the ball when it came to buying for the Mexicans," Anderau recalls. "Long before Lent, Babb ordered additional supplies of dried fruits, especially *pasas de uva* [raisins], *bellas* [candles] for their religious ceremonies, and untold amounts of their favorite

[16] Cartledge interview, October 17, 1965.
[17] Interview with Earl Anderau, Alpine, Texas, December 27, 1964.
[18] Ibid.

meat substitute, cheese."[19] But following Easter Sunday the store routine reverted to the pre-Lenten pattern and the pungent aroma of broiling *cabrito* again drifted across "the village of a thousand smells."[20]

The advent of automobile travel in the Big Bend in the 1930s added a new line of merchandise to the store inventory: automobile supplies and accessories. With improved roads in the mining district, automobile ownership became practical and the sale of gasoline at the Chisos Store became a profitable item. "Even at the minimum wages paid the Mexicans," according to Anderau, "almost everyone had some kind of a car and tried to drive it." Anderau continues:

> The company hauled the gasoline from Alpine in fifty-gallon drums and a store clerk hand-pumped it from these drums into a five-gallon spout can, from which he poured it into the tank of the automobile. Later the gasoline was deposited in an underground storage tank and we installed a more modern hand-pumped gravity tank with the overhead glass bowl. Gasoline that sold for 20¢ a gallon in Alpine would cost 35¢ at Terlingua and if they complained about the price, the stock reply was, "It's the transportation."[21]

Since the company store inventory fluctuated in accordance with customers' demands, this institution became a sensitive social and cultural barometer in the mining district. The earliest stock of merchandise "consisted of little else than the barest necessities of working men and their families—food, work clothing, a few bolts of calico, blue denim and shirting, and a few very cheap notions of a miscellaneous character."[22] The company's purchase orders to a growing list of vendors, however, document the changing buying habits of the company employees, as well as indicate the broadening demands of customers who patronized "the biggest store" from beyond the village periphery.

Vendor: The Studebaker Corporation of America, Dallas, Texas
Date: June 22, 1915
    1   #2904 Park Wagon with Two Seats . . . . Retail at $175.00
          (Wholesale to store, less 55%)

Vendor: Haymon Krupp & Company, El Paso, Texas
Date: (none)

---

[19] Ibid.
[20] Interview with Hattie Grace Elliot, Alpine, Texas, December 29, 1964.
[21] Anderau interview.
[22] Hawley, "Life Along the Border," p. 20.

```
1   Doz. Men's Pants . . . . . . . . . . . . . . . . . . . . . . .   $ 16.50
½   Doz. Coats . . . . . . . . . . . . . . . . . . . . . . . . . . .   $ 10.50
8   Cord Coats . . . . . . . . . . . . . . . . . . . . . . . . . . .   $ 22.50
```

Vendor: Texas Tanning & Manufacturing Company, Yoakum, Texas
Date: December 9, 1932

```
1   No. 14 Boy's Saddle . . . . . . . . . . . . . . . . . . . . .   $ 12.00
1   Doz. No. 100 Collars, 14" . . . . . . . . . . . . . . . .   $  7.25
½   Doz. #113 Bridles . . . . . . . . . . . . . . . . . . . . . .   $ 19.00
¼   Doz. ⅞" Riding Reins . . . . . . . . . . . . . . . . . . .   $  7.00
½   Doz. #82, 27" × 27" Saddle Pads . . . . . . . . . .   $  9.25
```

Vendor: Radford Grocery Company, Alpine, Texas
Date: September 4, 1941

```
1   case, Screw Worm Killer, Red Cow Brand
5   ctn LLF Cigarette Papers
5   ctn Wheat Straw Papers
5   bx Mexican Commerce Cigars
```

Vendor: Cooper Dry Goods Company, Waco, Texas
Date: September 4, 1941

```
1   Doz. Shirts, Assorted colors to retail for $1.25, Size 14
6   Rolls of oil cloth
1   Bolt 8 oz. white duckin
2   Doz. Lipstick, dark colors
2   Doz. Rouge for Mex trade[23]
```

The following undated grocery list, while emphasizing a masculine, utilitarian lifestyle, contains one item—the Victrola needles —which indicates that luxury buying had invaded the remote corners of the Big Bend.

```
1 — pair blue pants 32 × 30
1 —   "     "     "    31 × 31
1 —   "     "     "    33 × 30
1 —   "     "     "    33 × 31
2 — suit of underwear (Union Suit Heavy) size 40
1 — 2 piece suit of underwear, heavy, size 40
2 — shirts, size 15½
3 — small cans mustard
1 — bag potatoes
3 — cartons of vermicello
2 — yards of canvas, used to make water bags out of
1 — box of white thread, size 40
```

[23] Chisos Mining Papers.

1 — box gum
cigs, victrola needles, pancake flour, Valet R. Blades[24]

By the late 1920s company employees were apparently becoming attuned to the magic world of music and entertainment, as the Chisos purchasing agent received a steady stream of promotional literature touting current record releases and the combination "Victrola-Radiolas," on which customers were encouraged to play the records.

News events, famous personalities, and sports figures commanded consumer attention even along the west bank of Terlingua Creek. Less than two months following Charles A. Lindbergh's epoch-making flight to France, the Chisos company store joined the "Lindy" popularity binge and offered a choice of four discs by as many Victor recording artists, each bearing the Lone Eagle's charismatic name. A June 6, 1927, promotional tract states:

> The following Lindbergh specials "hot off the press" offer Victor dealers a wonderful opportunity and we are pleased to announce them for early shipment.
> "The Flight of Lucky Lindbergh" .... Ernest Rogers
> with guitar
> "Lindbergh The Eagle of the U.S.A." .... Vernon Dalhart
> with orch.
> "Like an Angel You Flew Into
> Everyone's Heart" .... Vaughn DeLeath
> with orch.
> "Lucky Lindy," Fox Trot .... Nat Shilkret and the
> Victor Orchestra[25]

The concurrent hawking of the newly developed combination radio-record players heralded a new era of home entertainment; Terlingua was not excepted. The Victor distributor, soliciting the Chisos Store trade for radios ranging in price from $135 to $1,000, claimed: "The Radio season is opening up. . . . increasing orders being sent us for the combination Victrola-Radiola types." And "The approach of the Tunney-Dempsey fight at Chicago September 22nd is giving added impetus to combination sales."[26]

While many of the company employees acquired hand-cranked phonographs, only a few administrative personnel could afford radios. In 1927 Chisos mechanic J. E. Anderau acquired one of the

24 Ibid.
25 Ibid.
26 Ibid.

first radios in Terlingua. An Atwater Kent brand, it had no cabinet and the tubes and working parts were exposed to the listener and Terlingua dust. Three tuning dials selected the station, while the broadcast signal came through both a large gooseneck speaker adjacent to the set and two pairs of headphones. This early radio served both luxury and utilitarian purposes: while providing entertainment for Anderau's family and neighbors, Anderau, a securities investor, returned home each day at lunch to listen to the noon stock quotations from station WOAI in San Antonio, more than three hundred miles away.[27]

Although distribution of luxury merchandise was highly restricted in Terlingua, the Chisos Store nevertheless introduced a whole new complex of cultural artifacts to a large segment of the community. If economic limitations denied them immediate possession, they at least responded to the availability. Through the commercial channels of mass distribution new cultural images and social ideals were being transferred from one society to another, and the tastemakers of Manhattan, Chicago, Alpine, or Yoakum were providing the material culture that lured a foreign people toward a new lifestyle. And while fruition still awaited in a distant and unpredictable future, the Chisos merchandise had opened the doors to a strange and inviting new world.

While some aspects of the acculturation process encountered involuntary resistance, others were readily accepted. A perplexed Robert Cartledge wrote Albert Mathis and Company in El Paso concerning receipt of an unordered shipment of men's underwear:

> Replying to your letter of March 15th . . . we beg to advise that we received a box containing ½ Doz of some kind of New-Fangled kind of Drawers. . . . One of our girls here took a fancy to these mens drawers, or whatever they are called nowdays, and we let her have them.
>
> We presume the package we received was shipped in error for 4 Doz #1200 P & B Ladies hose. . . . We presume we will now soon receive these ladies hose.[28]

Two factors account largely for the success of the Chisos Store: monopoly and cash sales. In addition to the natural monopoly isolation afforded, Perry applied the principle of economic overkill by eliminating local competition and virtually coercing company employees to trade at the company store. While the peripheral cus-

---

[27] Anderau interview.
[28] Chisos Mining Papers.

tomers had the choice of the retail facilities at Lajitas, Castolon, and Johnson's Farm, none maintained stocks of merchandise comparable to the Chisos Store. And even if company employees had access to these outlying stores, Chisos payroll and monetary systems discouraged, if not eliminated, their patronage.

Cash sales simplified bookkeeping procedures and assured the company maximum return on its investment. While Anglo mine employees, other mining companies, and some local ranchers frequently received credit (never the Mexican employees), the latter's accounts were sometimes difficult to collect. Writing to Alpine rancher T. J. Roberts who had made a note to the mining company for a delinquent account, Cartledge explains:

> Your note became due on May 1st. However to date we have failed to here from you with remitance. As I told you . . . we would at least expect Interest and parcial payment on this note.
>
> Tho we would like to help you all possible, at same time, we feel that we have helped you considerable by carring account so long, before we had you make note. At time we allowed credit, we . . . did same thru wanting to help out all posible ranch men.
>
> Hoping to recieve at least interest and part on note by early mail, we beg to remain.[29]

The conciliatory tone of Cartledge's letter to Roberts is totally lacking in the company's commercial relations with the Mexican laborers. This was strictly a "cash on the barrelhead" transaction, and while the Chisos Mining Company—as well as other operators of company stores—may be criticized for their ultraconservative mercantilist politics, practical social, economic, and monetary forces frequently lay at the crux of the matter.

Money substitutes—scrip, coupons, tokens, chits, foreign currency—were key factors in the economic structure of many company towns, and while this system of exchange was frequently linked to company exploitation of the employees, its widespread use also stemmed from economic necessity. Allen's research on the Western company town reveals that scrip served a variety of purposes. Other than a method of credit, it was a "convenient bookkeeping method for companies which extended credit, for they did not have to keep track of individual items purchased. . . . [Also] the discounting of scrip by other merchants, and sometimes even by the company, for cash was an additional profit-making technique."[30]

---

[29] Cartledge to T. J. Roberts, May 9, 1925, Chisos Mining Papers.
[30] Allen, *Company Town*, pp. 134–135.

Although the Chisos Mining Company used scrip extensively in Terlingua, there is no record of its ever being discounted by the company, nor is there any evidence of widespread employee dissatisfaction because scrip restricted their purchases to the Chisos Store. This acceptance was not always the case, however, in other mining districts; a group of Pennsylvania coal miners struck in 1842 in protest of being paid in scrip redeemable only at the company store.[31]

Adapting predominantly urban economic principles and practices to isolated mining camps is an important chapter in the history of Western mining. Substitution, innovation, and adaptation offered varying solutions. Maintaining a circulating medium in Terlingua without banking facilities posed some real problems to the Chisos management, which during its almost four decades of mining operations attacked this deficiency in various ways. Importing large cash reserves by stage from Marfa and Alpine, while not fulfilling all operational needs, was one step toward solving the company's monetary problems. Citing the difficulties of this operation and the risk involved, Hawley writes: "Once or twice a month, the Mexican stage drivers would bring money in sums of $1,000 to $5,000 from the bank in town. This was necessary because the flow of ready cash was always away from the mines instead of to them. Shipments of cash in silver and bills came in by express in small canvas bags in the custody of the stage drivers. No officer of the law accompanied these shipments."[32]

Temporary cash reserves, while a necessity in any industrial operation of the magnitude of the Chisos quicksilver operation, did not solve the company's total monetary needs. Exploring another approach to this problem, Perry wrote the Chisos management in 1907 that he believed that American Express Company money orders "would float out as currency from hand to hand. In buying these I would fill in our signature at upper left-hand corner, then when we paid one of them out we would fill in the name of the party in the middle space and countersign the same signature as at upper left-hand corner."[33] This method, no doubt, precluded extensive use of the money orders as no further reference to these items was found in the Chisos company papers.

The daily payroll was the company's major monetary problem.

[31] Wayne G. Broehl, Jr., *The Molly Maguires*, p. 96.
[32] Hawley, "Life Along the Border," p. 62.
[33] Perry to Chisos management, November 28, 1907, Chisos Mining Papers.

As the cash paid out at the end of each shift averaged from $5,000 to $7,000 monthly, the Chisos management settled primarily on two methods of payment designed to eliminate the need for large cash reserves in American money, as well as to streamline front office procedures. Mexican pesos, which flowed in abundance across the Texas-Mexico border, and company scrip became the accepted Chisos circulating mediums during most of Perry's tenure at Terlingua. The pesos system was inaugurated first.

While the Mexican silver currency may have been unique to the Terlingua operation, the use of foreign currency in company towns was not original with the Chisos management. Many New England manufacturing companies computed employee earnings in British shillings well into the present century. In his comprehensive history of the Whitin Machine Works at Whitinsville, Massachusetts, Thomas R. Navin writes:

> One reason for the continuation may have been the convenience with which daily rates quoted in shillings could be translated into weekly wages payable in dollars. Since there were six full working days in a week, and six shillings in a dollar, a daily rate in shillings was always the same as a week's pay in dollars. Ten shillings a day amounted to ten dollars a week.[34]

The Whitin management followed a tradition based on economic expediency, while Perry, on the other hand, found in the Mexican pesos a technique ideally suited to the company's largely Mexican national staff, as well as to the closed economy enjoyed by the mining company.

Perry, no doubt, reasoned logically that a two-lane access to profits is better than one. Frank Parker, a latter-day Terlingua citizen, provides one explanation. In 1910, while employed by Crombie and Company, an El Paso wholesale house, Parker processed mail orders from the Chisos Mining Company. He recalls:

> There was quite a talk over the country about Perry using Mexican money in that [Terlingua] camp. People never could figure it out. Now Mexican money [a peso] was worth about forty-eight cents on the dollar most of the time. They said that he sold stuff out of the store, taking in a peso for a dollar. In other words, he would sell a pair of shoes that sold for $5—that would cost $5 in American money —he would sell them to you for five pesos. He did that for years and years. Well, that kept all the business from going out to Alpine and Marathon and Marfa. They couldn't figure out how he did it.[35]

34 Navin, *Whitin Machine Works*, p. 67.
35 Interview with Frank Parker, Austin, Texas, May 19, 1972.

In later years while working at Terlingua for the Esperado Mining Company, the Chisos successor, Parker discovered the key to Perry's economic maneuver. Parker found in an abandoned warehouse a letter from Perry to the mine superintendent that

> set this business up of taking in this Mexican money. He told them to get $40,000 or $50,000 worth of pesos and die stamp them with a large "C" so they could recognize them as the ones they owned. He also told them to watch and see if any additional amount [of Mexican money] came in, for if it did, then they would stop it. Well, where he made his money was in the payroll. He paid them in pesos [worth forty-eight cents] instead of dollars. He got labor for forty-eight cents on the dollar and that was a big piece of money for Mr. Perry.[36]

Based on quicksilver sales income in American money, this exchange rate cited by Parker reduced the Chisos staff expense more than one-half while restricting the trade to the company store. Parker explained further that even in 1943 the conditions that made the exchange system work for Perry in 1910 still prevailed. When Parker forgot to order a supply of change from the bank in Alpine, the silver currency that he previously paid out "was coming back to the commissary as there was no place for it to go. It didn't matter if it was pesos, silver dollars, or whatever you wanted to use, the fund was just revolving. You could have even used rocks if you wanted to."[37]

Perry apparently found the Mexican peso suited the company's monetary needs, as the system was still in use in 1934. An agent for the National Recovery Administration reported in September of that year that the company still paid the "Miners in MEXICAN SILVER PESOS instead of the lawful currency, and at the rate of two pesos for one dollar, instead of the market rate of 3.53 pesos for $1.00, [which] made it necessary for the miners and employees to trade at the company store . . . practically required as a condition of employment."[38] When the exchange rate dropped to five pesos to the

---

36 Ibid.

37 Ibid.

38 Memorandum from T. U. Purcell to L. J. Martin, Chief, Compliance Division, National Recovery Administration, San Antonio, Texas, September 25, 1934. Record Group 9, National Archives, Washington, D.C. The use of Mexican pesos as a medium of exchange was practiced universally by all Big Bend border stores trading with Mexico. Elmo Johnson reported that when he opened his store below Castolon, he "brought in a water bucket full of dobie dollars" to use in the Mexican trade. Since this trade was largely barter, the

dollar, Perry reportedly continued to pay his men at the rate of two pesos per dollar, and the store accepted them at the same rate.[39]

The use of Mexican pesos as a circulating medium in Terlingua, therefore, yielded triple benefits to the Chisos management—only one was based on economic expediency. First, the unofficial rate of exchange recognized by them reduced further the already substandard wage scale; two, the same exchange rate reinsured the trade of individuals at the company store who were already captive customers; and finally, the system simplified the company bookkeeping system and reduced the amount of American money needed to maintain company operations.

Although two decades passed without this system stirring the somnolent social consciousness of Americans, such was not the case with the Mexican government. Materialistic concern, however, rather than human welfare, apparently altered Perry's monetary exchange system. Sometime prior to 1912 Chisos bookkeeper and store manager C. A. Hawley reported to the Mexican consul that the Chisos Mining Company was defacing large amounts of Mexican currency, according to former Big Bend citizen Guadalupe Hernandez. The consul, in turn, brought suit against the Chisos Mining Company forcing cessation of this practice. Hernandez maintains the die marking was stopped; Robert Cartledge, who joined the Chisos in 1911, did not recall the incident.[40] Mexican currency, however, remained a major factor in the Terlingua economy as long as the Chisos continued operating. A memorandum from the mine office in Terlingua to the Portland office, dated November 15, 1934, states, "Mr. Cartledge sent Radford Gro. Co quite an amount of mexican money to apply on our account, which they had discounted at the current rate of exchange, and applied the balance to our credit."[41]

The use of company-issued scrip as mediums of exchange in

pesos were used to identify merchandise values. As the merchandise value always equalled that of the trade goods, "they remained on the store counter as long as I was in the Big Bend" (interview with Elmo Johnson, Sonora, Texas, June 3, 1966). Jim and Stanley Casner operated a store at their farm at Chianti, near Presidio. Jim Casner reported that when he carried 3,000 "dobie dollars" to El Paso one time, "they were so heavy they broke the handle off the suitcase" he carried them in (interview with Jim Casner, Alpine, Texas, June 6, 1966).

[39] Virginia Madison, *The Big Bend Country of Texas*, p. 189.
[40] Interview with Guadalupe Hernandez, Marathon, Texas, June 4, 1966.
[41] V. T. B. (unidentified) to Beulah Patterson, November 15, 1934, Chisos Mining Papers.

Western mining towns had wider distribution than foreign currency and spans a greater period of time. Western historian James B. Allen reports that in 1846 the Sonora Exploring and Mining Company at Tubac, Arizona,

> dominated the economic life of its workers. . . . it paid [them] in "boletas," or scrip issued by the company and accepted by the merchants. This strange money was made of cardboard about two and one-half inches by four inches in size on which were printed pictures of animals (apparently to help illiterate Mexican workers identify its value). The various denominations were: 12.5 cents (a pig); 25 cents (a calf); 50 cents (a rooster); one dollar (a horse); five dollars (a bull); and ten dollars (a lion).[42]

While the Tubac system was unique, scrip issued by mining companies was by no means uniform in design and denomination. Allen continues: "Most often it [scrip] was composed of detachable coupons sold in books of various denominations from five to twenty dollars. Each book contained coupons ranging in value from one cent to one dollar. Sometimes metal tokens were used, and in other cases tokens were made of cardboard or other materials."[43]

The logical assumption that the Mexican government's opposition to the Chisos' die marking the silver pesos prompted the inauguration of the coupon system remains undocumented; however, the chain of ensuing events appears to support this hypothesis. Hawley, who reportedly "blew the whistle" on Perry, left the Chisos employment about one year following Robert Cartledge's employment in 1911. Cartledge claims that one of his "first jobs was to set up the coupon system for paying the Mexican employees."[44] A Master Coupon Book, dated 1911, now in the Archives Division of the Texas State Library supports Cartledge's statement.

Printed on 16½" × 26⅝" paper stock and headed, "The Chisos Mining Company Register of Coupon Books," this book contains a record of scrip transactions. After receiving the payroll coupon books in varying denominations corresponding to the daily wage scale—$1.25, $1.50, $1.75, $2.00, $2.50, and $2.75—the employees made their purchases and a store clerk cancelled coupon amounts— $0.05, $0.10, $0.25, $0.50, and $1.00—corresponding to their purchases. When the value of each coupon book had been spent, the

[42] Allen, *Company Town*, p. 10.
[43] Ibid., p. 137.
[44] Cartledge interview, January 22, 1966.

clerk returned the coupon book to the bookkeeper, who recorded the transaction on the register.

Each page of the register contains three columns: (1) consecutive numbers that corresponded to numbers on the individual coupon books; (2) date issued; and (3) date returned. This book provided the company with a daily record of scrip transactions: amount of payroll scrip issued, amount returned in purchases, and amount unrendered. For example, Book No. 21001 was issued on February 23, 1911, and returned for either cash or merchandise on February 27, 1911. The issued and returned dates were entered with a rubber stamp; neither the employee's name nor number appears on the book, only the number of the individual coupon book. The elapsed time between the issued and returned dates of other books varies from the following day to a week or ten days. Each register recorded coupons of only one denomination.[45]

Twice each month unrendered coupons were redeemable in cash at the company store, according to former Chisos engineer Harry Fovargue:

> [The miners] were paid daily as they came off shift, in scrip good only at the company store. Which sounds harsh to the modern world's workers. But there was an angle in Perry's favor perhaps, that never has seen print. On the first and fifteenth of each month any scrip that a miner might have left could be traded in for cash, which he could spend anywhere. They spent some of their dimes in later years to see movies in a theatre that Perry built.[46]

Both exchange mediums—scrip and Mexican currency—continued for many years as financial mainstays in Perry's monetary system, and on August 21, 1921, Cartledge responded favorably to Perry's suggestion that the dual system be maintained: "Paying part cash with coupons may make men feel easier and spend there cash but I am afraid that they are going to save there cash. If they did spend there cash it certainly would make it look better for me in the cash sale end. . . . but what I want to do is whatever is better for the Company." Cartledge, unlike Perry, was closer to the problem and understood the Mexican rationale. Realizing that fear and suspicion were motivating factors in the Mexican-Anglo relationship, Cartledge foresaw the morale benefits inherent in cash payments. He reasoned to Perry: ". . . if we give the appearance that we want to keep them from having any cash that it only stimulates

45 Chisos Mining Papers.
46 San Angelo *Standard-Times*, June 8, 1958.

The Chisos Mining Company's quicksilver complex at Terlingua, Texas, 1922. Furnace and condensing installations dominate the foreground; the towering headframe marks the site of the No. 8 shaft, source of much of the company's richest cinnabar ore. Courtesy, Smithers Collection, Humanities Research Center, University of Texas at Austin.

Howard E. Perry, founder and president of the Chisos Mining Company, 1898. When this wedding photograph was made, Perry, then forty years old, was still employed by a Chicago-based manufacturing company. Courtesy, Walter K. Bailey.

From such primitive beginnings a great mine grew. W. A. Goad with Laurance and Drury Phillips at the collar of the Chisos No. 4 shaft in 1906. A dagger and lechuguilla leaf ramada protects Drury Phillips, operating a hand-driven windlass hoist, from the desert sun. Courtesy, Adkins Collection, Humanities Research Center, University of Texas at Austin.

Chisos No. 1 shaft. In 1906 the company was "in good ore bodies at a depth of 200 feet . . . and running about 15 tons per day through the smelter." Courtesy, Adkins Collection, Humanities Research Center, University of Texas at Austin.

Perry (*right*) and noted geologist Dr. Johan August Udden on porch of Perry's residence at Terlingua. Perry's success with the Chisos resulted largely from Udden's knowledge of the Big Bend geology. Courtesy, Smithers Collection, Humanities Research Center, University of Texas at Austin.

Robert L. Cartledge joined the Chisos Mining Company in 1911 as store clerk and advanced to general manager in 1923. Until Perry mysteriously fired him in 1940, Cartledge was Perry's most trusted employee and personal confidant. Courtesy, the Cartledge family.

Wayne R. Cartledge joined the Chisos Mining Company in 1909 as bookkeeper, advanced to general manager in 1913, and resigned in 1919 to establish a border trading post with Perry at Castolon. Courtesy, the Cartledge family.

Eugene Cartledge, Austin attorney and land specialist whose exacting legal research gave Perry title to the disputed Section 295 and the richest cinnabar ore in the Terlingua district. His two sons occupied key administrative posts at the Chisos. Courtesy, the Cartledge family.

Furnace employee "working" soot saturated with quicksilver. During periodic cleanups, the furnace was allowed to cool and the silver-laden soot was collected. Liquid metal in tilted bin is flowing from soot into bucket at right. Courtesy, Marfa Junior Historians Photographic Collection.

Furnace employees at Colquitt-Tigner mine (under lease to the Chisos) charging a retort with cinnabar ore. Quicksilver can be recovered from high-grade ore by retorting, a relatively simple roasting and condensing process. Courtesy, Marfa Junior Historians Photographic Collection.

Quicksilver flasks stacked beside a Terlingua district furnace await transfer to railroad for shipment to eastern market. Each container weighed eighty-eight pounds when filled. The MM on the flask top probably indicates these flasks were originally the property of the Marfa and Mariposa mine. Courtesy, Smithers Collection, Humanities Research Center, University of Texas at Austin.

The Terlingua stage, a two-seated hack with a canopy top, provided the Terlingua Quicksilver District with its only passenger and mail service for almost two decades. The torturous two-day trip to Marfa cost $9, including overnight accommodations. Courtesy, Smithers Collection, Humanities Research Center, University of Texas at Austin.

The Chisos Hotel at Terlingua, center of Anglo social activity and temporary residence of visitors to the quicksilver district. Built around 1910, this two-storied frame building with a corrugated iron roof replaced the original facility, a church tent stretched over a plank floor. Courtesy, Hattie Grace Elliot.

The Chisos company store, erected in 1908, "sold everything from a spool of thread to a Studebaker wagon." It also housed the Chisos Mining Company general manager's office, the Terlingua post office, and the only telephone in the village. The store remained profitable long after the quicksilver mine ceased to produce. Courtesy, Adkins Collection, Humanities Research Center, University of Texas at Austin.

Mule-drawn freight wagons, Terlingua's first link with the outside world. Quicksilver flasks were the outbound cargo; the wagons returned with everything required to sustain life and operate the mine. Courtesy, Smithers Collection, Humanities Research Center, University of Texas at Austin.

In the desert environment of the Terlingua Quicksilver District water remained an acute problem for many years. This Mexican boy and his burro laden with two wooden ten-gallon kegs provided one temporary solution. Courtesy, Smithers Collection, Humanities Research Center, University of Texas at Austin.

The Terlingua water works in 1938. From this storage tank adjacent to the Chisos Store, Mexican families carried water to their homes. This wagoner carried water to the Mexican homes and charged a delivery fee. As early as the mid-1920s water was piped to most Anglo homes. Perry's Terlingua mansion is in the background. Courtesy, Smithers Collection, Humanities Research Center, University of Texas at Austin.

The Mexican section of Terlingua seen from the Chisos Store and the Terlingua Post Office. Terlingua was segregated, both socially and geographically: Mexicans lived north and east of the company store, Anglos to the west. Courtesy, Smithers Collection, Humanities Research Center, University of Texas at Austin.

The Terlingua jail, symbol of law and order in the Terlingua Quicksilver District. It is reported that only Mexicans were incarcerated in this small stone facility, and when the violators exceeded its capacity, they were chained to iron stakes outside. Courtesy, Dr. Ross A. Maxwell.

Disputed area in the Chisos underground invasion law suit. From the workings off the Chisos No. 9 shaft, in the background, Perry's diggers invaded the area beneath the two rock monuments (*foreground*), held by the court to be the property of the Rainbow Mining Company. Courtesy, Wilbur L. Matthews.

Rainbow staff and survey crew discover the northeast corner of Section 70, a key position in the Chisos underground invasion suit. The two men in the center are believed to be surveyors White and Dod. John Briscoe Brown stands at the right. Courtesy, Wilbur L. Matthews.

Site of the Chisos underground invasion. The Chisos headframe (*center background*) stands in close proximity to the Rainbow workings (*left foreground*). Courtesy, C. T. Armstrong.

"Stornaway," Perry's mansion overlooking Casco Bay near Portland, Maine. Called "the finest in the state between Beverly, Mass. and Bar Harbor," the estate included a multi-car garage, servant's quarters, extensive gardens, and a private yacht harbor. Courtesy, Walter K. Bailey.

Perry at the helm of one of his yachts. When asked how he could afford such craft, Perry replied that if he had not wanted the yachts, he would never have made so much money in the first place. Courtesy, Walter K. Bailey.

Perry gazes out of the window of his Terlingua mansion and ponders an
uncertain future. By the mid-1930s, with the Chisos in decline and all
indicators suggesting caution and restraint, Perry embarked on a search
for another bonanza that ended in bankruptcy. Courtesy, Mrs. G. E. Babb.

Ruins of Perry School at Terlingua. The four-room stucco and tile building was erected in 1930 by the Chisos Mining Company to promote employee stability and to generate rental income for the company, but many immigrant Mexican children began the American acculturation process here.

Terlingua today. Little remains of the once-bustling mining town. These crumbling adobe and rock dwellings that once housed Perry's Mexican miners stand in mute testimony to an age of abundance long past.

them to be afraid and make them save all the more. . . . And as you say that if you contemplate cutting wages, it would be better not to have them [coupons] made for awhile."[47]

Although the company continued to print the coupon books, Perry's contemplated wage cut stemmed from his concern for the company's economic well-being, which in 1921 was still recovering from the post-World War I slump. Earlier that summer Cartledge had written Perry that "the financial statement does not look very encouraging this last month. The store sales were small, production and everything else."[48] Production continued to decline the following year and the Chisos mine, now the district's only producer, reported only 3,172 flasks of quicksilver.[49] The company encountered so much water below the 800-foot level that during the summer of 1922 Perry suspended all mining operation for two-and-one-half months. The Chisos Store, however, remained open, and Cartledge, faced with requests for credit from the unemployed Mexican miners, solicited Perry's approval. His abrupt denial firmly established his position in the employee-employer relationship in Terlingua. Perry responded:

> My dear Robert: No, we are not going to give the Mexicans credit. They can get along while we are shut down. They will have to help one another. The company has practically supported that Big Bend Country for two years while it made no money, and now losing money. We are not going to let the Mexicans get into us. We not only want to save our money, but as soon as any Mexican or Mexicans have got into us then we will be obliged to hire them to get even. We are not going to have our hands tied in any such fashion.[50]

At the conclusion of the typed portion of his letter, Perry appended the following handwritten note: "About the matter of credit to Mexicans. Marcus [Hulings, a Chisos mining engineer] has a tender heart—and a sure fine thing. So have I a tender heart, as evidenced by what [I] have done for them [the] past two years—Now my heart is tender for the Company and so it will remain until we get out of our troubles."[51]

While evidence of Perry's self-proclaimed magnanimity during

[47] Chisos Mining Papers.
[48] James M. Day, "The Chisos Quicksilver Bonanza in the Big Bend of Texas," *The Southwestern Historical Quarterly* 64 (April 1961): 443.
[49] J. Harlan Johnson, "A History of Mercury Mining in the Terlingua District of Texas," *The Mines Magazine* 37 (March 1947): 28.
[50] Perry to Cartledge, July 13, 1922, Chisos Mining Papers.
[51] Ibid.

the "past two years" remains obscure, Cartledge's report to him on the following September 19th cites several recent developments that became dominant themes during Perry's ensuing years at Terlingua:

> Our statements have been showing over draft right along, but we always try to take into consideration deposit that is on road. This on account of accounts being behind and some of our creditors crowding us. However, I think . . . we are pretty well catching up.
>
> I do not believe that there . . . [are] many [coupons], if any at all that or outstanding from way back. In fact [I] think the closedown has caused them [the Mexican employees] to dig up all the old ones and spend them on the first of the closedown. I notice, that a few are still sending out to Sears Roebuck for merchandise.[52]

Creditors, overdrafts, declining production, unpaid bills, and threats to the store trade were topics that appeared with increasing frequency in the Cartledge-Perry correspondence, and in the end "signed thirty" to the Chisos story. Grappling almost daily with problems of finance, Cartledge attempted to form a human bulwark between Perry, who frequently ignored bad news, and a gathering host of agitators, who sought persistently to get the Chisos to resolve their unpaid accounts. Ultimately Cartledge failed and all parties were losers. But in the interim both Cartledge and Perry continued to look to the Chisos Store for consistent profits and moved with dispatch when anything interfered.

Mail-order catalogues and itinerant peddlers were two consistent threats to store income against which the company took appropriate measures. The first incursion into the Chisos monopoly appeared in the form of that great rural American merchandising medium—the mail-order catalogue. Perry's response was quick and forthright: he issued orders to destroy all incoming mail-order catalogues![53] With the Terlingua post office located in the Chisos Store and a company employee serving as postmaster, Perry could implement his instructions. And while no former employee admits participating in this illegal act, many agree that Perry's dictum was followed implicitly.

When questioned on the matter of Perry's "catalogue blockade," former Chisos employee Mack Waters responded hesitatingly in the affirmative, "I heard rumors to that effect."[54] H. C. Hernandez,

---

52 Cartledge to Perry, September 19, 1922, Chisos Mining Papers.

53 First reported to this writer by Dr. Ross A. Maxwell, to whom former Chisos employees related the catalogue episode while he was superintendent of the Big Bend National Park.

54 Waters interview.

Robert Cartledge's first justice court interpreter, was more emphatic in his reply. He claimed "to know for a fact that Perry didn't want the Mexican employees to see, much less order from, a mail-order catalogue."[55]

Even if some item penetrated the blockade, possession of a catalogue did not necessarily enable the holder to make mail-order purchases. Again Perry and the Chisos Mining Company exercised total control: the use of the catalogue necessitated purchasing a money order, available in Terlingua only at the company-operated post office.

As long as Terlingua remained an urban oasis in a land of scarcity, the company maintained a trade monopoly with little effort. But always the threat from "the outside" persisted. Perry's challenge of the "catalogue blockade" coincided significantly with the establishment of public school education in the mining district. Again Perry's nemesis was Mrs. Elizabeth Bledsoe, the Big Bend's dedicated school teacher and voice of independence. Mrs. Bledsoe states that it was general knowledge that Perry did not want any mail-order catalogues delivered to the mining camp. "But some of the Mexicans got the catalogues," she recalls, "and I told them, 'You bring your catalogues to school and I'm gonna' show you how to make out those orders.'" She noted that the items they ordered were "shoes, underwear, and stuff like that. Sometimes they could buy it at the store and sometimes they couldn't."[56]

As usual, Robert Cartledge carried Perry's banners in Terlingua, and Mrs. Bledsoe, incensed at the restrictions the company placed on the Mexican employees, defied Perry while venting her anger verbally on Cartledge. She recalls:

> Now I always got along with Bob [Cartledge] all right and I liked him, but do you know I taught those kids how to make out those post money orders in school because it was something that they needed to know. After the store closed at Study Butte, the only way they had to get it [merchandise] was to send to Sears and Roebuck or Montgomery Wards. They didn't have a car to go over to Terlingua and buy stuff. So they would make out the order and I would mail it when I would come to town [Terlingua]. Then when the stuff would come in I would bring it over from Terlingua for'em. Well, he [Cartledge] just got madder'n a hatter at me because he said I was takin' business away from the store there. And Old Man Perry was worth a million dollars. And why should he worry about the little bit

[55] Interview with H. C. Hernandez, Alpine, Texas, October 23, 1972.
[56] Interview with Mrs. Elizabeth Bledsoe, Alpine, Texas, October 25, 1972.

o' stuff they bought? It didn't do Mr. Perry any good or any harm either. Well, I said to Bob one day, "I taught 'em how to do it [make out money orders] and I'm going to help 'em do it and do it right."[57]

Sustaining Mrs. Bledsoe's explanation of the company's anti-money-order policy, Hernandez recalls, "You could get them but they would raise a little fuss about it. He [Perry] didn't want any money going out of the camp. He wanted a monopoly on the business and he had it too. It was the biggest store, I guess, [in] eighty miles either way."[58]

Control of the Terlingua postal service, which enabled Perry to enforce his "catalogue blockade," also opened other avenues of profit. He reasoned, with some logic, that since the postmaster's duties were performed on company time by a company employee, his salary should be credited on the company ledger as income.[59] This arrangement apparently had the concurrence of the Terlingua postmaster and store manager, Ed Babb (not that he had a choice), but an extremely delicate situation arose in the mid-1930s when the postmaster's account, still listed on the company books under Babb's name, had grown to $12,355.50.[60] Apparently fearing an investigation by the Post Office Department,[61] Cartledge hastened to have the account removed from the company books; however, aware that transferring that large amount might attract undue attention, he solicited the advice of Perry's Portland secretary:

> I realize as you do that to charge off such a big amount is likely to attrack the intention of some Inspector. In fact this is the reason I have been bringing the matter up for the past six years or more. I felt that it should [have] been charged off in some of the years that we were showing a loss before the account got so big. Then it would not be so noticable.
>
> About the only suggestion I can make to get same off of [the] books without our entry attracting attention, would be to have Mr. Perry increase my deposit by a certain amount each month. . . . [I could then] make out checks payable to G. E. Babb, Postmaster, and then intern have Babb indorse these checks and I send them back to

57 Ibid.

58 H. C. Hernandez interview.

59 Cartledge interview, April 29, 1965.

60 Cartledge to Perry, September 8, 1936.

61 Mrs. G. E. Babb was interviewed in Sanderson, Texas, on December 27, 1964, but Mr. Babb was unwilling to discuss his job assignment with the Chisos Mining Company.

Mr. Perry and he cash them and use the money on some of the many expenses that he has.[62]

Cartledge's urgency apparently went unnoticed by the Portland office for almost two years, when his recommendation was finally accepted. Acknowledging the change to the Portland secretary, he also cited other fiscal matters left too long unattended by the master of the Chisos mine. "This account," according to Cartledge, "should have been handled this way all the time but it was impossible for me to get Mr. Perry to agree to it. . . . We also have a good many accounts on our books that should be charged off to profit and loss, as they are now non-existent."[63]

Perry's continued steps to coerce company employees to patronize his store, exclusively, reflect the views of a man living well behind his appropriate time. While Perry remained philosophically immobile, something called social progress slowly passed him by, and the increased contacts with "the outside" gave company employees the power of awareness that neither Perry nor Cartledge could fully comprehend. The Mexican employee's determination to patronize the mail-order establishments is a case in point. Cartledge reported to Perry in 1921: "We have been selling . . . quite a few money orders the last few days. Will try and make you up a list of the purchasers of them by next mail."[64] Fourteen years later, despite the company's obstructionist tactics, Cartledge was still reporting the increasing outside flow of money, "Purchases thru mail-order houses seems to be growing right along."[65]

Although Perry and Cartledge accepted defeat at the hands of the mail-order establishments, they posted a better record with the Big Bend itinerant peddlers. With increased automobile travel to the Big Bend in the 1930s, peddlers from Alpine and Marfa frequented that area offering a wide variety of merchandise to farmers, ranchers, and mining company employees—all except those employed by the Chisos. Cartledge had standing orders from Perry to eject anyone found selling merchandise in Terlingua. He fulfilled this commitment when he apprehended a peddler in camp with the rear seat of a Buick automobile loaded with dry goods. Cartledge reported to Perry that "I ran him out. . . . I am wondering if we

[62] Cartledge to Allen P. Stevens, January 11, 1937.
[63] Cartledge to Beulah Patterson, December 11, 1939.
[64] Cartledge to Perry, July 5, 1921, Chisos Mining Papers.
[65] Cartledge to Perry, January 22, 1935, Chisos Mining Papers.

should not besides run them out of camp when we catch them, discharge or run out of camp any family that allows them to come to there house."[66] Perry, who relished the good news, responded, "I am glad you caught that peddler and ran him out of camp, and do so every time. . . . One reason I bought that section [of land] down near our road to Mariposa . . . was so no one could set a store or wagon near that road."[67]

Even those salesmen who offered noncompetitive merchandise and requested permission through proper channels to show their wares were unsuccessful. J. F. Reeves, an Alpine Wear-Ever Aluminum salesman, wrote Cartledge he "had thought about running down to Terlingua for one or more demonstrations, depending of course, on consent of officials with your company and interest shown."[68] Cartledge's response authorized Reeves to demonstrate his merchandise "to the Americans here. However, I am sure that your trip here would not pay you, as only a few American families live here." Referring to the Mexicans, Cartledge explains that they "are not able to buy this ware nor would they be able to understand a demonstration. In fact on account of this and other reasons, we would not be willing for you to work the Mexican population."[69] Implicit in Cartledge's rejection was the basic company premise that, while the small Anglo population was privileged to spend their money wherever they chose, funds spent by the Mexicans were destined only for the cash drawers of the company store.

Noting breaches in the company's merchandising blockade, Hernandez, however, explains that "in those days there was what we called roving merchants that made the rounds down there [the Big Bend] but they would never come in to the company. They would park down along Terlingua Creek and people would go down there and buy off them." Hernandez adds that local farmers were also permitted to peddle fresh meat and vegetables in the village, two items the company store never stocked. *Cabrito* [goat] and corn were priority items, as corn especially was a necessary staple for the Mexicans living in Terlingua. Hernandez continues: "They would soak it in lime and then that skin would come off after they washed it two or three times. Then they would grind it to make tortillas, make tamales, or something like a Frito now. Cut it in strips, of course. They didn't fry it like a Frito, but they cooked

66 Cartledge to Perry, January 24, 1934, Chisos Mining Papers.
67 Perry to Cartledge, December 31, 1934, Chisos Mining Papers.
68 Reeves to Cartledge, September 8, 1938, Chisos Mining Papers.
69 Cartledge to Reeves, undated, Chisos Mining Papers.

it in the fire until it got hard. Whenever they couldn't get any corn, boy, that was trouble right then."[70]

While neither the roving vendors that parked along Terlingua Creek nor the farmer-peddlers from Terlingua Abaja made any appreciable dent in the Chisos Store receipts, other areas of financial concern attracted Cartledge's and Perry's attention. Following the 1922 shutdown, Perry reopened the mine late in the year, although recovery operations were still hampered by flooding. Production for that year reached 2,954 flasks of quicksilver, followed by a gradual three-year decline: 1923, 2,027 flasks; 1924, 1,469 flasks; and in 1925, 985 flasks, the lowest annual recovery in more than twenty-five years. Perry, however, found the production reports for the next three years more to his liking. Although much of the 1926 production, 1,039 flasks, came from condenser cleanup operations, the new ore bodies discovered in the deeper workings are reflected in the 1927 production: 2,318 flasks marketed at $118.16. The following year production dropped to 1,679 flasks but a stronger market boosted quicksilver prices to $123.51 per flask.[71]

Although Perry could generate renewed optimism from the Terlingua production reports as the 1920s drew to a close, the new decade held little in store to indicate a return to the pre–World War I bonanza years. The depressed economy of the early 1930s is readily apparent in the extant fragmentary company records for that period. For the months of July, November, and December, 1931, the company posted net losses of $3,588.61, $288.10, and $1,126.71 respectively.[72] The quicksilver market for the ensuing years was equally dismal: 1932, $57.93 a flask; 1933, $59.23; 1934, $73.87; 1935, $72.00; 1936, $79.92;[73] and then on October 20, 1937, with poor Chisos production reports coming in the wake of another major break in the stock market, Perry again ordered the mine closed. For all practical purposes this event marks the end of the Chisos Mining Company. But Perry was not nearly ready to give up. He still believed implicitly that the Chisos mine was rich in undiscovered ore and that his greatest bonanza lay somewhere ahead in an unpredictable future.

Meanwhile, even as the depression deepened, both Cartledge and Perry were encouraged that the store continued to show a profit. In the three months in 1931 cited previously, when the com-

[70] Hernandez interview.
[71] Johnson, "Mercury Mining," 37 (March 1947): 30.
[72] Monthly Statement, Chisos Mining Papers.
[73] Johnson, "Mercury Mining," 37 (July 1947): 21–22.

pany reported a combined net loss of $5,003.42, the store posted regular monthly profits of $1,301.68, $1,225.98, and $1,278.44, on average monthly sales of $7,127.54. The end-of-the-year inventory totalled $34,587.65.[74]

With the mine operating at a loss, protecting the company store trade from outside competition became more important than ever. On May 16, 1932, Perry, apprising Cartledge of depressed market conditions, writes, "It is almost impossible to sell quicksilver. The two days I spent in New York were devoted to this kind of thing but not any sales." The thrust of this communication lies in devising methods to keep the mine employees from "buying outside and elsewhere." Without revealing his contemplated measures he confides to Cartledge, "there are two other steps . . . which I am considering. One of these steps will stop the outside buying completely." Perry then appends the following note in pencil, "We will begin to post names on the Bulletin Board—Soon as known quite fully about the offenders."[75]

While no record survives of what new measures Perry applied to stop "the outside buying," if he contemplated making company store trade a condition of employment at the Chisos mine—or expulsion, if they failed to comply—his method lacked originality. Wayne G. Broehl, Jr., for instance, cites a case in *The Molly Maguires* in which a Pennsylvania coal mine operator posted the following notice: "You will, therefore, please understand, one and all, that from this time henceforth we shall take particular note who deals at the store and who does not. And as the time is near at hand when we shall reduce the number of our men, just such men as have no account at the store will be dropped."[76]

When Perry's plans to eliminate outside patronage apparently came to naught, the Terlingua-Portland correspondence began to follow a new and milder line of reasoning. At the insistence of the United States government and the National Recovery Administration, Perry had involuntarily entered phase three of his monetary policy and in 1934 began paying American money to a mobile staff of Mexican employees who looked to a new merchandising Mecca ninety miles to the north. Since competition yields benefits, both Cartledge and Perry began restudying their sales methods at Terlingua, and the ensuing correspondence reflects the changing views of both men. Allusions to coercion are noticeably absent as interest

[74] Monthly Statement, Chisos Mining Papers.
[75] Chisos Mining Papers.
[76] Broehl, *Molly Maguires*, p. 84.

grows in introducing merchandising methods observed in other stores. Perry first suggests they consider "some kind of a bonus system, that is if you can pick up a lot of inexpensive trinkets or attractive stuff to give away with sizeable orders."[77] He also looked with racist envy on the successful Jewish shopkeepers in the Mexican districts of Austin and San Antonio. "It would pay you," he wrote Cartledge, "to spend fully a half day in San Antonio down in the Mexican store district talking with those Jew store people that run the most attractive stores. . . . There is no use for any white man to think he can hold his own with a Jew when it comes to merchandising."[78]

When Cartledge outlined a credit system to attract the Mexican employees to the company store, Perry vetoed the suggestion by return mail. He countered instead with a proposal to make the Chisos Store more attractive. "We will get more trade," Perry reasoned, "as soon as we paint the store on the inside and put in a new ceiling." This suggestion, no doubt, was in response to competition "on the outside." "For one thing," he continues, "our Mexicans, when they go to town and there are two or three nice stores down at the railroad, our place looks inferior to them."[79]

A new coat of white paint on the company store was the only tangible result from this brisk three-year correspondence, but even this failed because the white paint did not cover the original red paint. By the mid-1930s, when Perry finally realized that a new social and economic climate prevailed in the Big Bend, he was too much of a blind optimist to comprehend fully the impact of progress. "We no longer own our trade," he acknowledged to Cartledge. "The competition of the Alpine stores and the money order business," he continues, "is such we have got to get ready to do real merchandising and fight for the trade. And fight to persuade the Mexicans to buy things they really have no thought of buying. We will have to work them just like the Jews do."[80]

Cartledge agreed with the absentee mine owner. "There is no question that we do not have the trade grabbed as formerly." Citing a recent example, he continues:

> [I met a] Mexican and Waters on the road coming from a trip up town [Alpine]. One Mexican had a bed and springs on his car. I am

[77] Perry to Cartledge, August 28, 1934.
[78] Perry to Cartledge, January 22, 1935.
[79] Perry to Cartledge, February 1, 1935.
[80] Perry to Cartledge, July 23, 1935.

sure that he had no better bed and springs than we have and that he bought them no cheaper up there than he could down here. Maybe we are to blame in not knowing how to show him our beds.

Attempting to rationalize the loss of local trade, Cartledge continues, "Regardless of how cheap we sell or how we display our merchandise it is naturally with our men [Anglos] running to town, for them [Mexicans] to buy up there. A whole lots is humanature and a lots a matter of curiosity." Then illogically equating social progress with economic loss, he concludes:

> At time[s] it has appeared that Henry [Waters, an Anglo] even helps the Mexicans send off orders. Also we have mexicans in the district that have learned to read, and write a little that apparently go around and encourage other Mexicans to send out orders. . . . with the automobile and some of our people becoming educated things have gradually grown worse.[81]

Although some of Perry's customers abandoned him to trade in Alpine and the traffic in mail-order sales increased, store receipts failed to justify his concern for outside competition. In June, 1937, when the company reported a $10,291.90 loss, the store yielded $1,560.00 profit on gross sales of $7,169.50.[82] And much to Perry's surprise, in the face of declining population, store sales remained strong throughout the year. Cartledge explained that exploration crews working at Study Butte were trading at the store; also, "most men that were let out [following the closedown] left there families here and likely most of there savings to live on."[83] In addition, Cartledge notes that some of the men who had found work elsewhere were sending money back to their families. And one year later, with production still declining, the trade volume at the store held relatively firm. "I certainly was surprised," Perry wrote Cartledge, "to see our sales on the day before Christmas [1938] were up to $328.17."[84]

The last extant data on the Chisos operation is the September, 1941, monthly statement. Although the quicksilver operation continued to fail, showing a $4,961.59 loss for the month, the store business remained relatively stable, yielding a $1,036.45 profit on sales of $5,031.29. The significant statistic, however, is the inventory,

[81] Cartledge to Perry, October 19, 1935.
[82] Chisos Mining Papers.
[83] Cartledge to Perry, December 20, 1937.
[84] Perry to Cartledge, January 3, 1938.

which by September 1, 1941, had dropped to "approximately $7,802.36."[85]

One year later the Chisos Mining Company was bankrupt, but the store remained open and continued to serve the mining district as it had for almost a half-century. The five o'clock rush to the store was only a fond memory and most of the once vast stock of merchandise—"anything from a spool of thread to a windmill"—had all but disappeared. Gone too were those familiar names and faces—Cartledge, Babb, Ekdahl, and Anderau—who had helped build it into "the biggest store between Del Rio and El Paso." On December 11, 1942, one Emma Calanhan, apparently the new store manager, wrote to a Mr. Jacobs, presumably representing a Marfa wholesale firm, soliciting his advice in operating the store. The letter that follows is a sad and humiliating obituary to a once thriving institution.

> Quite frankly I do not know how to order any of the things we need. If you could supply us with a price list which would include quantity buying it would be greatly appreciated.
>
> At present we must conduct our business on a cash basis since the Chisos Mining Company is now bankrupt. Therefore send all supplies C.O.D. . . .
>
> Please send the following order:
> 1 doz bottles ink blue-black
> 2 doz notebook covers 8 × 11
> 3 doz boxes crayons to be sold for 5¢
> 3 doz spiral composition books No. 87 ruled
> 2 doz fountain pens to be sold for 25¢
>
> Any advice or information you may be able to give will be greatly appreciated.[86]

In his long and arduous struggle to eliminate competition with his company store, Perry was fighting an illusion. He enjoyed a natural near-monopoly, not because of company policy or the monetary system, but simply because of location. The Chisos employees had no other convenient place to trade in the Big Bend area, and the Chisos Store would have continued to make money for Perry, even after the mine closed, had it not been a part of the defunct mining company. Monopoly, even a natural monopoly, often offends the sensibilities of the victims and carries with it an appealing challenge. The Chisos employees ordered from mail-order catalogues,

---

[85] Monthly Statement, September, 1941, Chisos Mining Papers.
[86] Chisos Mining Papers.

probably not entirely for their economy, but largely because of the variety of choice and the visual appeal the forbidden mail-order catalogues afforded. Likewise, Robert Cartledge explained that the Chisos mark-up on merchandise was in line with that charged by the Alpine stores: groceries, 15 percent; dry goods, 50 percent; hardware, 50 percent; shoes, 60 percent; and jewelry, 100 percent.[87] H. C. Hernandez, the former Terlingua resident who traded at the company store, agrees with Cartledge. "There was a slight charge [for transportation] but it was reasonable," he explains.[88] Mack Waters, however, remarks, "A lot of people claimed they could get groceries cheaper in Alpine."[89]

Cartledge's and Hernandez' recollections, rather than Waters', are consistent with Allen's findings in his study of company stores in Western mining towns. He concludes that "even though most companies did not lose anything through the operation of stores, such profits as occurred usually were not out of order."[90]

Perry saw Terlingua for what it was, a company town, and the company belonged to him. Since he owned 99 percent of the Chisos Mining Company stock, which included the company store, he saw no justification to share anything. Monopoly, in Perry's nineteenth-century world of free and unbridled enterprise, was the one key to commercial success, and anything short of monopoly was a challenge to business management. But at Terlingua Perry became a victim of his own avarice: what he had he could have kept without the struggle, and no merchandising skill could have diverted the small percentage of the business that went elsewhere. Thus the challenge was always present to glean every possible peso from the people who lived in his town and worked in his mine. And the agony of partial success motivated Perry to continue the struggle.

Because Perry could not comprehend the phenomenon of social progress, the struggle was destined to continue. By the mid-1930s his once-captive customers were now aware of what the outside world had to offer. Terlingua Creek, Fresno Canyon, and the Rio Grande no longer circumscribed their habitat. They now knew of the wonders that awaited them ninety miles to the north: motion picture theaters, restaurants, parlors that served nothing but candies and ice cream, but most important of all, they discovered store after store where they could feast their eyes on marvels unknown at the

[87] Cartledge interview, October 17, 1965.
[88] Hernandez interview.
[89] Waters interview.
[90] Allen, *Company Town*, p. 134.

Chisos Store. When this new world of discovery led to a gradual change in the buying habits of the Mexicans, store receipts reflected the trickle of cash away from Terlingua.

The roots of the problem ran far deeper than Perry could ever comprehend, and the tide of human progress could not be stopped by simply offering "inexpensive trinkets or attractive stuff." If the reasons people ceased to trade at the Chisos Store escaped Perry, Cartledge's grassroots knowledge of the Mexican psyche touched closer to the crux of the matter. Merchandising was only part of the problem. The real problem was change, symbolized by education, curiosity, and the automobile. And if from Cartledge's refracted vision "things had gradually grown worse," the Mexicans viewed the present from a different perspective. For them things had grown gradually better. Perry had helped them leave the primitive rural life of Coahuila and Chihuahua and had introduced them to an urban life of sorts at Terlingua. With the benefits of Chisos employment they altered their manner of dress, added variety to their diet, cultivated new forms of entertainment, avoided disease, and lived healthier, while their children learned to read and write.

And whatever debt they owed Perry had long since been paid. Although the price was high of his own choosing, what they received in return, they liked. Slowly the account was being settled as they began sharing in a life that, a quarter-century before, was as remote to them as was Perry's belief that a Mexican could ever share the same plane of human dignity as he. This was the irony of change. As Perry stripped the region of its one item of value, quicksilver, he unknowingly and totally without human concern opened the paths to social betterment for a people whom he despised. Now they no longer accepted his brand of civilization. They knew there was something better.

CHAPTER 9

The Rainbow Mine Episode

HOWARD E. PERRY'S honesty—frequently questioned but never disproven—received its severest test in the Rainbow mine episode. Ironically this event, sometimes referred to as Perry's greatest indiscretion, came to light because of a breach in his "wall of secrecy."

It all began late in 1927 when John Briscoe Brown visited the Terlingua Quicksilver District. Although the reason for this trip has been lost with the passing of time, probably the rising price of quicksilver lured him to the Big Bend. With expanded industrial usage the liquid metal was selling at an annual average of $118.16 per flask, its highest market level since the end of World War I.[1] If Brown was there to investigate the prospects for quicksilver investment, he found the news encouraging: the Chisos production was triple that of the previous year, the Oakes interests continued prospecting on Section 39 and installed a rotary furnace, William D. Burcham reopened the Mariscal property, and the Waldron mine engaged a new engineer and planned extensive developmental work. It was a time of increased activity.

As a person to gather information in an area that thrived on secrecy, Brown was in his element. After graduating from the Colorado School of Mines, he had worked in Mexico for Standard Oil of New Jersey as an oil scout; as a procurer of involuntary data, he was a professional.[2] C. T. Armstrong, who knew him there and later joined him at Terlingua, recalls: "Brown would go to bars, and drink with the employees of other companies. He would just sit there like an owl, never take out a pencil, but would be recording mentally every word he heard. He would then go back to camp and write down what he had heard verbatim. Had a memory like an elephant." Armstrong then paused momentarily and a sly grin crossed his face. The expression must have revealed a savoring of

---

[1] J. Harlan Johnson, "A History of Mercury Mining in the Terlingua District of Texas," *The Mines Magazine* 37 (March 1947): 30.

[2] Interview with C. T. Armstrong, San Antonio, Texas, April 3, 1968.

pleasant and exciting experiences, long past. Then he spoke, "What a capacity he had for learning other people's business."[3]

Others remember Brown as a man with a chameleon personality who could adapt to any social environment. While mining was his profession, he had a dilettante's interest in economic philosophy, and with those who were not of the mining fraternity, his conversation turned from the mundane vernacular of physical science. Attorney Wilbur Matthews explains: "At times he gave indication of intellectual greatness. He wrote a book, had it published, and gave me a copy. It was an inscrutible piece that seemed to say that quicksilver should be adopted as a basis for United States currency instead of gold and silver. But as a mining engineer he was a man of ability."[4]

Brown's knowledge, as well as his indigenous charm, enabled him to establish rapport with all whom he met, both the intellectual and the illiterate. However, it was in the company of those who discussed faults, fractures, stopes, and veins that he functioned best. They remember his favorite environment—and the source of much of his information—was the barroom, because, whatever redeeming virtues John Briscoe Brown may have possessed, they did not include abstinence from drink.

While the nation—and Robert Cartledge—struggled with the temporary stigma of prohibition, the flow of sotol from across the Rio Grande was scarcely hindered. Mexican bootleggers flourished and the availability of alcohol in Terlingua gave a zest and a plausibility of an existence that, in the cold reality of sobriety, even the hardiest natives chose to escape. Life was hard, compensation pitifully small, and pleasures rare. Thus through the consumption of illicit liquors could the Mexicans escape the ordeal of oppression, enjoy the fellowship of those who shared their fate, and engage in casual conversation. In this environment Brown gained the information he was seeking. The results would prove remunerative.

These new-found friendships yielded two bits of information that sparked his imagination. He learned that the Chisos mine was removing extremely high quality ore from westward-trending drifts off the No. 9 shaft. Two decades later geologists would report that "the Chisos development on the 600 level encountered the Del Rio–Edwards limestone contact along the Chisos-Rainbow northwest

3 Ibid.

4 Interview with Wilbur L. Matthews, San Antonio, Texas, August 24, 1966.

fault" and surfaced a high grade ore that boosted 1927 production
to 477 flasks in August and 493 in September. In that year the company marketed 2,318 flasks that netted over a quarter-million
dollars.[5]

But what interested Brown even more was that the Mexican
muckers believed that the Chisos tunneling, which worked the Mc-
Kinney-Parker claims, had penetrated the adjacent Rainbow property. Brown knew that the Rainbow mine, virtually dormant since
1918 and totally undeveloped, was available for lease, and the
probability of such excellent ore in that close proximity increased
its appeal. And the fact that the Chisos holdings bordered the Rainbow claims on three sides heightened the plausibility of this underground deviation. To substantiate further these rumors, the Mexicans heard that Marcus Hulings, the Chisos mine superintendent,
also believed they were removing Rainbow ore and wanted to stop
the westward drifts until a surface survey could be made. However,
the mine management forbade it and instructed him to continue the
work.[6] The farther west they tunneled, the Mexicans told Brown,
the richer the ore became.

If Brown had not already made his decision at this point to
acquire a quicksilver property, then the apparent existence of this
great ore body spurred him to action. While his personal espionage
enabled him to gain this information, his decision to acquire the
Rainbow property was based on his knowledge of the Terlingua
geology. Armstrong sustains his former colleague's logic:

> Brown knew that cinnabar deposits in the Terlingua district oc-
> curred in cracks in the limestone formation. He also knew that these
> fractures occur at frequent intervals. Therefore, he reasoned that if
> the Chisos had good ore on the west side of Section 295, it probably
> continued on Section 70. History proved him correct. North of the
> Chisos Number Nine shaft, we discovered parallel fractures that
> contained some of the finest cinnabar in the district.[7]

The Rainbow Mining Company properties consisted of five
mining claims, whose record as a quicksilver producer was undis-
tinguished prior to 1927. The history of this company dates from
February 8, 1909, when J. M. Wattenbarger filed application for
the first mining claim on Survey 70, Block G-12.[8] The record of this

---

[5] Johnson, "Mercury Mining," 37 (March 1947): 30.

[6] Interview with Robert L. Cartledge, Austin, Texas, February 24, 1968.

[7] Armstrong interview.

[8] M. G. Whitlock et al. vs. Chisos Mining Company, "The Question of

survey was filed in the General Land Office on February 16, 1909, and the claim identified as G4-1. It was on this claim that the rumored invasion had occurred.

Wattenbarger did very little developmental work; however, by 1913 he had extended his holdings to include two additional claims, and on April 22 of that year, he acquired two more leases from A. C. Chisholm.[9] United States' entry into World War I stimulated activity in the district, and on November 15, 1917, Wattenbarger leased the property to J. S. Darnell for one-sixth of the minerals produced. On June 12, 1918, Darnell in turn transferred his lease contract to the Rainbow Mining Company. This lease was made in trust for L. T. Millican, M. B. Whitlock, Wayne Darnell, M. W. Davenport, and J. S. Darnell, "who formed a corporation known as Rainbow Mining Company." Limited developmental work was undertaken, and later in the year their shaft operations had only reached the 250-foot level.[10]

During the early 1920s, while the Chisos reported the district's only recovery, the Rainbow interests enacted various organizational changes. On May 12, 1920, the stockholders executed an agreement that pooled the lands and minerals and specified that they be "handled as a consolidated mining property under the management and control of five trustees."[11] These were M. B. Whitlock, L. T. Millican, J. M. Wattenbarger, J. S. Darnell, and J. W. Darling, "who should administer the trust in the interest of the owners, who agreed to associate themselves together for the purpose of forming an unincorporated association under a declaration of trust in the name of Rainbow Mining Company." The capital stock was $1,250,000.[12]

The capitalization of the Rainbow Mining Company represented an optimism not based on actual production, and while the property languished unwanted amid reports of a general postwar economic resurgence, the stockholders again decided to restructure the company for the purpose of liquidation. On May 22, 1924, they executed an instrument that empowered M. B. Whitlock to dispose of the five mining claims.[13] This was not a prophetic gesture, because four years elapsed before the reorganization plan netted a

---

Title," Attorney's file in possession of Wilbur L. Matthews, San Antonio, Texas. Hereafter cited as Attorney's file.

[9] Ibid.
[10] Ibid.
[11] Ibid.
[12] Ibid.
[13] Ibid.

lease, which was negotiated by "their agent and attorney in fact, Whitlock . . . to W. W. Devine and J. B. Brown, dated November 8, 1927."[14] The lease provided for a payment of $60,000 and 15 percent royalty on all quicksilver recovered. Although the price was high, Brown's vigilance was beginning to yield rewards; they would increase with the passage of time.

Perry, no doubt, found the news of the Rainbow lease doubly upsetting. He, probably more than anyone else, knew the Rainbow's vast potential and would have liked to acquire it himself. But that opportunity was now long past. The same lease for which Brown had paid $60,000 had been offered to Perry for $9,000, and against the strong insistence of Wayne Cartledge, he had rejected it. Many years later Cartledge recalled, "If he hadn't been such a tight-wad he could have had it all for himself."[15]

But probably what troubled Perry even more was that his "gospel of secrecy" had been violated by his own employees. For Perry secrecy was an occupational necessity—a tenet by which one lived and survived—and in Terlingua secrecy was synonymous with mystery and distrust. This indescribable aura of obscurity seemed to pervade every facet of the Chisos operation. The knowledge of this indiscretion, however, was heightened by the fact that his company was removing ore from an area that Perry undoubtedly believed was the property of the new owners, one of whom, Brown, was staying at the Chisos Hotel.

Perry first became suspicious of Brown's presence in Terlingua soon after he learned he had leased the Rainbow property. Not knowing the character of the man with whom he was dealing, and certainly not his reason for remaining there, Perry naively attempted to resolve the issue by letter. He wrote Brown:

> [Terlingua] is not a public place as of course you know, but entirely privately owned. One has a right to come up here to the post office, but nothing further unless mutually agreeable. Anyone stopping at our hotel is really there as our guest; that is, a paying guest. If you and your associates are entirely friendly we greet you, but if otherwise, you will of course understand that you have no place here.[16]

While one of the Mexicans who aided Brown had been dis-

---

[14] Ibid. This transaction was recorded on March 13, 1928, vol. 3, p. 40, Bill of Sale Records, Brewster County, Alpine, Texas.

[15] Interview with Wayne Cartledge, Marfa, Texas, December 29, 1964.

[16] Perry to Cartledge, December 24, 1927, Chisos Mining Company Papers, Archives of the Texas State Library, Austin.

missed by the Chisos management and forced to leave Terlingua, the fact that Brown continued to reside at the Chisos Hotel was a source of growing anxiety. Now confident that his intentions were not "entirely friendly," Perry submitted to Robert Cartledge a proposal for Brown's removal.

> Please let me know whether Brown . . . stay[s] at the hotel any more. . . . If I had known when at the mine, or *before we drove that Mexican out*, that this fellow had those *Palmbutter Firearms* [probably code words for some bits of incriminating evidence Brown was carrying] I would have made a move to have him embroiled in some kind of disturbance, and of course had him searched; that would have been fine business. In case we catch any more such fellows I would like to have this done. We will take away from them anything they have that would be incriminating until we have straightened them out.[17]

Former associates do not recall when Brown left the Chisos premises, or if he did so voluntarily, but whatever plans Perry may have had for "straightening out" Brown, they were never implemented. At this point Brown knew—but Perry did not—that his knowledge of the supposed invasion was founded solely on rumor and suspicion. While he felt this was sufficient evidence to acquire the lease, it was still far from conclusive, at least not a matter upon which to base a law suit. The obstacles were great and the undertaking fraught with occupational uncertainties; however, the prospect of financial returns held high promise of a venture worth embarking on. It was solely upon this premise that Brown acquired the Rainbow.

While no longer welcome at the Chisos Hotel, Brown was not one to allow Perry's inhospitality silence his flow of information. He still traded at the Chisos Store, got his mail there, and continued to engage the Mexicans in conversation. His hope for the future lay in increasing his knowledge, not only of where the Chisos company was finding ore, but where it was in relationship to his property. While he may not have had the freedom he enjoyed when he first arrived at Terlingua, he nevertheless continued his clandestine visits; little by little he pieced together the fragmentary reports of what was occurring far beneath the barren hills of the Terlingua country.

Armstrong believes that Brown's continued success in the face of increased Chisos surveillance was because "Brown knew the

[17] Ibid.

Mexicans, spoke their language, identified with them, and achieved a rapport the Chisos management never understood." By gaining their confidence, he undermined the very principles of secrecy and distrust with which Perry had endowed his company and persuaded the Mexicans to reveal to him information that had been Perry's well-kept secrets for over a quarter-century. Armstrong concludes, "He was indeed a professional. He soon knew more about the workings in the Chisos than Perry did."[18]

With each bit of evidence Brown ferreted out he became increasingly certain that the Rainbow property had been invaded. While he knew he still lacked the evidence needed to engage Perry in the courts, he foresaw the possibility of confronting him with the information already in hand. A negotiated settlement might possibly yield sufficient cash to enable Brown to develop more fully the Rainbow mine's potential. However, since dealing with Perry was too great an undertaking for him alone, the Rainbow organization was reconstructed again. On March 12, 1928, Brown and Devine transferred their interests in the property to Rainbow Mines, Inc., and applied for a charter in that name.[19] This reorganization increased the company's financial potential for confronting Perry with their evidence of invasion. Such an undertaking, Brown reasoned, could become costly.

This maneuver drew into the active management of the company a man who, though lacking Brown's technical knowledge, possessed a shrewd tenacity that counterbalanced the blatant confidence and stoic self-righteousness of Perry. M. B. Whitlock, attorney-in-fact and experienced in the hazardous melee of the Texas wildcat oil industry, was endowed with the qualities of survival that placed him on a competitive level with Perry. If Perry was never to accept Whitlock as his social and intellectual equal, this association would prove to be a costly encounter that he would long remember. On March 28, 1928, in the company of John Briscoe Brown, they met in the Holland Hotel in Alpine.

Following the introduction by Brown, Perry quickly took the initiative and accused Whitlock of "making some nasty remarks about [me] and [my] company." Whitlock recalled: "I told him I had not and he said that he understood that I had said that he had trespassed on our property. I told him that I believed he had and

[18] Armstrong interview.
[19] "Question of Title," Attorney's file; Deed Records, Brewster County, Alpine, Texas, vol. 60, p. 306.

he says, 'Well, I haven't.' "[20] Although they discussed other matters, the meeting was brief, and Perry quickly annulled whatever hopes Whitlock and Brown may have entertained for an early settlement.

If they agreed to concede the first round to Perry, they were not ready to accept total defeat. Two days later, with the aid of Alpine attorney John Perkins, Whitlock drafted a letter to Perry, the first of eight exchanges of correspondence over the following five months. While little, if anything, resulted from these communications, the opponents appeared to be evaluating each other's knowledge, appraising their relative strengths, and maneuvering for an advantage, when and if it became necessary to settle the issue in court.

The first letter, dated March 30, 1928, and addressed to Chisos Mining Company, Terlingua, Texas, began with the cool formality of "Gentlemen." While not referring to their meeting two days earlier, Whitlock repeats his claim "that there has been a trespass on the property known as the Rainbow Mining Property by tunneling beneath the surface and removing ore therefrom. My information is that this trespass has been committed by the Chisos Mining Company."[21]

The tone of the communication was conciliatory and simply raised the question of a possible invasion. He suggests that the matter be investigated: "If there has been a trespass, the owners of the Rainbow Mining properties desire to be reimbursed for actual damages. . . . If there has been no trespass, these owners, of course have no cause of complaint." To resolve the question, he proposes that "the privilege of examination and survey of the underground works adjoining the Rainbow Mining Company, be permitted by you. If there has been no trespass this survey will have the effect of completely clearing up the suspicion which we now have; likewise if there has been a trespass this survey will determine the extent of such trespass and will give us a basis for discussion." Then, turning to a more cordial attitude, Whitlock concludes, "In writing this letter, I wish you to understand that I am doing so in a friendly state of mind with the desire of amicably disposing of the question which I am here raising."[22]

Two weeks later, on April 13, 1929, Perry responded from New

[20] Oral testimony, *Cause No. 1252, M. B. Whitlock et al.* vs. *Chisos Mining Company*, Attorney's file.
[21] M. B. Whitlock to Chisos Mining Company, March 30, 1928, Attorney's file.
[22] Ibid.

York. Considering that the original letter was directed to the mine at Terlingua and had to be forwarded to him in Portland, Perry lost little time in preparing an answer. The tone and brevity of the reply indicated Perry had sought legal advice. While the letter exuded cordiality, the response was firm: "Possibly you are not aware of the practice among many mining companies, and which has always been followed by our Company from the beginning, viz: not to allow anyone to go down into their mines except those in their employ, and no outsider has ever been in our mine. To reverse this rule solely of the suspicion which you mention would of course not seem reasonable to anyone."[23]

Significantly, there is no hint of denial of any wrongdoing on the part of Perry and "the suspicion which you mention" is, as far as Perry wishes to admit, the extent of Whitlock's evidence. However, Whitlock's letter undoubtedly caused grave concern for the little mine owner. He now suspected the Rainbow owners knew something and their knowledge was sufficient to justify a written statement of their claims. Just how much they knew was still a mystery, but Perry's closing sentences—"Consideration of any facts in your possession regarding the question will of course be given"[24]—was obviously phrased to evoke an answer that might reveal the extent of the opposition's knowledge.

But whatever Perry hoped to learn from Whitlock, the information was not forthcoming. Yet Whitlock's letter dated April 30, 1928, must have conveyed some indication of the strength and resiliency of Perry's opponent. The firmness of his opening sentence—"your answer is unsatisfactory"—undoubtedly revealed to Perry that more than verbal or written exchanges would be required to silence Whitlock. He continues by assuring Perry that "it has always been my policy to adjust any matter of a controversial nature in a friendly spirit," and repeats his request for a joint inspection of the disputed property. "This survey would," he reasons, "settle the persistent rumors . . . that the Chisos Mining Co. has trespassed on the property of the Rainbow." While Whitlock still alluded to "persistent rumors," he obviously intended to tantalize Perry with the knowledge that his accusation had sound documentation. "I wish to assure you," he concludes, "that this request is *not made* out of idle curiosity."[25]

Perry's second letter, dated May 23, 1928, again ignores Whit-

23 Perry to Whitlock, April 13, 1928, Attorney's file.
24 Ibid.
25 Whitlock to Perry, April 30, 1928, Attorney's file.

lock's suggestion of an inspection and continues to emphasize the fact that "the matter which is causing you concern is based only on rumors." But again Perry is careful not to close the door to further negotiations and suggests that another meeting might be beneficial. Still hoping to learn the true extent of Whitlock's information, he suggests: "Possibly there are some things in your mind which it has not occurred to you to tell me. . . . Under these circumstances perhaps you would think it better for us to sit down together and talk the matter over."[26]

Obviously becoming impatient with Perry's hesitancy, Whitlock responded on June 6 with firmness in his claim that, "I have information that the property of the Rainbow Company was invaded and a trespass committed on it during the early part of June, 1927, and continued for several months."[27] Whitlock, playing a longshot, based his claim on two questionable sources of information. While Brown continued to glean information from the Chisos employees, he had a surface survey made in an attempt to define the location and extent of the penetration. Neither matter would convey much argumentative weight when negotiating with Perry, especially the latter, as this occurred under the close scrutiny of Robert Cartledge, who assessed the reliability of the work by noting the condition of the personnel. On May 8 he wrote Perry, "they have been surveying around up on the hill [the] last few days. . . . I understand they are drunk most of the time."[28]

Although still uncertain of the extent of Whitlock's knowledge, Perry nevertheless succumbed to his accusation and in essence divulged the point of the invasion. He writes, "I suppose you think the trespass occurred on the line of our Maud claim." While committing himself on the location, Perry knew other factors were working in his favor. He had learned in three decades that Big Bend land surveys were flexible and uncertain entities, and on this point he continues, "I am wondering whether you have in mind that last spring when a surveyor acting for the Rainbow interests was running his line, Eugene Cartledge . . . stated to your surveyor that he could not recognize the line surveyed at that time."[29] At this juncture it is doubtful if Perry realized it, but the question of the Chisos-Rainbow boundary line would become the crux of all future negotiations and, in the end, would be the issue that would decide

[26] Perry to Whitlock, May 23, 1928, Attorney's file.
[27] Whitlock to Perry, June 6, 1928, Attorney's file.
[28] Cartledge to Perry, May 8, 1928.
[29] Perry to Whitlock, July 3, 1928, Attorney's file.

the outcome of the controversy. But for the present it was an issue that Perry could use to forestall a settlement. Time was becoming a factor.

Realizing that he was slowly becoming entangled in Whitlock's verbal maneuvering, Perry attempted to inject a new defensive element into the negotiations. L. T. Millican had been a member of the Rainbow group since 1918, but recently friction had developed between him and Whitlock over the latter's management of the property. When Perry learned of this internal dissension, he began negotiating to purchase Millican's interest in the Rainbow property. What Perry did not realize was that by surrendering his power of attorney to Whitlock, Millican had invalidated his authority to act independently. With well-calculated craft Perry reminded Whitlock, "You know of my deal with Mr. Millican, and I have heard something of the differences between you and him." Perry emphasized that any settlement must include a provision for carrying out the agreement made between him and Millican for his purchase of the Rainbow property. He added that until this matter is clarified, he should not be expected "to enter into consideration of the friendly adjustment which you suggest. I consider, at least, Mr. Millican's share as really belonging to me."[30]

As Whitlock gathered the import of Perry's attempted treachery, he quickly retaliated. On July 13 he wrote Perry: "The Rainbow Company is not and cannot become interested in any private deal or contract between you and Mr. L. T. Millican. Those having authority never authorized Mr. Millican to sell the Rainbow property. . . . I have heard Mr. Millican undertook to do something he had no authority to do, and could not do."[31]

In his futile attempt to negotiate with Millican, Perry revealed his desperation to avoid a legal entanglement. But while Whitlock undoubtedly felt he had halted this maneuver, the victory was only partial and temporary. He was certain an invasion existed, he believed that ore, legally Rainbow's, was being removed daily by the Chisos, and he felt that little progress was being made on Section 70 to develop the Rainbow holdings. The immediate obstacle was lack of cash. Again he appealed to Perry for a settlement but the response duplicated earlier letters. "Unable to see how immediate action could be expected from us," Perry replied, focusing on the problem of the difference in the boundary lines that each company

30 Ibid.
31 Whitlock to Perry, July 13, 1928, Attorney's file.

maintained was correct. Responding in legal vernacular, he explained,

> it will probably be necessary to determine the surface lines before any one can possibly make any representations to you with regard to any possible claim against us. . . . we do not want to get into a long litigation either with regard to boundaries, powers of attorney, or any other question.[32]

Cunningly claiming that "the writer does not claim any familiarity with the legal question," he concludes with a firmness that suggested that the matter had reached a stalemate, "our rights are such that we do not see how we could forego them." Then, attempting to eliminate the possibility of another meeting with Whitlock, Perry resorted to a legal obstructive technique that he would use with increased frequency. He claims, "My doctors would not wish me to risk the trip to Texas at this time."[33]

Written on August 8, 1928, this letter is the last known correspondence exchanged between Perry and Whitlock. Reluctantly Perry set the stage for litigation. His contemporaries in the Big Bend still speculate on the reasons for his stubborn unwillingness to permit an inspection and resolve the issue. Perry represented one point of view in his first letter to Whitlock—"no outsider has ever been in our mine."[34] His reverence for secrecy was part of the history of mining carried to inordinate extremes, and Perry could not rescind a policy that represented to him not only a manner of doing business but a way of life. And as long as he could maintain the accusation was merely a rumor, his legal advisors could sustain his position.

Also, the premise that Perry's stubbornness was a subterfuge for wholesale thievery[35] is challenged by those who cite Eugene Cartledge's advice as the keystone of his Rainbow policy. Perry had high regard for Eugene Cartledge, who convinced him that they could legally sustain their claim for a boundary line that included the disputed land.[36] Perry had every reason to trust Cartledge, for it had been on his advice that in 1901 Perry had challenged the records of the Texas General Land Office and gained possession of the most valuable quicksilver property in the district.

[32] Perry to Whitlock, August 8, 1928, Attorney's file.
[33] Ibid.
[34] Perry to Whitlock, April 13, 1928, Attorney's file.
[35] Interview with Hunter Metcalf, Marfa, Texas, June 6, 1966.
[36] Cartledge interview, February 28, 1968.

Another factor influencing Perry's position was that news from the Rainbow camp indicated Brown and Whitlock probably would not press the issue beyond negotiation. Perry's counterespionage revealed the apparent demise of Brown's earlier optimism.

> The snakes [Rainbow] have shipped 20 flasks Qs . . . and are making about ½ flasks per day, but they would be making less each day, as they are burning there richest ore first. According to Joe [White] they should have gotten more Qs. from the first rich ore . . . but claims they lost lots of it on account of not know how to run a retort. From Joe it would seem that the bunch over there are drinking considerable. Joe says he met Devine and Birkman going to Lajitas a few days ago, evidently for Booz. He says that the old bunch is mad at Brown and that Brown wants to sell out and get away.[37]

As negotiations ceased a calm settled over the district and for the next year little was heard about the supposed invasion. The Chisos company continued to find good ore and by the end of the year had marketed 1,679 flasks of quicksilver at an annual average of $123.51 per flask.[38] Rainbow prospecting was less remunerative, as Perry's informants had correctly reported. This was not the calm of peace, however, but a calm of preparation. Secretly the two adversaries sought means of strengthening their claim for who legally owned the blood-red cinnabar that the Chisos No. 9 shaft continued to disgorge.

Eugene Cartledge, returning to Austin, focused his search on the records in the General Land Office. There he discovered an error in an early land survey that he believed would strengthen Perry's position. It is ironic that this error resulted from an attempt to give order to the greatly confused issue of Big Bend land titles. The Texas legislature, on act of April 19, 1907, authorized the land commissioner to resurvey all school lands lying between the Pecos and the Rio Grande. The emergency clause stated that "the land lines and surveys in the mineral-bearing school land territory is in great confusion, uncertain and of doubtful location."[39] Brown was not the first miner to experience the frustration of disputed ownership. The Alpine *Times* reports in 1904 that "Ed Lindsey is having a hard time over the surveying of his mining claims. . . . One survey puts them on state land and then the next puts them on railroad

---

[37] Cartledge to Perry, February 8, 1929, Chisos Mining Papers.
[38] Johnson, "Mercury Mining," 37 (March 1947): 30.
[39] "Question of Title," Attorney's file.

land. According to the last survey, Mr. Lindsey will lose them entirely."[40]

However, the high purpose of the law was not fulfilled. State Surveyor R. L. Dod resurveyed Survey 70 and all adjoiners in 1909 and 1910 and, discovering a discrepancy in his calculations, attempted to reconcile it by a mathematical relocation that created a strip of unclaimed land—at least on paper. Eugene Cartledge discovered Dod's error and on September 27, 1928, Robert Cartledge, acting in Perry's stead, filed a mineral claim embracing 8.63 acres identified as "Junior." Ninety-four varas wide, it encompassed the area of the alleged trespass. Eugene Cartledge assured Perry that this filing would silence the Rainbow claims; for the present, he appeared to be correct.[41]

While Cartledge cited the error-ridden 1909–1910 resurvey to sustain the Chisos position, the Rainbow staff concentrated on locating on the ground the original corner of a 1900 survey. Their hopes rested on the memory of J. J. Dawson, who in June, 1900, had filed the first mining claim on Section 70. Prior to filing, however, he had served as chain bearer for Brewster County Surveyor W. M. Harmon in running the lines. On October 22, 1928, almost one month after Robert Cartledge filed his "Junior" claim, Dawson located the all-important corner. His affidavit reads: "in the presence of J. B. Brown, J. M. F. Burkman, W. W. Devine . . . and M. B. Whitlock, [I did] locate and identify a certain rock mound marked 9, as being the same monument that I, in company with and presence of W. M. Harmon [did] erect upon the spot designated by him as the true and correct S. W. Corner of Mining Claim G4-4. The aforesaid monument . . . was located and erected in the summer of 1900."[42] To authenticate their discovery, both Brown and Whitlock photographed the rock mound.

Later they would look back on this event as an important milestone in their fight with Perry, and a new survey made from this corner would further sustain this view. By the end of 1928 Brown could turn to the future with a more optimistic perspective, as a legal confrontation with Perry appeared increasingly desirable. Now in his own mind he was certain there was a trespass, but before he dared challenge Perry in the courts, he needed more evidence.

[40] Alpine *Times*, July 20, 1904.
[41] Cartledge interview, February 28, 1968; also, "Identification of Land in Suit," Attorney's file.
[42] "Identification of Land in Suit," Attorney's file.

The new year brought with it two events that encouraged Brown. On January 9, 1929, the Texas land commissioner rejected Robert Cartledge's "Junior" claim,[43] and soon thereafter Brown and Whitlock added to their group a man whose skill and knowledge would greatly aid them in the forthcoming encounter with Perry. Brown first met C. T. Armstrong in Pachuca, Mexico, where both were employed in a silver mine. A graduate of Columbia University School of Engineering, Armstrong was in San Antonio on sick leave when Brown approached him with a proposition to join the Rainbow staff. He confided to Armstrong that he desperately needed additional financing to develop the mine. What Armstrong remembers most graphically about this meeting, however, was Brown's confidence in the Rainbow's potential and his relative certainty of the Chisos invasion.[44]

Brown's proposal appealed to Armstrong, who arranged to purchase Wayne Darnell's stock interest in the company. Replacing Birkman as secretary-treasurer, he departed immediately for the Big Bend and took residence in Alpine. "When I arrived at the mine," Armstrong recalls, "production was low, the labor force was small, and the methods of recovery quite primitive. But that was soon changed."[45]

Their first program of work provided for a series of drifts leading toward the area where Brown had been told the rich ore deposits lay. As they drove eastward, the sounds of the Rainbow muckers were punctuated by the distant hammering of air drills as the Chisos workmen pursued an opposite course. "We knew they were working our claim," Armstrong adds, "but we couldn't prove it. But we came up with an idea that could have only been concocted in the sometimes brilliant, frequently psychotic, mind of John Briscoe Brown."[46]

From his knowledge of mathematics and engineering Brown rationalized a plan to locate the exact point of the Chisos penetration. While the sounds of the workmen were clearly audible, the human ear lacked the directional qualities necessary to orient their point of origin. But Brown found a solution. Sometime prior to Armstrong's joining the Rainbow staff, Brown had acquired a World War I sensory device called a geophone. Although designed for use in European land mining operation to detect underground move-

[43] "Analysis of Title," Attorney's file.
[44] Armstrong interview, August 24, 1966.
[45] Ibid.
[46] Ibid.

ments, this piece of war surplus equipment was destined to help alter the legal fortunes of Howard Perry.

The basic units of the geophone were two large tripod-mounted, copper-looking bowls covered with sensitive diaphragms that operated on the same principle as a mammoth stethoscope. These were highly directional and a single ear piece ran from each unit through which the operator listened. As Armstrong explains, "The diaphragms were attached to the walls of the mine shaft and we would listen through one unit and then the other to determine which gave the strongest reading."[47] After determining the general direction of the sound, the diaphragms were then moved to another shaft and the process repeated until a directional fix was established. Armstrong recalls asking Brown, "How do you know how accurate the sounds are you are listening to?" Whereupon Brown replied, "Try it yourself and you will see." "Well, I put the headphones on," Armstrong continues, "and when you faced south you got nothing, but when you turned the geophone diaphragms toward the east [toward the Chisos workings] it would practically knock your head off."[48] Brown was able to follow the progress of the drift, Armstrong continues, "triangulate its location, and establish the exact location of the Chisos workings." Pausing as if to recall the delight of their achievement, he concludes, "It was quite a gadget. Simple, yet quite a gadget."[49]

The final readings were taken on March 28–29,[50] and the sounds Brown and Armstrong heard represented the most blatant invasion of Perry's "wall of secrecy." While Brown listened to the geophone deep in the Rainbow workings, a Chisos foreman tapped a prearranged signal on the western face of the No. 9 shaft.[51] With reasonable certainty, Brown plotted the position of the most westward drift of the Chisos workings; he now had the information he sought. The people whom Perry trusted most had exposed to the prying ears of an unseen enemy the information that would help redirect him on a course that would lead to legal defeat, disappointment, and eventual bankruptcy.

To reconfirm his findings Brown transferred the data gathered by the geophone soundings to a map. Results were conclusive; the origin of the sounds was "some 200 feet from the East line of G4-1,

---

[47] Ibid.
[48] Ibid., November 15, 1973.
[49] Ibid., August 24, 1966.
[50] Oral testimony, *Cause No. 1252*, Attorney's file.
[51] Armstrong interview.

well within the Rainbow property," and originated from a depth of about 600 feet, about the same level from which Brown and Armstrong had monitored the sounds.[52] When Brown apprised Whitlock of this new information, the latter arranged a second conference with Perry. This meeting, also held at the Holland Hotel, yielded nothing. Again Perry took the initiative and "started out by saying that he understood I was going to file suit against him. I told him that I probably was. He deplored the fact and said that he didn't want to be sued." Whitlock explained to Perry that he hoped to avoid legal action, but Perry still refused an inspection, stubbornly maintaining that "he had opened up a structure that he thought was his and nobody else had a right to see it."[53]

The talks, extending over three or four days, ended on May 18 with negative results. Now desperately in need of money and fully convinced that only by suing Perry could the issue be resolved, the Rainbow management embarked for San Antonio to consult an attorney. At Armstrong's suggestion they contacted Will Kennon, who referred them to Judge Howard Templeton of Templeton, Brooks, Napier, and Brown, one of the city's leading legal firms.

On the morning of May 25 a trio from the Rainbow—Armstrong, Brown, and Whitlock—presented their evidence and suspicions to Judge Templeton and a young attorney with his firm. After discussing the matter at length, Judge Templeton advised that they lacked sufficient admissible evidence to make a case against the Chisos. However, he turned to Wilbur L. Matthews, the young member of the firm and said, "It's a long way out there and much time will be spent traveling. We'll let Matthews take it and see what he can do with it."[54]

While youth and distance greatly influenced Matthews' selection, Templeton's conclusion failed to dampen his enthusiasm. "I told Judge Templeton I couldn't think of anything I'd rather do," Matthews explains. "I based the whole plea on information and belief and prayed for a large amount of money. As I recall, Judge Brooks [a member of the firm] went out to Terlingua only one time. It was all mine."[55] Matthews was just twenty-six years old, had practiced law for three years, and had gained his only experience representing defendants in damage suits. This, the first suit he

---

[52] Ibid.
[53] Oral testimony, Attorney's file.
[54] Matthews interview.
[55] Ibid.

had ever handled for a plaintiff, was the greatest opportunity of his career; he would prove himself equal to the challenge.

The magnitude of Matthews' task in developing his case revealed the source of Judge Templeton's pessimism. No matter how certain Armstrong and Brown were of the invasion, their evidence, when equated with the difficulty of proving it in court, was the basis for Templeton's reticence. Probably only because of Matthews' youth, inexperience, and the prospects of an undertaking that promised high adventure was the Rainbow successful in stopping Perry.

From the confused and contradictory maze of Big Bend land surveys, Matthews first had to define and locate on the ground the legal boundary between Sections 70 and 295. Herein lay the crux of the problem. Second, he had to prove that Perry had committed a trespass over 600 feet underground by following an ore deposit discovered on Survey 295, across the legal boundary to Section 70. This would not be easy. Third, once he had proven the Chisos guilty of invasion, Matthews had to establish precedent for an underground inspection to gather evidence that an invasion had occurred and something of value had been removed. Both were without precedent in Texas. And last, if Matthews was successful in proving his case, there still remained the formidable task of forcing Perry to accept the verdict. Legal tradition in the Big Bend had too long been synonymous with the best interests of the Chisos Mining Company, and a favorable verdict from a Brewster County jury appeared remote.

Matthews acted with dispatch. Writing Whitlock the results of his first efforts, he states: I "spent Thursday and Friday in Alpine. The suit against the Chisos Mining Company was filed and all necessary citations, notices and subpoenas were issued and placed in the hands of the Sheriff."[56] The suit, identified as Plaintiffs' Petition and Application for Temporary Injunction, was filed on May 31, 1929, as *Case No. 1252, M. B. Whitlock et al.* vs. *Chisos Mining Company.* The document states: "That defendant acted recklessly, wilfully and intentionally in so mining and removing said ores of plaintiffs. . . . [The] defendant has wilfully and intentionally concealed the facts. . . . [and the] plaintiffs pray that defendant be required by proper order of the court to forthwith state and file herein an account . . . of all ore taken by it each day from any such tunnels since the second of June, 1927."[57]

---

[56] Matthews to Whitlock, June 4, 1929, Attorney's file.

[57] Plaintiff's Petition and Application for Temporary Injunction, filed May 31, 1929, Attorney's file.

Continuing, Matthews pleaded that the Chisos Mining Company be required to keep and preserve for presentation at trial all records of its operation in the disputed area. It further emphasized that

> the plaintiffs fear, and they believe with good reason, that the defendant will not voluntarily preserve said records. . . . WHEREFORE plaintiffs sue . . . and pray for temporary injunction enjoining and restraining defendant from continuing to trespass on their said premises . . . and on final hearing the plaintiffs pray for judgment for the title and possession of their said premises and for writ of restitution, and for their said damages, with legal interest thereon.[58]

By filing this petition Matthews removed the matter from the partisan environment of the Holland Hotel and placed it in the legal jurisdiction of the district court. Beyond this point the procedural steps would be determined by law, and Perry's whims and idiosyncrasies would play a lesser role in the developments. Yet his influence, so firmly entrenched in the Big Bend, would still make itself felt.

Matthews first encountered the overpowering aura of Howard E. Perry when he attempted to select a local attorney to aid in the case. Whitlock recommended John Perkins of Alpine, who had previously helped him compose his original letter to Perry. Perkins, however, was reluctant to accept the assignment. He told Matthews that "Perry has got this country pretty well sewed up out here and he didn't feel it was to his advantage to oppose Perry in something as big and incriminating as the Rainbow suit appeared to be."[59] But apparently the prospect of an attractive fee altered his view, because on June 5 he wrote Matthews, "I am very much pleased to be associated with your firm and appreciate the fee arrangement of twenty percent."[60]

Sensing that Perry might attempt to manipulate L. T. Millican in another attempt to thwart the negotiations, Matthews moved to increase Whitlock's rights and obligations in the Rainbow corporate structure. His foresight was prophetic. Whitlock responded, "I will fail to secure the signiature of L. T. Millican to the Adenda to the power of attorney. . . . because there was a possabilaty that he could cause some trouble."[61] But whatever hopes Perry and Millican

[58] Ibid.
[59] Matthews interview.
[60] Perkins to Matthews, June 5, 1929, Attorney's file.
[61] Whitlock to Matthews, June 5, 1929, Attorney's file.

may have entertained to "cause some trouble," they were circumscribed with the execution of the document on June 1, 1929. Matthews was carefully removing all obstacles from the paths of litigation; those that remained were still formidable.

The Chisos response to the Rainbow's application for a temporary injunction focused on that matter that would occupy Matthews' attention for the next six months and test his legal skill to its untried limits. Perry's attorneys claimed that "the real issue in controversy in this suit is as to the location of the boundary line between the property owned by Plaintiffs and by this Defendant; that if the boundary line of Plaintiffs' land is located as claimed by this Defendant, *there has been no trespass.*"[62]

Implicit in proving a trespass was establishing a legal boundary favorable to the Rainbow's interest. Matthews understood his success rested in large part upon this premise, and in his first trial brief he cites the problem as locating G4-1 of Survey 70 on the ground. "According to the original field notes," Matthews writes, "the North line of Survey 70 was located where mining claim G4-1 . . . would include the land in controversy, but the re-surveys locate that line about 300 varas North of the original survey line . . . [and] would not include the land in controversy."[63]

Matthews, therefore, built his case on the belief that he could establish the priority of the original survey, and his success would rest on the validity of the original corner located by J. J. Dawson. Soon after filing suit, Whitlock explained this to Dawson: ". . . we will stand on the survey made by Mr. White . . . from the corner marked #9 which you pointed out to us last fall."[64] However, time, research, and investigation would be required to establish legally the location of G4-1; for the present, Matthews turned to other matters.

When Matthews failed to get the restraints he sought in filing the plaintiffs' petition, he chose the alternative of taking oral depositions. Through the issuance of subpoenas this legal technique would enable him to question Chisos executives under oath and require them to produce maps and records of work performed in the disputed area. At this juncture this seemed the only course to follow in order to determine if an illegal invasion existed. On July 3, 1929, Matthews wrote Perkins that he had decided to proceed with ex-

---

[62] Chisos response to Plaintiff's Petition and Application for Temporary Injunction, June 14, 1929, Attorney's file.

[63] "Question of Title," Attorney's file.

[64] Whitlock to Dawson, June 5, 1929, Attorney's file.

amining the Chisos employees involved in exploration and recovery. The subpoenas cited Frank E. Lewis, Marcus Hulings, and E. J. Dahlgren. Matthews emphasized that despite its inaccessibility, Terlingua would be the preferable site to take depositions, as that is "where the records of the Chisos Mining Company are kept."[65]

Later that same day Matthews wrote Brown requesting additional names of Chisos employees whose testimony might be valuable to their cause. Brown's reply revealed that his espionage skills were still acute. Referring to the specific location of machinery underground at the Chisos mine, he had learned that mechanic J. E. Anderau had supervised the installation of an air blower at the 600-foot level, "near our boundary. This ventilator station was moved some distance to the east, from its original location, and it is thought that the latter was within our property." Brown knew well that while he spied on the Chisos operation, he was also being kept under its close surveillance. With this in mind he cautioned Matthews, "The receipt of a telegram cannot be depended upon."[66]

When Judge Sutton denied the plaintiff's plea for a temporary injunction, the need for the deposition increased. Time was of the essence and the brisk exchange of telegrams on July 16, 1929, revealed the exigency of the matter.

Perkins to Matthews, 9:54 A.M.

VAN SICKLE [Chisos attorney] AGREES TO THURSDAY JULY TWENTY FIFTH FOR DEPOSITIONS.

Matthews to Perkins, Later same day

THURSDAY JULY TWENTY FIFTH SATISFACTORY. . . . WE INTEND TO REQUIRE PRODUCTION OF BOOKS AND RECORDS AND THEREFORE MUST ISSUE SUBPOENAS TO APPEAR AT PARTICULAR PLACE.[67]

The "particular place" was the mine office in Terlingua. While the decor lacked the staid elegance of a courtroom, the procedures followed were clearly defined by statute. Each witness was properly sworn by Edith Hopson, the court reporter, gave his testimony under oath, was questioned by the attorney for the plaintiff, and was cross-examined by the attorney for the defendant. Each attorney was entitled to enter in the record his objections to the admission of any evidence he deemed inadmissible or detrimental to the interests of his client. As the witnesses testified, this maneuver would

65 Matthews to Perkins, July 3, 1929, Attorney's file.
66 Brown to Matthews, July 9, 1929, Attorney's file.
67 Attorney's file.

be employed with monotonous regularity; there was much to be objected to.

The key witness was Marcus Hulings, superintendent of mining at the Chisos camp. Capable, personable, and well liked, he had been graduated from Rensselaer Polytechnic Institute in 1903 and had engaged in various mining activities in Mexico prior to joining the Chisos mine as an engineer in 1916. Two years later Perry had advanced him to superintendent. With full authority over all underground work, Hulings was the only person who possessed the information sought by the plaintiff. His testimony, however, was not always enlightening.

Q. You keep a record of ore mined from each section of these drifts and the quality of ore from each place do you not?
A. No, sir.
Q. You don't?
A. No, sir.
Q. Don't you know, can't you tell from your books, how much ore has been mined from this territory in here?
A. No, sir. It could be arrived at with some work, but there is no record taken off.
Q. Your books are so kept that it could be arrived at?
A. No, sir, not books.
Q. Or records?
A. No, sir.[68]

Hulings' vagueness aroused Armstrong's suspicion; himself a mining engineer, he became convinced that Hulings was withholding evidence. He would later write Matthews: "According to Mr. Hulings the Chisos did not keep a map showing the relative position of underground workings to surface boundaries. . . . This is decidedly not in accord with the usual mining practice."[69]

When questioned about a stope, Hulings stated that the work was performed in June of 1927 and the material removed was "very nice ore." However, he denied that the amount removed was ever recorded. Hulings further denied that the Chisos management kept a record of the ore hoisted from the No. 9 shaft. When asked, "A while ago you said that you kept a record of the ore hoisted from Shaft No. 9, and then it was carried on and mingled," he responded, "I haven't any record"; when asked who in the company kept that record, he stated, "I don't know."[70]

[68] Oral testimony, Marcus Hulings, Attorney's file.
[69] Armstrong to Matthews, August 3, 1929, Attorney's file.
[70] Oral testimony, Marcus Hulings, Attorney's file.

While most of the testimony related to the location, quality, and amount of ore surfaced, the critical question of survey boundaries was thoroughly explored. Questioning Whitlock about the northeast corner of Survey 70, Matthews struck at the crux of the dispute. The location of that site would eventually resolve the matter. Knowing this, he focused his questioning accordingly:

Q. There is not much dispute about the location of that boundary line between 70 and 295 on the ground is there?

A. I don't know about that. Our claims don't go to the line of 70.

Q. They don't go to it but they call for the corner of 70, don't they?

A. I think it does. I think it calls for so many varas each way from the N. E. Corner of 70.

Q. Your petition alleges that the N. E. Corner of 70 is the beginning point and that they lie west of 295.

A. I think so.[71]

Under cross examination C. E. Mead, attorney for the Chisos mine, also alluded to that corner, and in so doing defined that disputed strip of land in which over six hundred feet down lay the valuable cinnabar deposit that both companies claimed. He asked Whitlock:

Q. You called for the N. E. Corner of Survey 70. Isn't there another condition, also, that even if it is down where you say it was, G-1, that still there is 84 varas clear between Survey 295 and your line?

A. Mr. Perry claims there was.

Q. You claim less than 84.

A. I made no claim as to where it was other than the *survey made by Mr. White* [based on the 1900 Harmon-Dawson corner].

Q. That is the controversy, is it not?

A. It probably is.

Q. And that controversy has got to be settled?

A. I don't know how it will be settled.

Q. That is only about 7 varas wide?

A. 7.1 varas, I understand.

Q. You don't think we have taken our $1,000,000 worth of ore from a 21 foot strip there, do you?

A. No, I think they came in 200 feet, which is a lot more than 20 feet.[72]

Thus Mead redefined the problem: each company claimed a

---

[71] Oral testimony, M. B. Whitlock, Attorney's file.
[72] Ibid.

different boundary, and the one that stood would yield the successful litigant huge financial returns. While each counsellor would base his claim on existing land records, obviously both could not be held correct. The solution was yet to be found.

The boundary alone would not resolve the dispute; Matthews had to know the extent of the Chisos workings off the No. 9 shaft and the amount and value of the ore removed. This information Hulings was not going to reveal, and his assistant, Frank E. Lewis, was "away on vacation." On this matter Armstrong recalls, "Hulings hemmed-and-hawed. Lewis was not a 'company man,' and he probably would have told the truth. We heard that Hulings sent Lewis on vacation. Wanted to get him the hell out of there. He was straight-forward. He would have told the truth."[73]

Matthews' plan to take the depositions in the mine office at Terlingua was not without foresight. For in that room the Chisos Mining Company kept the maps and charts on which Hulings and Lewis plotted the entire underground workings. Matthews knew that if he could force Hulings to produce an accurate rendering of the disputed area, the extent of the penetration and the amount of ore surfaced could be approximated. To the accompaniment of the steady drone of Mead's objections, Matthews demanded that Hulings show, and interpret for the court record, the pertinent documents:

> *Defendant objected* to it because there is no proper cause for this. . . . *Defendant objected* to any notations being made on the map as it is not offered in evidence. . . . *Defendant objected* to the map being offered in evidence because it is a record of the office and is unauthorized to be taken away and it is not right and proper to be taken from the defendant in this suit. . . . *Defendant objected* to the requests, because the subpoena asked that the maps be brought here, and not that they might be copied. . . . *Defendant objected.* . . . *Defendant objected.* . . . *Defendant objected* [emphasis added].[74]

Against the verbal background of Matthews' questions, Hulings' responses, and Mead's objections, John Briscoe Brown and C. T. Armstrong remained, as always, acutely alert. As each map was produced, Brown and Armstrong took notes on the locations, elevations, and directions of the Chisos drifts and compared this data with the calculations on their map. Matthews recalled that they knew *"for the first time,* the Chisos had penetrated the Rainbow property, and

[73] Armstrong interview, April 3, 1968.
[74] Oral testimony, Attorney's file.

even farther than they had originally suspected."[75] The young attorney's persistence was producing results.

The Rainbow staff left Terlingua that night elated with the success of the depositions. Going to the Rainbow camp they held an impromptu celebration and toasted Matthews and each other repeatedly. In the absence of a vintage beverage the local supply of "imported tax free" tequila sufficed. Many years later Armstrong remembered that "the quantity of the staff compensated for what it lacked in quality. Flush with success and full of fiery tequila, Brown toasted Matthews and claimed, 'I'll own old man Perry's yacht yet.'" By that time, Armstrong added, the last two words— "yacht yet"—posed a pronunciation problem for Brown.[76]

Matthews returned to San Antonio the following day and reported the results of the hearing to Whitlock. Tempering his enthusiasm with a lawyer's cool logic, he explained the necessity of "securing an inspection of the maps of the underground workings."[77] More than anyone else, Matthews realized that final success rested on the Rainbow's boundary claim. He wrote:

> there has been a substantial invasion of the Rainbow properties, provided the boundary questions are determined in accordance with Mr. White's survey. It does not appear improbable to us that the amount of ore taken from the property will amount to over One Half Million Dollars. . . . We intend to immediately enter into a thorough investigation and compilation of data with reference to the surface boundaries.[78]

The site of Matthews' "thorough investigation" was the General Land Office in Austin. Realizing that he needed additional data before calling the matter to trial, he and Judge Templeton went to Austin, examined the records, and "secured some data which we think will assist us materially in determining the conflict with respect to the boundaries on the surface."[79] With the new information Matthews decided to return to Terlingua and with a surveyor establish the surface lines of the Rainbow claim. He wrote Perkins, "I feel that the theory of our case . . . will have to be worked out from the monuments located by Mr. Dawson."[80]

As Matthews gathered the survey data, he received some start-

75 Matthews interview.
76 Armstrong interview, April 3, 1968.
77 Matthews to Whitlock, July 29, 1929, Attorney's file.
78 Ibid.
79 Matthews to Whitlock, August 3, 1929, Attorney's file.
80 Matthews to Perkins, October 4, 1929, Attorney's file.

ling news from Armstrong that caused him to alter his plans temporarily. "Mr. Brown is in receipt of some reliable information," Armstrong wrote, "that is highly important. . . . there has been considerable work done not shown on the maps we saw."[81] While the Terlingua hearings had yielded important data, if the Chisos management was withholding information, the only solution that remained was an underground inspection. Matthews realized that forcing Perry to permit such an inspection would be difficult and could be achieved only by lengthy court action. But even more challenging was the fact that this procedure was without precedent in Texas. However, he wrote Armstrong that "during the August term of the Court, it is our intention to present an application for the inspection and survey of the . . . Chisos Mining Company. We are now engaged in the preparation of a brief on the subject for the submission to the Court.[82]

Matthews' brief is a monumental achievement. In twenty-five pages of well-researched analysis he cited precedents dating from 1799, when the Courts of Chancery of England "made orders of inspection and survey of mines in suits involving underground trespass and taking of minerals or ores."[83] The Circuit Court of the District of Nevada had established the United States precedent in 1867, when "the plaintiff commenced an action . . . to restrain an adjoining owner from mining the lode which it was alleged had been done."[84] With this memorandum of authorities, Matthews sent Perkins an application and motion for inspection and survey which he filed on August 16, 1929. Matthews expected a repercussion; it was not long in coming. The two telegrams read as follows:

Perkins to Matthews, August 21, 1929, 4:06 P.M.:
ATTORNEYS FOR DEFENDANT ARE OBJECTING CLAIMING WANT TIME TO FILE BRIEF AND URGING OVERWORK STOP COURT IS WILLING TO HEAR AT ANY REASONABLE TIME

Matthews to Perkins, later the same day:
WE SEE NO BASIS IN JUSTICE FOR DEFENDANTS OBJECTION TO EXAMINATION OF THEIR MINE AND NO BASIS IN LAW FOR THEM TO RESIST A COURT ORDER ALLOWING INSPECTION AND SURVEY[85]

[81] Armstrong to Matthews, August 8, 1929, Attorney's file.
[82] Matthews to Armstrong, August 10, 1929, Attorney's file.
[83] *Cause No. 1252*, "Memorandum Relative to the Right of Inspection and Surveying in Mining Cases," Attorney's file.
[84] Ibid.
[85] Attorney's file.

If Matthews failed to gain access to the Chisos shafts by court mandate, Brown knew that one other approach to the Chisos workings remained. That was through the ground. While this method lacked the sanction of the court, Matthews concurred that a delay in the court hearing "could be taken advantage of by pushing the work of making a ground connection through our shaft."[86] Brown reassured Matthews:

> As soon as we connect with the Chisos workings, we will accurately determine the position of these workings with reference to the boundary. . . . It is possible that an attempt may be made to remove all traces of high grade ore. By waiting until we connect through, it is almost certain that we can expose ore, adjacent to that removed by them, which would give an excellent basis for establishing . . . by comparison, the grade of ore removed.[87]

Armstrong, recalling the urgency of that effort, said: "We knew we were working against time. We worked two shifts, eight to four, and four to twelve. Put as many men in there as could work. We really tore a hole in the ground."[88]

While Brown and Armstrong drove the Rainbow shaft eastward toward the Chisos workings, Matthews turned again to the matter of a surface survey. On October 21, 1929, he arrived in Alpine, and in the company of two surveyors, a Mr. White and James P. Dod, county surveyor of Presidio County, embarked for Terlingua to resurvey the disputed land for the last time.

The following day the Rainbow crew began surveying the two properties. Matthews recalls that on the second day another group of engineers arrived and began a survey for the Chisos legal staff. As the Rainbow crew measured the distances in the area of the northeast corner of Section 70, they found a mound of rocks concealed in a thorny bush. Being near the assumed location of that critical site, Matthews hollered to the surveyor, "Don't touch it!"[89] After a hasty examination, he was certain they had found the original corner of Survey 70.

Calling to the Chisos crew working nearby, Matthews suggested that a representative from each staff examine the rocks to determine if that was the original monument. Together they carefully removed each stone and inspected it. Finally, near the bottom of the pile they

86 Matthews to Brown, August 28, 1929, Attorney's file.
87 Brown to Matthews, August 20, 1929, Attorney's file.
88 Armstrong interview, April 3, 1968.
89 Matthews interview.

discovered a large flat rock bearing the conclusive chisel mark—
N.E. 70—JJD—placed there by surveyor J. J. Dawson in 1900
when he made the initial survey. Matthews recalls with just pride:
"Right then and there, they [Chisos] said, 'This is it.' It was on the
Chisos side of the fence. We measured the distance and as I recall
there was approximately a two hundred foot difference between the
rock mound and the fence that the Chisos claimed was the correct
boundary. Seven hundred feet down, that was a valuable two
hundred feet. That's when they began to talk settlement."[90]

The pride of achievement was evident in a letter to Perkins.
Matthews reported: "We made a very satisfactory survey . . . and
were successful in finding upon the ground the old Northeast corner
of Section 70. It seems to me that our work will result in substantially settling all disputes with respect to the boundaries."[91] By
locating that critical boundary, Matthews established the legal
premise for his suit. Up to this point his argument was based solely
on belief, but now the suspected invasion was an established reality.
For the first time since that meeting in Judge Templeton's office
Matthews had reason to share the enthusiasm of success. Later that
same day he wrote Judge Ike White in Austin [not to be confused
with surveyor White, previously mentioned], Perry's chief legal
representative in Texas, with a firmness and conviction never before
expressed. He could now state that he had "definitely established
what our clients will contend is the correct boundary between
Mining Claim G4 #1 and Mining Claim 'Maude.'"[92]

While Perry was not ready to admit total defeat, the correspondence reaching Matthews' office conveyed a more conciliatory
attitude. On October 29 Judge White consented to a supervised
inspection of the disputed area when the Rainbow breakthrough
occurred.[93] While careful to admit nothing, he specified "that the
Plaintiffs would have the right to inspect the workings to a point
2400-feet East of the Southwest corner of Mining Claim G4-4 as
pointed out . . . by Mr. Dawson."[94] He further agreed to make a

[90] Ibid.

[91] Matthews to Perkins, October 29, 1929, Attorney's file.

[92] Matthews to White, October 29, 1929, Attorney's file.

[93] White's action was involuntary. Judge Sutton had ruled previously that
the Rainbow staff had a right to an examination of the underground workings
of the land involved and that the counsel was to work out the details. The
judge's order was never prepared, as in the interim the boundary dispute was
resolved and Ike White agreed to a joint inspection.

[94] White to Matthews, October 28, 1929, Attorney's file.

complete statement and disclosure showing the amount and value of ore taken from the ground claimed by the plaintiff.

Six days later this achievement was climaxed by the breakthrough into the Chisos No. 9 shaft. Armstrong well remembers the late-night drama of the event. "Can't recall the time. I remember wondering who was calling me at that time of night. But when I first heard Brown's voice I knew what he was going to say. 'We've busted through. Came through pretty high but *we can see the ore!'* " Armstrong, continuing to reminisce about the events that preceded the final joining of the shafts, remarks:

> We worked to the 295 line and stopped. We were at the 600-foot level and we could hear them working below us. We were damn careful not to invade their property. When we arrived at their line, we sank a winze down and plowed into 'em. Apparently the Chisos knew we were coming, because when we broke into their workings, they had abandoned the drift and had erected a barrier at the property line. That was admission of guilt, if nothing else was proven.[95]

The following morning Armstrong called Matthews to share the good news. The latter, in turn, called Judge White in Austin. Matthews "advised him that we would like to make an immediate inspection of the underground workings of the Company with someone representing his clients." White concurred and told Matthews he would leave Austin on November 19, 1929. He also added that he hoped both legal staffs would have complete representation at the inspection and would stay "until all investigation was complete and all arrangements reached which the parties can possibly come to."[96]

While Matthews engaged Judge White in a legal battle that promised to reward the winner richly, their confrontations were confined to the legal rhetoric of their correspondence. As the case progressed a warm personal friendship developed between the two men. On the trip to the Big Bend for the inspection, Matthews joined Judge White in the diner of the Southern Pacific passenger train. Matthews laughingly recalled that the judge liked to have a drink now and then, but since it was during prohibition, he concealed his whiskey in a medicine bottle. Matthews added that legal

---

[95] Armstrong interview, August 24, 1966.
[96] Matthews to Brown, November 5, 1929, Attorney's file.

maneuvers were temporarily forgotten as "we ate dinner together and drank from Ike's medicine bottle."[97]

The respite, however, was only temporary, because the evidence they encountered at the Chisos mine finally revealed the full extent of Perry's transgressions. Arriving at the mine, representatives of both companies went to the entrance of the No. 9 shaft. In small groups they huddled together in the ore car and began their slow descent into the darkened cavity of the shaft. As they moved quietly past successive strata of the earth's structure, Matthews sensed the inner satisfaction of final achievement; it had required much litigation and the surfacing of vast amounts of Rainbow earth to force Perry to allow people whom he had never seen to enter the almost sacred inner reaches of his mine.[98]

Matthews said that his first impressions of the long-forbidden workings of the Chisos mine, when they finally bumped to an abrupt stop at the base of the shaft, remain graphic in his memory. "The heat in the shaft was intense," he recalls, "and the sweating semi-nude bodies of the Mexican miners appeared like gleaming savages in the subdued light of the carbide lanterns."[99] Armstrong, long accustomed to the underground workings of a mine, was equally impressed by the spectacular showing of rich ore. He recalls: "[William D.] Burcham was there when I arrived. He was sitting on a pile of ore. He said to me, 'I've never seen such ore as this.' It looked like you had killed a cow. Red everywhere. [The richness of cinnabar ore is partially determined by the redness of its color.] In the artificial light it looked like a room full of rubies. Worth a helluva lot more." Armstrong added that the geological reports state that the ore found in the area of the Chisos-Rainbow contact was some of the richest in the district.[100]

[97] Matthews interview.

[98] It is probable that, until the Rainbow litigation, Dr. J. A. Udden, Perry's consulting geologist, was the only person not employed at the Chisos mine permitted entry to the Chisos workings.

[99] Matthews interview.

[100] Armstrong interview. William D. Burcham's role in the Rainbow mine episode remains a matter of disagreement. Virginia Madison, citing Burcham as her source, writes: "Neither side would go into the disputed territory without his [Burcham's] presence. Perry's engineers were authorized to show Burcham that the Chisos Mining Company had removed no more than $211,000 worth of Rainbow ore, which information Burcham was permitted to transmit to the Rainbow people by word of mouth, but not written" (*The Big Bend Country of Texas*, p. 186). Mrs. Burcham [writing for Burcham] states: "A representa-

While their entry into the Chisos mine had been costly, the prospects for their reward were good; therefore, Brown and Armstrong chose to make their own calculations. With the aid of the Chisos stope maps they measured the drifts and noted the directions. Matthews adds: "These maps showed the dimensions, foot by foot, showed the assay of the ore removed, and the dates it was surfaced. When they completed their survey they found the Chisos calculations were correct."[101]

Matthews left Terlingua armed with data that he believed would soon lead to a negotiated settlement. The correspondence that followed between him and Judge White supported this view. But then in early January what had seemed so near suddenly appeared remote and far away; a new psychology seemed to pervade all communications from the defendant. On January 9, 1930, White wrote Matthews: "You have suggested that you claim all the ore West of a line 75.9 vrs. from the West line of Survey #295; and we beg to say now that if that is your attitude, it would be a waste of time to discuss a settlement. We might as well get ready to try the lawsuit."[102] Left with no alternative, Matthews informed Perkins on January 20, "We are making every preparation to try the case at the February term."[103]

What obstructive forces were at work can only be conjectured, but instructions from Perry probably altered the progress of the settlement. Robert Cartledge states that Perry's legal counsel enjoyed little unanimity. Firms in Marfa, Alpine, Austin, and New York made up his legal staff and this composite resulted in varying opinions. Cartledge, for instance, says that his father, Eugene Cartledge, would write to Perry his opinion on a matter, and Perry would respond that Daniel Burke in New York disagreed. As a result Perry made many decisions that should have rested with his attorneys.[104]

tive of each side asked him [Burcham], as a qualified engineer to referee the case. He was beholden to neither side for his opinion and report on the case, which office he performed for no monetary remuneration" (Rubye and W. D. Burcham to Kenneth B. Ragsdale, April 4, 1968). Matthews, however, writes that "the statement that W. D. Burcham was chosen as a mediator by both sides in the dispute is absolutely untrue. He came into the case as a witness for Perry as to the value of the ore taken from the invasion area" (Matthews to Ragsdale, June 5, 1968).

101 Matthews interview.
102 White to Matthews, January 9, 1930, Attorney's file.
103 Matthews to Perkins, January 20, 1930, Attorney's file.
104 Cartledge interview.

However, if White's threat was a tactical maneuver, Perkins countered by filing a supplementary petition to the original styled *Cause No. 1252*. Scheduled for hearing in the District Court of Brewster County, Texas, the instrument claimed the suit "is in the form of a trespass to try title, it is also a boundary suit and a suit for compensation for ores of quicksilver taken from plaintiffs patented mining claim G4-1."[105] It further claimed that during the year 1927–1928 the Chisos Mining Company removed ore valued at $75,000.

The response was almost immediate. On January 27 a conference was held in White's office in Austin to discuss the joint use of records, papers, and instruments relative to the boundary line question in the forthcoming trial. However, during the meeting White appeared "to be very anxious to discuss the question of settlement."[106] Matthews assured him that the Rainbow staff was not disposed to accept any amount less than $100,000. When White scoffed at that figure, Matthews countered by suggesting that they should continue the trial preparations.

Matthews again had the initiative. He was trying to maneuver White for the Rainbow's advantage, and the prospects suddenly appeared brighter. He wrote Brown:

> Just before I left his office he said that he would take the matter of settlement up with his people and would make us an offer within the next two weeks. I told him that we would be glad to consider any offer that they wished to make, but that we wanted it strictly understood that the case would be tried in February. . . . It is my notion that Judge White will submit an offer of settlement of something over $50,000.00 between now and the middle of February.[107]

Matthews was prophetic. The brisk exchange of telegrams reflects the accuracy of his prediction:

White to Matthews, February 6, 1930:
REFERRING MATTER OF SETTLEMENT ADVISE LEAST CASH YOUR CLIENTS WILL ACCEPT

Matthews to Brown & Whitlock, February 6, 1930:
WOULD LIKE TO HAVE CONFERENCE WITH YOU IN SAN ANTONIO SATURDAY FEBRUARY EIGHT VERY IMPORTANT THAT YOU BE HERE

[105] *Cause No. 1252*, Attorney's file.
[106] Matthews to Brown, January 28, 1930, Attorney's file.
[107] Ibid.

Matthews to Perkins, February 7, 1930:

CONSIDERING HERE TOMORROW SETTLEMENT CASE FOR SEVENTY FIVE
THOUSAND DOLLARS, WHICH WE WILL RECOMMEND[108]

On Saturday, February 8, 1930, Matthews, Whitlock, Armstrong, and Brown met again in the same conference room they had
occupied almost ten months earlier. However, a different atmosphere pervaded the discussions: the Chisos management was proposing the settlement. Matthews had been careful to allow White
to take the initiative in naming a figure, but when he suggested
$75,000 to settle the matter out of court, Matthews thought the
figure was too small. He explains:

> As I recall, that was just about a net figure of the ore removed. Prob-
> ably four times that amount—$300,000—would have been a more
> realistic figure. I said to them, "boys this is what they have offered.
> You know about what they have taken. Their offer represents money
> lying on the table." The fact that the Rainbow was in good ore and
> needed money for a furnace helped influence the settlement. They
> finally said, "Let's take it."[109]

All parties to the suit agreed to meet a week from the following
Monday, February 17, 1930, in the Holland Hotel in Alpine to con-
clude the agreement. What appeared to be a routine legal procedure quickly erupted into another emotional confrontation between Perry and the Rainbow staff: he stubbornly refused to sign
the document and threatened to go to trial. This position was taken,
no doubt, in response to the exhortation of Eugene Cartledge, who
continued to the very end to assure Perry that his boundary claim
could be sustained in court. Tensions mounted.

To add comic relief to a melodrama that suffered too much
from dramatic extremes, the controversial John Briscoe Brown
usually provided a line well calculated to get a laugh. When Perry
hesitated to settle, Brown continued to tantalize him: "Now, Mr.
Perry, if you haven't got the money, we'll be glad to take your
yacht!"[110] Whether on this occasion or at another hotel conference
with Perry, no one seems to recall, but they all remember that
Brown, saturated, came down the stairs to the lobby of the Holland
Hotel stark naked!

In response to either Brown's low comedy or the insistence of
Ike White, Perry altered his position and later that day agreed to

108 Attorney's file.
109 Matthews interview.
110 Armstrong interview, August 24, 1966.

the settlement. Hunter Metcalf, partner of the Marfa law firm of Mead and Metcalf, who was present at the meeting, recalls: "I remember Perry wrote a check for $75,000 and with a broad well-calculated gesture tossed it across the table to Wilbur Matthews. The little rascal tried to be quite a showman."[111] That afternoon Perkins went to the Brewster County courthouse and filed a motion to dismiss his earlier motion for inspection and survey. A judgment giving the Rainbow Mining Company, Inc., title and possession of mining claim G4-1 and $75,000 damages bears the same date.[112]

Matthews and Whitlock departed immediately for San Antonio and the following day distributed the funds received from the Chisos Mining Company. After Whitlock returned to Denton, Matthews inquired about the settlement with Millican. Whitlock replied: "Mr. Millican has received his part of the Judgment . . . [and] I received from him all that I requested. . . . So everything is settled to my satisfaction."[113]

While the Rainbow mine episode was not a monumental occurrence in the Big Bend—the Alpine *Avalanche* failed to report it —the event, nevertheless, seemed to mark a turning point in the lives and the fortunes of the participants. For Wilbur L. Matthews the episode at Terlingua was an important achievement in his early career. Other than winning a sizable cash reward for his clients, his scholarly research and his skillful handling of the negotiations established him with a firm that two decades later listed him as the senior member.

The legal settlement also marks the beginning of a period of success for the Rainbow Mining Company. It immediately began surfacing the rich ore from the disputed area and soon discovered another rich deposit in a parallel fracture a few yards north of the breakthrough. With the Chisos settlement money the Rainbow owners purchased a new Herrshoff furnace and "pretty soon we were making one hundred flasks a month. That was higher," Armstrong adds, "after we began working that second stope. I guess we must have recovered over ten thousand flasks. Got in the big class. Started shipping by car load lots."[114]

But the man probably most responsible for the Rainbow mine's

---

[111] Metcalf interview.

[112] Sara Pugh, county and district clerk, Brewster County, to Kenneth B. Ragsdale, April 3, 1968, in possession of author.

[113] Whitlock to Matthews, March 3, 1930, Attorney's file.

[114] Armstrong interview, August 24, 1966.

legal triumph over Perry received little of the benefits. John Briscoe Brown left the company soon after the settlement. The reason was simple: alcohol. Armstrong explains: "They picked him up in the middle of a street in San Antonio looking at the moon. Said he was taking a sight on the moon. Trying to figure out where he was, I guess." He added that Brown had invested heavily in the futures market and after being placed in jail decided to dispose of his holdings. When the jail authorities refused him permission to call his broker he complained to Armstrong, "These guys are crazy. I'm the only sane person in the whole damn jail."[115]

Brown attempted a literary career, and in 1933 wrote a manuscript on economic philosophy based on the theory that "other forms of matter have a 'time lag' in the response to the pull of gravity, while mercury does not."[116] In this work Brown advanced a plan based on this theory for ending the enonomic depression of the 1930s. Equally confused was the publisher who rejected his manuscript, who writes: "It may be that you have found a new theory which will eventually be accepted. At the present time, however, we feel certain that it will be disputed by many scientists. . . . So many books offering cures for the depression have been published during the last year or two that the market has been fairly well saturated."[117] Brown, however, rejected the advice and in 1934 pubished the book, entitled *Rescue from Chaos*, himself. Wilbur Matthews received a copy that bore the following inscription, "To W. L. Matthews with compliments of J. B. Brown, March 30, 1936." That was the last anyone in Terlingua heard of John Briscoe Brown.

Armstrong is succinct in documenting the demise of Brown's successor. "I made Starr superintendent after Brown left. A good man. Kept good records. Later had the V. B. Oil Company in Alpine. Went broke. Had a race horse at the mine. It never ran."[118] Starr's former adversary, Howard E. Perry, contributed to his failure in business. When the Chisos Mining Company admitted insolvency in 1942, the company owed the V. B. Oil Company $472.30.[119]

The Rainbow mine's days of glory were numbered and when recovery began to decline in the mid-1930s, Armstrong moved seven miles east to Study Butte. Explaining this change, he says, "I

---

[115] Ibid.

[116] J. B. Brown, *Rescue from Chaos*, p. ix.

[117] Ibid., p. xvi.

[118] Armstrong interview, August 24, 1966.

[119] Bankruptcy file, *Cause No. 688*, Federal Records Center, Fort Worth, Texas.

thought Study Butte had a hell of a lot of low grade ore that no one else had."[120] Also, Bill Burcham's Brewster Quicksilver Properties, Inc., was in financial difficulties, so with two investors from San Antonio, Alvin C. Hope and J. V. Rowan, he formed Southwest Mines, Inc., which assumed Burcham's liabilities.

Neither Burcham nor Armstrong benefited from the transaction. Burcham was soon dismissed by the new owners, and when the price of quicksilver declined, Armstrong, discovering that he and his partners were "simply swapping dollars," also left the firm. In 1943 Burcham would return to the district to work for the Esperado Mining Company in its costly attempt to resurrect the Chisos mine. In 1939 Armstrong joined Standard Oil of New Jersey, which assigned him to an oil exploration team in the Sinai Desert. Following the outbreak of World War II, when repeated Italian bombings made exploration work there hazardous, the company transferred him to South America where he continued exploratory work in Colombia and Venezuela until he returned to Texas in 1952. Choosing a comfortable retirement in San Antonio, Armstrong still dreams of a bonanza yet undiscovered in the Terlingua Quicksilver District.

The tragic figure of the Rainbow mine episode was Marcus Hulings. Regarded by his contemporaries as honest and forthright, he played a role in the negotiations, they recall, that was totally incongruous with his high moral principles. It was Hulings who first confided to Robert Cartledge that he feared the Chisos shaft had penetrated Rainbow property. When Perry refused his request for a surface survey, he reluctantly continued the westward drift.[121]

The Rainbow staff evidenced much skepticism as they listened to his testimony at the Terlingua hearings. During the underground inspection he showed Burcham where the Chisos operation had removed Rainbow ore at the 650-foot level. In response to Burcham's inquiry if that was the only level in dispute, Hulings replied in the affirmative. However, when Burcham later worked the Chisos property, "he discovered that on another level, the 300-foot level, the Chisos Mining Company had also extracted ore on the Rainbow side of the line." Burcham still believes, "knowing his sense of integrity," that Hulings testified under duress.[122]

Those who knew Hulings best believe that this event—for whatever part he played in the drama—preyed heavily on his mind.

[120] Armstrong interview, August 24, 1966.

[121] Cartledge interview, February 28, 1968.

[122] W. D. Burcham, Alpine, Texas to Kenneth B. Ragsdale, April 4, 1968, in possession of author.

Already saddened by the death of his wife and greatly disturbed over a spurious Arizona mining venture in which several close friends lost heavily, he was unable to cope with this apparent deception. On the afternoon of June 15, 1931, Hulings left Alpine and drove east on the Fort Stockton road. About eighteen miles from Alpine he parked his car, removed his hat and glasses, and placed a .38-caliber revolver to his right temple. A few minutes later a passing motorist found his body lying by the side of the car.[123]

For a man who seldom exhibited genuine concern for other human beings, Perry admitted a great personal loss with Hulings' death. He wrote Robert Cartledge: "I am just too sad and broken up—broken hearted to write scarcely at all. . . . What was it that caused him to do it?"[124] Many feel they know the answer, but the real truth died with Marcus Hulings.

For Howard E. Perry, the impact of the Rainbow encounter was neither instantaneous nor decisive, but he would long remember those Holland Hotel meetings. He could view the event with mixed emotions, as in his estimation, he had both won and lost. Soon after the settlement, Harris Smith, owner of the Fresno mine, asked Perry, "Mr. Perry, didn't it hurt you pretty bad to have to give the Rainbow $75,000?" Perry replied: "Now Harris, would you mind giving away $75,000 when you had already taken in $250,000? After all, I was paying them with their own money!"[125] Perry is reported to have told this story many times, always followed by explosions of delighted laughter.

Yet in assessing the extent of Perry's transgression, he should not be removed from the context of the age that produced him. The historian Edward Kirkland, commenting on late nineteenth-century business methods, states, "Piracy, no longer found on the high seas, was domesticated on land; it invaded business"[126] and became a part of the technique of fortune expansion. To Perry, his invasion of the Rainbow property, irrespective of the boundary issue, did not raise a moral issue. He was simply caught with his hands in the Rainbow's pocket; he had violated a rule imposed by law and he chose to accept the penalty. The game could now continue. And the game did continue, though the coming decade brought only decline and disappointment. The year 1930 marked a point of change.

[123] Alpine *Avalanche*, June 19, 1931.
[124] Perry to Cartledge, June 16, 1931.
[125] Interview with Harris Smith, Austin, Texas, July 7, 1965.
[126] Edward Kirkland, *Business in the Gilded Age*, p. 15.

A former member of Perry's legal staff supported the Kirkland thesis. In his declining years Hunter Metcalf discussed the controversy with the author. At the beginning of the interview Metcalf spoke of the Rainbow litigation with extreme caution, attempting not to betray the professional obligation of the attorney to his client. However, as the discussion proceeded it became apparent that he questioned Perry's moral integrity. Finally, breaking into uproarious laughter, Metcalf asked, "How much can a man deviate? At 600 feet he put a digger on a lateral headin' west. He got all he could while he was getting it." Then attempting to regain the composure befitting a man of his profession, he smiled and added, "I reserve the right to claim I was misquoted."[127]

While partisan in its purpose, a Matthews office memorandum sustains Metcalf's view. Attempting to establish precedent for the ownership of minerals discovered in tunnels and shafts, Matthews concludes that "the only reason for any tunnel or shaft ever being dug by the Chisos Mining Company . . . in the vicinity of its [Rainbow's] properties was that an inspection of exposed strata disclosed the lodes or veins which they expected to develop through their work."[128]

On the other hand, Eugene Cartledge believed that Perry's position in the Rainbow matter was defensible and insisted that the issue be settled in court. Yet aside from the moral issue involved, Perry's never-ending demand for secrecy, his refusal to show Marcus Hulings the southwest corner of Section 295, his reluctance to permit an inspection of the disputed area, and his delight in commenting that "he paid them with their own money," all serve to endow him with a guilt that he may or may not deserve.

Although Perry liked to reflect on the Rainbow confrontation with flippant casualness, by the mid-1930s he was forced to look to the future with sobering uncertainty. The economic issues were clearly defined and all indicators forecast decline. Yet Perry's capacity for reason was too well insulated by his inherent optimism for him to respond to the news from Pennsylvania Avenue and Wall Street. Still true to his calling he believed implicitly that another bonanza lay waiting "behind the next rock." Thus, the drama was destined to continue, but only temporarily.

[127] Metcalf interview.
[128] Office memorandum, p. 10, Attorney's file.

CHAPTER 10

When the Blue Eagle
Screamed

THE decade of the 1930s marked an important milestone in American history. Ideas, concepts, principles, and attitudes long held as fundamental by many Americans, were entering an age of flux. Economic upheaval heralded the new decade and secondary shock waves, penetrating every facet of human endeavor, sent ancient idols tumbling. In politics a 1932 Democratic landslide wrested the government from conservative Republicans and placed it in the hands of crusading liberals. A festival of experimentation ensued and the functions of government invaded the sanctity of private business and altered the concepts of free enterprise.

Socially changes were equally vast as the whole tenor of American life underwent transformation. Urban America, infested with industry, sick and dying, became the focal point of national concern, rendering obsolete forever the historically persistent belief in the priority of an agrarian society. Economic realignment had long since dissolved that myth and with it went another casualty, the misguided belief in the self-made man. Horatio Alger had made way for a brigade of well-tailored young college men on the make.

Many of these factors, evident since before the turn of the century, were now magnified by a depressed economy. Recognition and acceptance of these changes were some of the forces that sent society into turmoil in the 1930s. Change was the theme of the new decade. America had rounded a corner, and while many tried to turn back the clock to a simpler yesteryear—real or fancied—a new era was at hand.

While Terlingua during the early years of the Depression may have seemed remote and geographically isolated from Washington, Pittsburgh, and the Dust Bowl, or the Salinas Valley, the same economic forces that strangled the nation also presaged change for the Chisos Mining Company. Although the Chisos management had grappled with the dual problems of increasing costs and declining revenues during the preceding decade, the economic exigencies and emergency legislation of the early 1930s aggravated further a malady that ate away at the vitals of Perry's once great quicksilver bonanza. For a mining company with declining reserves, whose once-critical product remained unsold in a depressed market, the issue of recovery legislation was largely academic.

But for Perry the matter was essentially personal. Every edict that emanated from Washington after 1933 was contrary to the basic principles upon which he operated his business. Howard E. Perry *was* the Chisos Mining Company. It was his property that he had created by his own initiative with his own capital. The philosophy of government-supervised cooperative enterprise, therefore, was the antithesis of everything in which he believed; one perspective looked to the future for economic recovery and social betterment, while the other longed for the past and the *status quo*. The latter was the "law of the land of Perry," the premise upon which his crumbling empire had been built. His was a closed corporation, with no margin for exterior interference or peripheral dictates. He was the sole master of his domain; to storm his desert bastion would be a breach of procedural ethics, a challenge to the system that he felt had made America great. Perry was among the last of his breed, a dinosaur facing an encroaching ice age, refusing to cope and unable to adapt.

As the 1920s drew to a close, Perry unknowingly was approaching the twilight of his great quicksilver bonanza. While the events of Thursday, October 24, 1929, hastened the evolving drama that would be played out during the ensuing decade, Perry misread the vast ramifications of the Wall Street disaster and wrote Cartedge as follows: "You have heard of course about the great smash we have had in the stock market. There was no excuse for such an awful smash. The New York banks are to blame mostly; though it is true that some stocks were too high. I have had to change over some of my investment stocks into stronger issues. The thing is all over now I think, though the market will not come up high again for some time."[1]

From the rising chorus of voices that responded to the Wall Street catastrophe, Herbert Hoover's flaccid monotone had a particular appeal for Perry, as the nation's president articulated his continued belief "in modified *laissez faire*, the gold standard, individual enterprise and the profit motive as the mainsprings of progress, and in savings and self-denial as the essence of economic

---

[1] Perry to Cartledge, October 30, 1929, Chisos Mining Company Papers, Archives of the Texas State Library, Austin.

security."[2] If Hoover's advocacy of self-denial fell on Perry's inattentive ears, a later reference to rugged individualism struck a responsive chord. Real or fancied, Perry considered himself a self-made man, functioning in a climate of free and unbridled and enterprise. He had invested personal funds in a risk-prone venture, assumed total liability with the prospect of profit, had beaten the odds, and had won. Therefore, when Hoover sanctioned rugged individualism, Perry considered it a personal tribute. Unknowingly, both men had lost touch with the changing realities of twentieth-century America.

The 1932 presidential campaign cast in stark relief two divergent philosophies of government. While Hoover maintained that the government should protect the economic freedom of the individual, Roosevelt instead took a closer look at the individual's diminishing alternatives and addressed himself to a nation racked with fear and uncertainty. Speaking with confidence he proposed to shift the focus of governmental concern from the top of the social and economic spectrum to the bottom, and on April 7, 1932, the nation's radios transmitted a soon-to-be-familiar voice expressing "his solicitude for the under-privileged in a phrase: 'the forgotten man at the bottom of the economic pyramid.' "[3] The people concurred and November 8, 1932, marked a new departure in the American experience.

While it is doubtful if many Terlingua citizens could grasp the import of the emergency legislation the new administration was ushering through Congress, its mushrooming bureaucracy would soon find its way to the Big Bend and saturate the very foundations of the Chisos Mining Company with ideas and attitudes heretofore foreign to that operation. This hastily assembled recovery program, labeled the New Deal by Roosevelt, reflected the spirit of the 1930s, which the historian Dixon Wecter describes as follows:

> Under the New Deal, Capitol Hill and the White House replaced Manhattan and Wall Street as the nation's cerebral cortex. The government began to impinge upon the life of the citizen as never before . . . taxing, lending, spending, building, setting quotas in agriculture and conditions of employment in industry. . . . Awareness of this new relationship of government to the daily life of the citizen dawned for many with the advent early in the summer of 1933 of the National Recovery Administration. It was President Roosevelt's chief prescription for recovery.[4]

[2] Dixon Wecter, *The Age of the Great Depression, 1929–1941*, p. 42.
[3] Ibid., p. 52.
[4] Ibid., pp. 81–82.

Perry's world would never be the same. He would long remember that summer of 1933.

A blue eagle bearing the slogan, "We Do Our Part," became the symbol of the National Recovery Administration (NRA) legislative program whose purpose, Roosevelt claimed, was to put people back to work and which would in the process raise the purchasing power of labor by reducing hours and increasing wages. This new program, based on the concept of business-government partnership, began organizing industry through trade group cooperatives. The industrial code was the program's primary objective, which NRA administrator Hugh S. Johnson solicited through cooperation rather than administrative control. He claimed the "NRA's job was not to impose codes but to accept them."[5] Business appeared in a cooperative mood as "representatives of nearly 800 groups of fabricators and distributors of goods and services . . . thronged into Washington to get codes of their own. In the great stampede it seemed as if no industry wanted to be left out."[6]

Representatives of the depressed quicksilver industry appeared in the vanguard of the trek east. The National Quicksilver Producers Association (NQPA) was formed on July 18, 1933, and filed an application for a Code of Fair Competition on July 31. This association, with membership in Texas, Arizona, Arkansas, Alaska, Oregon, Washington, California, and Nevada, represented "thirty-five operators employing 450 men, but seventy-five mines are ready for production with a probable employment at the mines of above 1,000." Association documents show W. W. Klipstein (California), president; Irving Ballard (California), secretary; Howard Perry, vice-president; and W. D. Burcham, owner of the Brewster Quicksilver Consolidated at Study Butte, vice-president.[7] This group would attempt to chart the course of the quicksilver industry through the troubled waters of the depression years. The ensuing voyage would not be a placid one. Eight months of hearings, code revisions, and political maneuvering intervened between the original presentation of the code on July 31, 1933, and its final approval on March 21, 1934.

The quicksilver producers viewed tariff protection from foreign imports as an essential item in their code. This is expressed in Arti-

[5] Arthur M. Schlesinger, Jr., *The Age of Roosevelt: The Coming of the New Deal*, p. 110.

[6] Wecter, *Great Depression*, p. 84.

[7] National Recovery Administration, Record Group 9, National Archives, Washington, D.C.

cle IV, Foreign Competition, which states that since domestic production costs would increase with NRA code approval,

> it is inevitable that conformance with any code written in the spirit of the Act would make impossible operation at profit and tend to reduce labor employment through forced shutdown of domestic producing units, unless protection is given under the Act against low cost and low priced foreign competition. The industry therefore agrees that this code shall be effective only after adequate protection against foreign competition has been granted.[8]

The quicksilver producers attempted to secure the protection provision through political support, but hope for relief faded when they discovered that the maze of Washington's political roads all led eventually to the White House. On September 22, 1933, a Miss French, secretary to California Congressman Harry L. Englebright, wrote Presidential Secretary Stephen T. Early about

> the grave situation which the quicksilver industry is facing as a result of the enormous importation of the metal, and it is feared that many, if not all, of the domestic mines will close down for the next year or more, which will materially add to the unemployment question.
>
> I understand that the question has been brought up with the President on previous occasions, but great stress has been brought to bear that unless something is done to alleviate the matter, very shortly, dire results may be predicted.[9]

Three days later Early referred that matter to NRA Administrator Hugh S. Johnson, who wrote Miss French on September 27 that he was "asking Mr. Ryder who is in charge of our tariff division to give every consideration to the problem you present."[10] The NQPA administration must have realized it was getting the big run-around.

If Roosevelt's broadcast eloquence instilled confidence in the American people, many of those who responded to his promises of personal easement found their hopes unfulfilled. Association secretary Irving Ballard, tiring of the perennial Washington buck passing, also succumbed to the magic voice that made Americans believe that "all they had to fear was fear itself."[11] As a last ditch

[8] Ibid.

[9] Miss French to Stephen T. Early, September 22, 1933, R. G. 9, National Archives.

[10] Hugh S. Johnson to Miss French, September 27, 1933, R. G. 9, National Archives.

[11] Franklin D. Roosevelt's first inaugural address, March 4, 1933, Washington, D.C., B. D. Zevin, ed., *Nothing to Fear: The Selected Addresses of Franklin Delano Roosevelt, 1932–1945*, p. 12.

effort to obtain tariff relief through executive order, Ballard wrote Roosevelt on October 19, 1933. Beginning in a warm and friendly manner, Ballard states that "in some of your wonderful radio talks you invited anyone who did not seem to get their troubles properly attended to, to write you personally. This I am doing." He explained to the president that the matter of tariff protection had been referred to him and that Presidential Secretary Louis McHenry Howe reported that the president

> had taken the matter up with Gen. Johnson, and Gen. Johnson wrote us that he had put the matter up to Mr. Oscar B. Ryder, Liaison Officer of the NRA and the Tariff Commission. We have taken the matter up with Secretary [Daniel C.] Roper and Secretary [of State Cordell] Hull who both referred the matter back to Mr. Ryder who tells us very frankly that he can do nothing without an executive order from you.
>
> Unless this condition is not immediately taken care of the entire Quicksilver Industry will be put out of business. . . . To date this year there has been enough quicksilver imported into the United States to render practically inoperative your recovery programme in so far as this essential industry is concerned.[12]

Attaching a cover letter to Howe, Ballard directed the letter to the presidential secretary, who in turn, directed it to William W. Bardsley, code assistant, who wrote Ballard that "the code has been discussed with Mr. J. F. Gallbreath [secretary, American Mining Congress], . . . who we understand has the Association's authority for any necessary revision of the code."[13] Whatever authority Gallbreath possessed, it was never exercised, as the Code of Fair Competition for the Quicksilver Industry was approved on March 21, 1934, with the protective tariff provision omitted.

Although the tariff issue would become an argumentative club that Perry would wield repeatedly during the following months, the issue to which he took immediate exception was contained in Article IV, Section 1 of the code, which stated that "no employee shall be paid in any pay period less than at the rate of 42½¢ per hour; provided, however, that in the Southern District the minimum rate that shall be paid in any pay period shall be 30¢ per hour."[14]

[12] Irving Ballard to Franklin D. Roosevelt, October 19, 1933, R. G. 9, National Archives.

[13] William W. Bardsley to Irving Ballard, no date, R. G. 9, National Archives.

[14] Code of Fair Competition for the Quicksilver Industry, R. G. 9, National Archives.

For some unexplained reason Perry, who served originally as the National Quicksilver Producers Association vice-president, disassociated himself from that group some time during the code hearings. Extant data seem to indicate that the hourly pay rate expressed in the code prompted this move. Assuming an independent role in code making, Perry solicited the support of four other Terlingua District mine owners in forming a nonassociation splinter group of Texas quicksilver producers. On April 20, 1934, Rainbow Mine vice-president C. T. Armstrong and Brewster Quicksilver Consolidated president W. D. Burcham [also a former Association vice-president], sent telegrams to NQPA secretary Irving Ballard in San Francisco supporting Perry as their "unanimous choice as nonassociation member code authority for Texas quicksilver producers."[15] The same day Perry also alerted Ballard to expect receipt of the foregoing telegram "naming myself as the non-association member code authority for Texas for this position fits in with work doing in Washington and been doing ever since you were there."[16]

Under Perry's leadership the group moved with dispatch. Engaging two Washington attorneys, Paul M. Segal and George S. Smith, the Texas producers had prepared an Application for Exemption, Exception From, or Modification of the Code, which they directed to NRA administrator Johnson. Focusing entirely on the inefficiency of Mexican labor, the application states:

> Conditions in the quicksilver industry in the Terlingua District with regard to labor and payment, therefore, are entirely different from those obtaining in other portions of the country. . . . Ninety percent of all employees in the industry and a larger percentage of all laborers are Mexican people. Because of this, and the proximity of Mexico, the quicksilver industry is conducted in the Terlingua District from practically all standpoints as though said District were within Mexico. The needs and tastes of the laborers employed are simple and their wants are inexpensively satisfied. The prevailing wages in the District . . . range from 14½¢ to 25¢ an hour. Such labor is not highly productive, nor is it entirely continuous. The prevailing wages in other activities among the Mexicans, such as ranching, is room and board only. . . . The efficiency and productiveness of the labor is not such as to permit the payment of this minimum

15 C. T. Armstrong to Irving Ballard, April 20, 1934, R. G. 9, National Archives.
16 Howard E. Perry to Irving Ballard, April 20, 1934, R. G. 9, National Archives.

wage, and specifically in the cases of Mexicans performing unskilled labor of the type above described.[17]

This document concludes with a request for an exemption from the minimum wage provisions of the code and an authorization to pay a minimum rate of $0.20 per hour in the Terlingua district. The examining board hearing the application for exemption issued a favorable verdict, and on April 23, 1934, deputy administrator W. A. Janssen signed Order No. 351-3, Order Granting Temporary Exemption From Wage Provisions.[18]

This was a clear-cut victory for Perry. And while, no doubt, sound practical basis existed for the claim that "the efficiency and productiveness of the labor is not such as to permit the payment of this minimum wage," the key phrase of the NRA document referred to "the proximity of Mexico." For all practical purposes the Chisos mining business was conducted "as though said District were in Mexico." Remoteness and isolation were still Perry's greatest allies. With Washington almost 3,000 miles and a week away, and Alpine a hard day's drive to the north, the Terlingua district remained the "land of Perry." Socially, politically, and especially economically, he made unchallenged decisions, which a submissive staff enforced without question. But this condition was destined to change. The NRA legislation had so decreed it.

Less than two months following the approval of Exemption Order No. 351-3, NRA executive assistant T. U. Purcell entered the Chisos drama, and Perry met one of his most tenacious adversaries. While little is known of Purcell prior to 1934, a former associate describes him as "dedicated, sincere, with above average education. Tried to do a good job."[19] Purcell's work on the Chisos Mining Company case confirms his former colleague's appraisal, as he approached this assignment with the crusading fervor that characterized the work of many of the NRA adherents. While his motives and methods would be later challenged by the Terlingua district quicksilver producers, as well as the state NRA administrator, he attacked Perry and his business practices with a reformer's zeal that jarred the foundations of Perry's Big Bend empire. Both men would long remember this encounter.

[17] *Application for Exemption to, Exemption From or Modification of The Code*, R. G. 9, National Archives.
[18] R. G. 9, National Archives.
[19] Interview with Zola Avery, Dallas, Texas, November 28, 1973.

On June 29, 1934, J. R. Martin, El Paso district manager for the Department of Labor, and W. O. Hale, former Brewster County sheriff, reported to Purcell that the Chisos Mining Company was not complying with the NRA wage code and was paying employees in Mexican silver pesos instead of United States currency. Two weeks later Purcell went to Terlingua to investigate the complaints, whereupon Perry "freely admitted that he was working his miners nine hours per day, fifty-four per week, instead of the forty provided in the code, and that he was paying them from $1.00 to $2.75 United States currency per day, which is also under the code."[20] Perry also explained to Purcell "very emphatically, that the Quicksilver Code had been 'completely stayed' by the administration, and that his industry, therefore, was not subject in any way to the Quicksilver Industry Code."[21]

This encounter, which apparently ended amicably, revealed one of the basic weaknesses of the NRA administration—poor field communications. Purcell learned from Perry, instead of the NRA state headquarters, that Order No. 351-3, temporarily exempting the Terlingua district quicksilver producers from the $0.30 per hour minimum wage, had been issued more than two months previously. Delayed communications from Washington verified that Perry's claim that "he was the Code authority for Texas" and that the code had been "completely stayed" were both false.

Following the June 29 meeting, Perry telegraphed Janssen in Washington, noting Purcell's surprise that "We are not operating under code. . . . He will likely wire compliance board or yourself. Please wire me whether it has been possible yet to do anything about embargo or quota foreign. We are sweating blood."[22] Janssen responded on July 6 that "Compliance board have been advised of exemption granted to producers in Brewster County, Texas. It will be necessary for the code authority to make application for industry for protection against imports under Three E of the act."[23]

Considering the clear-cut wording of Order No. 351-3, Perry's statements to Janssen that "we are not operating under [the] code," and to Purcell that the code had been "completely stayed," were

[20] T. U. Purcell to L. J. Martin, March 5, 1935, R. G. 9, National Archives.
[21] T. U. Purcell, to H. P. Drought, September 19, 1934, R. G. 9, National Archives.
[22] Howard E. Perry to W. A. Janssen, June 29, 1934, R. G. 9, National Archives.
[23] W. A. Janssen to Howard E. Perry, July 6, 1934, R. G. 9, National Archives.

probably Perry's tactics calculated to create confusion, bring additional pressure on the NRA, and delay compliance. The temporary exemption order states clearly that the minimum wage in the Terlingua district was $0.20 per hour; all other code conditions—including the eight-hour day—remained unchanged. But Perry probably knew, as did NRA administrator Hugh S. Johnson, that bureaucratic confusion favored lax code enforcement. Alluding to this administrative weakness, Johnson explained later that the "NRA organization was new, hastily thrown together, uninstructed, inexperienced, and growing at the rate of 100 [people] a day. . . . I was trying to operate with a staff and executives all of whom I did not know by their first names."[24]

If Perry felt he could delay, or even avoid compliance because of misinterpretation or the embargo matter, he also knew that others in the business community shared his disdain for NRA bureaucratic incursions. According to Wecter,

> Violation of codes by those displaying the Blue Eagle became so manifold that after a few months the public began to wax cynical. . . . Policing, half-hearted and ill-financed, soon grew as lax as under national prohibition in its dying days; violent spasms of enforcement bore results equally demoralizing. . . . The most conspicuous rebel to defy the Blue Eagle, rugged Henry Ford, suffered no apparent loss in sales; indeed, under the swelling outcry against the NRA, he came to be hailed in conservative circles as a hero.[25]

The Big Bend was indeed about to witness a spasm of enforcement, as Purcell had already made his first offensive move. After encountering open hostility by Perry's staff during a third visit to Terlingua in July, Purcell nevertheless gathered sufficient data to file a complaint against the Chisos Mining Company for noncompliance. Reporting this episode to the NRA compliance director, H. P. Drought, in San Antonio, Purcell writes:

> The manager of the Chisos Mining Company [Robert Cartledge] . . . refused to make available to me such books as contained the time and amounts paid miners, but after a delay of twenty-four hours, during which time he took the matter up with Mr. Perry [apparently in Washington], he again made available to me the records. . . . I then prepared a summary of all of the statements which amounted in the aggregate to $16,206.78, of which approximately $13,130.00 represented back wages due miners, and that balance of $3,076.00

24 Hugh S. Johnson, *The Blue Eagle From Egg to Earth*, p. 248.
25 Wecter, *Great Depression*, p. 85.

represents payments due American office and mechanical employees, none of whom can be considered as executives under the Code.[26]

Purcell then cites specific violations of the quicksilver code by the Chisos Mining Company as follows:

1. *Maximum Hours.* Practically all employees were working "six days per week, nine hours per day, or a total of fifty-four hours per week during the entire time since the Code became effective." A number of employees had been required to work nine-hour shifts, seven days per week.

2. *Minimum Wages.* The $0.20 minimum wage provided for under the exemption order "was not paid employees from the effective date of such order, April 23rd, at anytime between that date and September 1, 1934. I find, in fact, that some employees had been paid as low as 75¢ for a nine hour working day and that the majority of the employees had been paid at the rate of $1.00 for nine hours work, or approximately 11.1¢ per hour."

3. *Overtime.* Purcell noted that "no overtime was paid any employee for hours worked in excess of 40 hours per week."

4. *Child Labor.* The report stated that "a number of underground employees and surface employees were under 18 years of age." Purcell also reported that "'Standards for Safety and Health' had been disregarded in that proper safety precautions are not used."

5. *Failure to Post Code.* The Chisos management had failed to post copies of the code, and "in fact the employees had been kept in ignorance of Code provisions and even of the existence of a Code."

6. *Coercion.* Purcell reported a violation of Section 7, Article 5, in that "employees had been practically required as a condition of employment to trade at the Chisos Mining Company store." This report also cited the payment of Mexican pesos "at the rate of two pesos for one dollar, instead of the market rate of 3.53 pesos for $1.00."

7. *Doctor's Fee.* All employees were required to permit deductions from their wages of a doctor's fee not to exceed $1.50 per month.

8. *Safety.* This report cited the Chisos coal mine as a "death trap."

9. *Oppression.* Purcell reports that the attached "affidavits show that the Chisos Mining Company have for many years . . . effectively practiced oppression of its employees in one of the most aggravated

[26] Purcell to Drought, September 19, 1934, R. G. 9, National Archives.

forms that has ever come to my attention. These affidavits likewise show numerous abuses on the part of R. L. Cartledge."

10. *Embargo.* Purcell explains further that Perry informed the Terlingua district quicksilver producers "that they would not be subject to the Code . . . until and unless the Administration granted them an embargo." Purcell notes that neither the code, nor Amendment 351-3 "contain any provision making it subject to . . . the granting of an embargo."

11. *Supporting Testimony.* Purcell supported his claims with specific complaints filed by several Chisos employees. The affidavit of Dr. J. O. Jeter of Alpine, for example, states that deficient medical attention "has resulted in a high mortality among children of Terlingua."[27]

Purcell's report drew the attention of NRA state director Drought, who telegraphed John Swope of the compliance division's field section in Washington: "Believe El Paso executive assistant has uncovered deplorable situation bordering on peonage. . . . Dispatching field adjusters August one to investigate."[28] Drought's telegram cued the closing chords of the overture; another Big Bend drama was about to begin.

What occurred during the next three weeks will remain a mystery forever, as documentation for this period is lacking and the participants are long-since deceased. Purcell's return to the district, however, drove tempers to the breaking point, and Perry, unable to secure embargo protection from Washington and chafing at the prospect of having to produce $16,206.78 in back wages, closed the Chisos mine on September 1 in apparent retaliation. This most recent confrontation with Purcell apparently dramatized for Perry that he once more was facing a formidable opponent in a struggle for survival: shades of Whitlock, Matthews, and the Rainbow mine episode. Perry again attempted to maneuver for a technical advantage but uncovered nothing in Purcell's background that would strengthen his position. "P[urcell] wants to make his spurs on this," Perry wrote Cartledge. "I have a Dunn [and Bradstreet] report on him—says 'Means small and seems pressed for money'—salary $3,500.00—slow pay—general reputation not bad at all."[29]

Perry must have realized at this point that the NRA compliance issues would not be easily resolved in his favor and that his greatest strength lay in creating public resentment against the agency cre-

27 Ibid. The doctor was probably D. O. Jester of Alpine.
28 H. P. Drought to John Swope, July 23, 1934, R. G. 9, National Archives.
29 Perry to Cartledge, September 1, 1934.

ated to reestablish economic solvency. After voluntarily ordering the
Chisos mine to cease operations, Perry instructed his attorney to
issue a totally false "story to the press to the effect that the mines
had been closed by NRA, thereby throwing hundreds of destitute
Mexicans on relief rolls."[30] The Alpine *Avalanche* picked up the
story and reports:

> The Chisos quicksilver mines at Terlingua . . . were closed down
> Monday of this week . . . by reason of the company's inability to
> meet governmental requirements.
> Howard E. Perry, president of the mining company, is now in
> Washington, D. C., attempting to iron out the differences with the
> authorities. . . . It is not known definitely at this time what the nature
> of the controversy between the Chisos mines and the government
> is. . . . The only thing given out thus far concerning the shutdown
> is to the effect that 'the mines are unable to meet certain require-
> ments of the government'. . . .
> The Chisos quicksilver mines employ some 150 miners, mostly
> Mexican.[31]

During much of Purcell's preliminary investigation at Terlingua, Perry had remained in Washington preoccupied with the
quicksilver embargo issue. Unable to secure relief through the
NQPA, Perry turned to the president of the United States. On July
20, 1934, he directed a six-page memorandum to the chief executive,
soliciting his aid in this matter. Perry wrote: "Although fully cog-
nizant of the multitudinous demands upon your time in connection
with the rehabilitation of American industries, nevertheless we are
constrained to ask your aid in our behalf."[32]

Perry then cited for the president the destructive nature of the
foreign imports and related this to a matter dear to Roosevelt's
heart—gold. Perry continued: "Incidentally, it appears rather incongruous that in one industry—namely, gold production—now
fostered by the administration, such industry should be dependent
almost entirely on imported quicksilver for recovery." He then made
his most direct thrust by quoting the provisions of Section 3-E of
the quicksilver code, which states:

> on his [the President's] own motion . . . that any article or articles
> are being imported into the United States in substantial quantities

[30] H. P. Drought to L. J. Martin, September 25, 1934, R. G. 9, National
Archives.

[31] Alpine *Avalanche*, September 7, 1934.

[32] Perry's memorandum to the president of the United States, July 20,
1934, R. G. 9, National Archives.

under such conditions as to render ineffective . . . the maintenance of any code or agreement under this title, *the President may cause an immediate investigation to be made, and if,* after such investigation *the President shall find* the existence of such facts, '*he shall in order to effectuate the policy of this title, direct that the article . . . shall be permitted entry in the United States only upon such terms and conditions and subject to the payment of such fees* [in order not to render] . . . *ineffective any code or agreement made under this title.*'[33]

Perry, however, weakened his argument by citing the divergent opinions issued by the Tariff Commission in response to an earlier request by the president. While the majority report opposed restrictive action on quicksilver imports, the minority report pointed to " 'the advisability of some action by the President under NIRA [National Industrial Recovery Act] to maintain this domestic industry on a more or less normal operating basis.' Time has shown the wisdom of the minority report." Perry concluded by requesting the president, on behalf of the officers of the National Quicksilver Producers Association and the Code Authority of the Industry, to "exercise the power vested in you and grant relief to the domestic industry by restricting importations of quicksilver by license or embargo."[34]

Whatever hope Perry may have entertained for presidential intervention in the embargo matter, it did not prevent him from seeking support elsewhere in Washington. The day following his dispatching the July 20 memorandum to Roosevelt, he wrote Cartledge that he was up to his "neck on the embargo on limitation. . . . Have finally retained a very high up Washington man—a Democratic National Committeeman—(and he comes high) we will see what that will do for us."[35]

Although Perry remained in Washington and Portland throughout most of July and August, he maintained constant communication with the mine administration and his attorneys in Alpine and Del Rio. On September 11 he was still in Portland working on a company wage structure that he believed the NRA Compliance Division would accept. "My general plan," he wrote Cartledge, "is to raise 75¢ men to $1.00; $1.00 to $1.25; $1.25 to $1.50 with further raises to come along as the business will stand it; but I do not want to raise any Mexican until we get our code and embargo placed."

[33] Ibid.
[34] Ibid.
[35] Perry to Cartledge, July 21, 1934.

While his statement regarding "our code"—whose provisions were well defined—is unclear, he no doubt was assuming an independent posture alongside many other of the nation's business leaders. He explains to Cartledge that because of the lack of uniform administration and enforcement, "the great majority of NRA concerns have not as yet raised wages."[36]

One week later Perry was back in Texas seeking a settlement with the state NRA compliance director in San Antonio. His first target: Purcell. Flanked by his attorneys,[37] he called on the state director, H. P. Drought, and demanded that Purcell be removed from the case. Perry placed the entire blame for the strife at Terlingua on the executive assistant, claiming "that as a result of his speech to the workers the Chisos Mining Company had found it necessary to close its mine for fear the workers would damage the property." Unyielding, Drought stated that "he would not take any action without the consent of Mr. Purcell . . . and that he would agree to a conference between the State Director, the mine owners and Mr. Purcell."[38]

Robert J. Smith, Drought's administrative assistant, telephoned Purcell in El Paso and arranged a joint meeting the following day in the San Antonio office. Purcell arrived early in the afternoon of September 18, and following a brief private conference with Drought, the two men joined the owners of the Chisos, Rainbow, and Brewster Quicksilver Consolidated mines in an adjacent office. Smith recalls that their hopes of resolving the issues that afternoon went unfulfilled, as the meeting ended in a near disaster. Drought opened the meeting by requesting Purcell to present a report of his investigation, but as a "consequence of dealing in personalities by Mr. Purcell, and despite repeated instructions from the State Director that he limit his report to the operations of the mine rather than of individuals, nothing was accomplished save to widen the breach between Mr. Purcell and the mine owners."[39]

While Purcell emerges as the center of controversy in the quicksilver code compliance issue, the major obstacle to settlement was

[36] Perry to Cartledge, September 11, 1934, Chisos Mining Papers.

[37] Probably Hal Browne of Del Rio and Wigfall Van Sickle of Alpine. Perry told C. T. Armstrong, Rainbow Mines, Inc., vice-president, that he never attended an important meeting without legal counsel (Armstrong interview, October 15, 1973).

[38] Robert J. Smith to L. J. Martin, March 20, 1935, R. G. 9, National Archives.

[39] Ibid.

apparently a personal conflict between Perry and Purcell, which Drought attempted to resolve through open discussion at the San Antonio meeting. When Perry charged without Purcell present that Purcell was aggravating the situation, Drought told Perry that the executive assistant "had charges to make against him, too, and I thought that the charges by each against the other could best be presented in the presence of all concerned."[40] Perry became less bold in the presence of the accused, who recalled that "Mr. Perry, who had tentatively preferred charges against me to the State Director, refused an opportunity to repeat those charges in my presence."[41]

After yielding little toward a settlement, the meeting finally adjourned later in the afternoon, and at seven o'clock that evening Purcell and Smith joined Drought at his home to attempt to prepare a basis for adjustment that would meet the provisions of the code. Drought recalled that they discussed the matter until well past midnight and that Purcell was again the obstructive force. "The basis for an adjustment was finally worked out," Drought recalls, "after Purcell had spent more than three hours talking about the violations with no suggestions to offer relative to an adjustment."[42] The resulting document, however, is surprisingly lenient in its conditions of compliance, especially coming in the wake of the emotion-packed meeting a few hours earlier. Also, it bears the concurrence of Purcell, Perry's prime antagonist, to whom Drought explained during the meeting: "This is your case: You have worked it up from the beginning and I do not intend to take any action in connection with it which you do not approve, nor will I suggest or accept any basis of adjustment which you do not feel is proper and just, and which you will not wholeheartedly approve."[43]

The nine-point document, entitled *Proposed Adjustment of Complaints Against Chisos Mining Company, Rainbow Mines Incorporated and Brewster Quicksilver Consolidated*, was completed the following day and contains these key provisions:

> 1. All mines that have been closed will reopen September 24, 1934, and operate to the same extent as heretofore, and they will be operated in strict conformity to the Code of Fair Competition. . . .

[40] Drought to Martin, September 25, 1934, R. G. 9, National Archives.

[41] Purcell to Martin, March 5, 1935, R. G. 9, National Archives.

[42] H. P. Drought to Robert J. Smith, March 18, 1935, R. G. 9, National Archives.

[43] Smith to Martin, March 20, 1935, R. G. 9, National Archives.

2. The State Director will recommend that restitution of back wages under the code and exemptions will be calculated not on the basis of the code, but on the basis of the exemptions from the effective date of the code.

3. The State Director will recommend that the payment of restitution may be withheld until such time as by proper action by the proper federal authority, some relief in the form of increased tariff duties or quota arrangement is granted to the Quicksilver Industry. . . .

4. The State Director will report these complaints in full to Washington . . . strongly recommending that relief . . . be granted immediately by proper federal authority.

5. Each of the above-named respondents will permit all properly authorized NRA officers access . . . to their properties, payrolls and other records of wages paid and hours worked, and will permit such officers to inform respondent's employees of their rights under the aforesaid code and will in all other respects lend cooperation to said officers.[44]

On September 20 the mine owners reconvened in Drought's office to consider the adjustment proposal. Purcell was also present, and either by accident or intent again pushed Perry to the brink of violence. After congratulating the Rainbow representatives for their past cooperative attitude, Purcell then made some uncomplimentary statement about the Chisos' antipathy in the compliance matter, and "Perry blew up."[45] Drought apparently salvaged the meeting by proposing "a basis of adjustment which was entirely satisfactory to the Rainbow and Brewster, and which I felt constrained to tell Perry he had to accept." But Perry, no doubt still chafing from Purcell's latest verbal barbs, took exception to the proposal but finally "accepted it at the last moment. The others accepted it upon its proposal."[46]

If the members of Drought's staff found satisfaction in the outcome of the conference, Perry failed to share their sense of achievement. He had lost a battle but was still at war with the bureaucratic establishment. And for a man who faced life daily through litigation, little had changed. Perry was still master of his Big Bend empire and if Purcell returned again, the meeting would be on Perry's terms. Time was in Perry's favor.

[44] *Proposed Adjustment of Complaints Against Chisos Mining Company, Rainbow Mines Incorporated and Brewster Quicksilver Consolidated*, R. G. 9, National Archives.

[45] Interview with C. T. Armstrong, San Antonio, Texas, August 24, 1966.

[46] Drought to Martin, September 25, 1934, R. G. 9, National Archives.

With this latest encounter with Purcell still fresh on his mind, Perry wrote Cartledge the following day from San Antonio about the outcome of the meeting. "We had a hell of a time here," he reports, "the like of which I never knew before. And I hated to increase wages but for the settlement of a very serious controversy with the government one has to give in on something."[47] The adjusted wage scale increased the $1.25 minimum to $1.40 per day, substantially above the Portland plan, which shows the person making $0.75 per day being raised to $1.00.

Although Perry's concession to the NRA was accepting the $0.20 hourly minimum, the matter of residual payments to the Mexicans offered the racist mine owner a possible two-lane escape route to freedom: he could victimize Purcell while avoiding the issue. Perry continues:

> Of course the Mexicans will want to know whether they are to get back payment on wages and overtime and I have thought a great deal about our answer, one that will not lead us into a lot of talk-a-talkee. I have advised with Hal Browne quite a bit over it, and we have decided as follows:—Just give a great big wave of your hand and say, "O—we do not know about that and add that Purcell is not to be trusted, because he down in his heart wants to get rid of them for white men. . . ."[48]

The Chisos mine resumed operations on September 24, as agreed in the adjustment of complaints, but increased wages coming in the face of declining recovery—no production in 1931, 1,711 flasks in 1932, and 1,178 in 1933—and a depressed market damaged Perry's hopes for the Chisos' future. Government bureaucracy and Purcell were the unquestioned culprits. "I may have to close—before long—," he wrote Cartledge, "I hope not but no one can tell— the CWA [NRA] and Purcell have about ruined us. Damn them."[49] Perry continued to exercise political pressure to have Purcell removed from the Chisos case, and by the end of September a gleam of Perry's former optimism returned to his correspondence as he reports to Cartledge: "I have been practically assured that that fanatic, Mr. P.[urcell] will not come there anymore, in case he does come . . . turn him down on anything he wants and . . . ask all our

[47] Perry to Cartledge, September 21, 1934.
[48] Ibid.
[49] Perry to Cartledge, September 28, 1934. Perry is probably referring to NRA. The Civilian Works Administration (CWA) was a concurrent government-administered public works program.

white men not to hold converse with him."[50] Perry was fortifying his
position.

Perry's optimism, however, was premature. As he attempted to
restrain Purcell through political pressure and victimize him
through character assassination, Purcell was also actively plotting
his attack on Perry's domain. The irony of this episode is that Perry's
"wall of secrecy" was again penetrated by Chisos employees, who
continued to transmit reports to Purcell on the company's failure to
comply with the memorandum of agreement. "Purcell called the
State Director by long distance telephone [from El Paso]," Smith
writes, "and . . . he directed Mr. Purcell to visit the Brewster County
mines and 'to make a routine inspection.' He was specifically instructed to take no action; make no demands. He was told particularly to take no guards or outside parties."[51] Purcell's crusading
spirit prevailed, however, as he categorically violated each of his
superior's clearly defined directives. The pendulum was about to
swing in the other direction.

Purcell arrived at Cartledge's office in Terlingua on the morning of November 8 accompanied by A. G. Brooks, a border patrol-
man, and Vernon M. Elwell, an employee of the regional office of
the National Reemployment Service in Alpine. He informed Cartledge that he was there "to investigate a number of complaints . . .
that his company had materially raised the price on various staple
articles . . . in their store," and requested a list showing both the
current prices of those articles, as well as those at the time of the
shutdown on September 1. At first Cartledge stated he would prepare the list, but later declined, claiming that "they had not increased their profits on anything." He then added: "You fellows
have just picked out the Chisos Company to give us the worst of it
and you are not doing anything with the balance of them." To
which Purcell replied: "You are wrong Mr. Cartledge. I am down
here to investigate all of the mining companies and have investigated several already on this trip."[52]

Failing to secure Cartledge's cooperation in the matter of store
prices, Purcell then asked permission to audit certain payroll records
that he had been told were not in accord with the compliance

---

[50] Perry to Cartledge, September 29, 1934.

[51] Smith to Martin, March 20, 1935, R. G. 9, National Archives.

[52] Purcell to Drought, November 10, 1934, R. G. 9, National Archives.
Purcell drafted two reports on the evening of November 9, but one is dated
November 10. One relates his encounter at the Chisos mine, the other with the
Brewster Quicksilver Consolidated staff at Study Butte.

agreement. Since these records were kept in the office of the mine superintendent, Cartledge referred Purcell to Frank Lewis. "He [Lewis] reluctantly, after considerable thought," Purcell reported, "handed me what he stated represented the payroll for the day of November 9th only." As Purcell was asking Lewis to explain an apparent discrepancy, Cartledge entered the office, interrupted Lewis, and addressed Purcell: "We will not bother explaining these details to you. We have told you that we are trying to comply with the Code. I wired Mr. Perry that you were here requesting information and he . . . [instructed] me to tell you that you have again overstepped your authority, and to refuse to show you any books whatever; and furthermore, to tell you to get off of this property."[53] Whereupon Purcell departed with his two escorts.

Of Drought's three directives to Purcell—take no action, make no demands, and take no outside parties—it was one down and two to go. But before Purcell left the district, he would card a perfect score, and his methods would hang him before the end of the year. The pendulum was swinging in Perry's direction.

Purcell committed his greatest indiscretion, not at Terlingua, but in his volatile confrontation with William Burcham and the Brewster Quicksilver Consolidated staff at Study Butte. After being refused access to company records by a bookkeeper named Bennett, Purcell, "naturally became suspicious and immediately went to several miners' houses and asked them what hours they were required to work." On learning that they were paid $1.00 in company scrip for a nine-hour day, he confronted Burcham, who, according to Purcell, "said he could not pay them [his employees] in U.S. currency and that, therefore, he would immediately shut down all operations." Purcell then took matters in his own hands. Addressing an assemblage of the Mexican employees, he told them he would "go to Alpine immediately and try to arrange with the FERA [Federal Emergency Relief Administration] director there to provide them with necessary relief."[54]

Still escorted by Brooks and Elwell, Purcell returned to Study Butte on the morning of November 9 with Mabel Hamilton, a case worker, and Nora B. Dibrell, an observation case worker, where he "arranged with the Mexican foreman to have all the miners congregate at one point . . . to furnish the case workers with the required information . . . and we left the Brewster mines at 1:42 P.M." Ap-

[53] Ibid.
[54] Ibid.

parently the executive assistant knew his actions of the last three days would be challenged by Drought, as well as the mining companies he had been sent to investigate. On returning to Alpine, Purcell had both reports—the Chisos report and the Brewster Quicksilver Consolidated report—signed by the persons accompanying him to the Terlingua district and notarized by Annie K. Turney, the Alpine postmistress. Elwell stated that he accompanied "the writer from Alpine to Terlingua between November 7th and November 9th . . . and hereby subscribe to the truth and accuracy of the occurrences and conversations."[55]

Truth alone was sufficient to bring the matter to a head. Drought's response was immediate. On November 14 he acknowledged receipt of the notarized reports but added that "the Chisos Mining Company has made certain representations to me, and the Brewster [Quicksilver] Consolidated has sent me affidavits by employees, which are in conflict with your reports. These companies charge that they are being made the victims of persecution and that your reports are false." Drought explains further that while his attitude toward the companies remains unchanged, it is "incumbent on me to order another investigation by men against whom the charges have not been made." The state director explains that he is "asking Hal J. Wright to visit these mines and to investigate thoroughly and report to me." Although copies of the complaints to Drought have not been preserved, evidently race relations lay at the crux of the conflict. Drought continues: "I am also asking Alfone Newton, Jr., to accompany him [Wright]. The latter . . . is not only partly of Spanish extraction, but as chief deputy and sheriff of this county for many years, he has dealt with Mexicans of all types and knows them as well as they can be known." The urgency of the matter is contained in Drought's concluding sentences: "Wright and Newton will go to the mines fully advised of all of the facts and circumstances which you have reported to me. I expect them to leave today."[56]

Purcell's departure from Study Butte on November 9 marks the end of his NRA field work in the Terlingua district. The convergence of circumstances had so willed. While Perry had restricted his effectiveness by forbidding company cooperation, Purcell had further ostracized himself by assuming a "personal interest in the matter . . . [and since] his attitude was not conducive to compliance

[55] Ibid.
[56] Drought to Purcell, November 14, 1934, R. G. 9, National Archives.

but was rather that of enforcement, with the result that relations between the mine owners and Mr. Purcell were always strained."[57] Drought had obviously attempted to be objective in staff matters concerning the executive assistant; however, Purcell's mishandling of the quicksilver code compliance matter had affected their personal relationship. Purcell later complained to L. J. Martin, chief of the Compliance Division in Washington, that Drought had not fulfilled his commitment to report to him fully on Wright's and Newton's reinvestigation: "The State Director promised me . . . that he would furnish me a complete report . . . but the only report that I have ever had . . . is a brief paragraph . . . advising me that these two men had returned to San Antonio, and had reported to him verbally that 'they had done a good job, and that everything was O. K. again.' "[58]

Drought's consideration for Purcell possibly accounts for his failure to fulfill his promise. The apparent success of Wright and Newton further highlights the extent of Purcell's failure. Recalling the incident, Robert Smith writes:

> When they reached the Brewster County mines they had immediate access to the records of the Chisos Mining Company and found no instance of non-compliance. When they visited the Brewster Quicksilver Consolidated mines, they found the workers on relief, the mine closed, with considerable damage to the mine because the mines operate below water level, and with the pumps stopped the mines were filling. . . . They did find that employees of mines other than the Brewster Quicksilver Consolidated were paid in cash every day; that the Brewster Quicksilver Consolidated settled with its men once each week. . . . However, as a result of the visit of Messrs. Wright and Newton, the mine was re-opened, the employees taken off relief rolls, and the mine instituted a policy of paying employees daily in cash.[59]

For all practical purposes this ended Perry's conflict with the NRA. But Purcell, refusing to allow the issue to die, became more critical of Drought following the state director's departure from the NRA. Still resenting being replaced by Wright and Newton in the quicksilver investigation, Purcell wrote Compliance Chief L. J. Martin in Washington insisting that the NRA should take "some action as might be necessary to correct this situation, which, in my

[57] Smith to Martin, March 20, 1935, R. G. 9, National Archives.
[58] Purcell to Martin, March 5, 1935, R. G. 9, National Archives.
[59] Smith to Martin, March 20, 1935, R. G. 9, National Archives.

opinion, is untenable from a compliance standpoint."[60] Before expediting the matter, J. R. Howell, Jr., executive assistant in the Houston NRA office, offered Drought the opportunity to make an additional report on the case. Drought, by then state director of the National Emergency Council, declined, referring the matter instead to Robert J. Smith with the explanation that "I think that you can take care of this better than I can, for you can deal with it from an impersonal standpoint, while I might be unable to do so."[61] Smith's six-page defense of Drought's actions did nothing to salvage Purcell's deflated ego. Smith concludes: "The State Director was obviously justified in acting on the later findings of these two NRA officials whose investigation was made at a later date than Mr. Purcell's, and whose statements are equally trust-worthy. Mr. Purcell goes entirely beyond the scope of his duty when he criticizes the action of the State Director."[62]

The Terlingua district compliance issue continued to smoulder throughout the spring of 1935, and on April 30, Fred C. Rogers of the El Paso NRA office wrote Perry concerning the pending adjustment of complaints. "I understand that investigations of your mines have already been made," Rogers explains, "and it has been disclosed that you have been in violation of your code."[63] Perry, of course, countered by maintaining that "'we are complying and will continue to comply,'" which, according to Rogers, was "unsatisfactory in view of our instructions to make a re-investigation."[64] The charges did indeed appear damning: six- and seven-day weeks with nine-hour days, wages below minimum, no overtime, and "underground and above ground, working boys under fifteen years old."[65]

On May 16 and 17, Joseph R. Shannon, an NRA field representative, accompanied by Rogers and a Mrs. Acreman, proceeded to Terlingua to reinvestigate code violations at the Chisos mine, but "Mr. Purcell decided not to accompany us. . . . We spent the entire day of May 17, mostly at the Chisos Mining Company, the results

[60] Purcell to Martin, March 5, 1935, R. G. 9, National Archives.

[61] Drought to Smith, March 18, 1935, R. G. 9, National Archives.

[62] Smith to Martin, March 20, 1935, R. G. 9, National Archives.

[63] Fred C. Rogers to Howard E. Perry, April 30, 1935, R. G. 9, National Archives.

[64] Fred C. Rogers to E. L. Tutt, May 11, 1935, R. G. 9, National Archives

[65] Labor complaint to the Chisos Mining Company, April 30, 1935, R. G. 9, National Archives.

of our investigation being covered by affidavits made by Mr. Rogers and myself [Shannon]."[66]

Whatever code violations the investigating team discovered, the issues remained unresolved. If Purcell believed that by remaining in Alpine, Shannon's team could operate unencumbered by latent animosities, his optimism went unrewarded, for Perry was also conducting his private war with Purcell and the NRA on another battlefield. Although six months had passed since Purcell had departed the Terlingua district, Perry's hatred for both the man and the system still burned deep. "Hal [Browne, Perry's attorney] and I," Perry wrote Cartledge on May 21, "are looking to get Perfidious [Purcell] fired from the Service—I am also taking quite a hand to push along in Washington the Senate move or bill to wipe out the NRA in nine months."[67] Fate was now with Perry, for one week later, on May 27, 1935, the Supreme Court by unanimous decision declared Section 3, Title I of the National Recovery Administration unconstitutional. When Cartledge gathered the import of the decision, he asked the jubilant mine owner, "Since the NRA is now dead should not we work more hours?"[68]

The NRA had come and gone, and neither Terlingua nor the American nation would ever be the same. While the program's total impact on the economy in particular and society in general would long be debated, the "festival of experimentation" altered the economic relationship between business and government, so that in its wake came vast changes in the total framework of American society. If remoteness helped cushion the impact on the quicksilver district, this experience dramatized more than any other single event that improved communications and transportation had at last drawn this frontier settlement into the mainstream of national experience.

Nevertheless, Perry remained master of his domain. Even though he had been forced to comply partially with the dictates of the self-imposed quicksilver code, many aspects of the Terlingua NRA experience had strikingly similar parallels in other areas of the nation. Compliance and enforcement were national problems and the NRA, according to Arthur M. Schlesinger, Jr., "had not always succeeded in holding people to its social objectives in wages,

[66] Joseph R. Shannon to E. L. Tutt, May 25, 1935, R. G. 9, National Archives.
[67] Perry to Cartledge, May 21, 1935.
[68] Cartledge to Perry, June 28, 1935.

hours, and labor organizations." Schlesinger explains further that a four-state survey made late in 1933 revealed that "codes were being violated, evaded, and ignored, and that NRA's apparatus of local compliance boards were lax and ineffectual."[69]

While the accusation that the business community exploited the huge reservoir of unemployed was also widespread, the exigencies of the emergency forced many people, both labor and management, to assume unbecoming roles. Whether one lived in a rural or urban environment in the mid-1930s meant little, as everyone everywhere grappled for survival. The rent-free dwellings that Perry provided his Mexican miners, who received less than $0.20 per hour, probably placed them in a more advantageous economic position than "a group of working girls in Chicago . . . [who were] toiling for less than twenty-five cents an hour," and some for less than ten cents. "Makers of ready-to-wear dresses, confectionery employees and cannery workers," according to Wecter, "were among the classes exploited most callously."[70]

The relative success or failure of the NRA programs varied with locale, and while NRA administrator Johnson claims his organization "did more to create employment than all other emergency agencies put together,"[71] Wecter maintains that because of "increased manufacturing costs and prices attributable to the NRA [it] slowed rather than speeded the [recovery] effort."[72] Perry would obviously stand by Wecter's statement, as code compliance increased his expenses while, according to Perry, the Chisos operation continued at a loss. "Mine labor increased in 1934 for two reason[s]," Cartledge reported in January, 1935. "Partly on account of raises in wages in the last three months of 1934 and partly on account of more extensive working at Mariposa."[73]

While the NRA actually helped generate employment in some industrial areas by shortening hours and consequently spreading work, this principle failed to apply to the Chisos Mining Company. Terlingua was a "labor island" with just so much work for just so many people. And as long as a cheap imported product flowed unrestricted into a depressed domestic market, the NRA brand of economic recovery remained a philosophical dream in which Perry never indulged.

[69] Schlesinger, Coming of the New Deal, pp. 120–121.
[70] Wecter, Great Depression, p. 18.
[71] Johnson, Blue Eagle, pp. ix–x.
[72] Wecter, Great Depression, p. 86.
[73] Cartledge to Perry, January 24, 1935.

The NRA left enduring achievements, but these lay in the areas of social rather than economic reforms. Explaining these changes, Schlesinger writes: "Here NRA accomplished a fantastic series of reforms, any one of which would have staggered the nation a few years earlier. It established the principle of maximum hours and minimum wages on a national basis. It abolished child labor. It dealt a fatal blow to sweatshops. It made collective bargaining a national policy and thereby transformed the position of organized labor."[74]

In the case of these major reforms Terlingua remained on the periphery of the Roosevelt administration's social legislation. Perry dictated a labor policy based on an ever-abundant supply of workers from south of the Rio Grande and regarded hunger as his greatest ally. "A hungry Mexican," he reveled to Cartledge, "is a good Mexican. . . . One of the good things about a depression is that labor can not huckleberry to us."[75] Schlesinger's reference to the demise of child labor also failed to apply to the Chisos. The passage of the Social Security Act on August 14, 1935, concerned Cartledge, as he reported to Perry that "we now have a good many boys working for us that are under age but who are filing for numbers under the Social Security Act."[76]

Of all the sweeping changes to emanate from the reforming zeal of the New Deal, the strengthening of organized labor had probably the most far-reaching impact on American society. In Terlingua it went virtually unnoticed. The distribution of skills and the cultural and ethnic structure of the labor force determined this failure, not Perry. The Anglo employees occupying administrative and supervisory posts who could comprehend the provisions of labor legislation were excluded from the benefits, and those to whom union protection was being extended were, because of the language barrier, probably unaware of its existence. But the social legislation of the 1930s was not spent uselessly on Terlingua's citizens. There were lasting benefits. If teenage boys were aware of social security benefits, so must have been their parents. And whatever was his motivation and his method, Purcell brought awareness to the Mexican employees of the Terlingua district that "the forgotten man at the bottom of the economic pyramid"[77] had not been abandoned during some of the nation's darkest hours. Emphasizing the psychological reward of the NRA programs, Schlesinger states that the

---

[74] Schlesinger, *Coming of the New Deal*, p. 174.
[75] Perry to Cartledge, February 7, 1937.
[76] Cartledge to Perry, December 1, 1936.
[77] Wecter, *Great Depression*, p. 84.

"Blue Eagle campaign changed the popular mood from despair to affirmation and activity. . . . 'It was,' said [Secretary of Labor] Frances Perkins, 'as though the community rose from the dead.' "[78]

Although the high floating clouds in the Big Bend sky may have acquired the proverbial silver lining, the lifestyle in Terlingua continued much the same following the demise of the NRA. Employees' raises effected under the emergency program remained, but Perry was already planning ways to retrieve the cash for the company coffers. Some time late in 1936 he erected adjacent to the company store the Chisos Theatre, a motion picture facility, and the Oasis, a confectionery shop, which began siphoning off some of the excess cash that remained in the village.

While with the passing of the NRA Perry had lost his once great antagonist—Purcell—he had no problem in filling the mental void formerly occupied by the executive assistant. Perry had defeated Purcell, the Supreme Court had annihilated the NRA, and the master of the Chisos mine had both won and lost his skirmish with bureaucracy. But Perry's greatest battle still lay ahead: his battle for survival.

[78] Schlesinger, *Coming of the New Deal*, p. 175.

CHAPTER 11

Events That Hastened the Fall

AS Howard E. Perry unknowingly approached the twilight of his great quicksilver bonanza, he was beset with a multiplicity of problems other than encroaching government bureaucracy: declining recovery, a nonproducing silver mine at Sierra Blanca, a spurious gold mining venture in Canada, continuing administrative blunders, failure of obsolete equipment, lawsuits, and ever-constant hounding by a growing list of creditors.

Whether Perry actually comprehended the seriousness of his financial condition until the very end is debatable, because throughout a decade of decline he maintained the image of a successful absentee mine owner. As production dwindled and expenses mounted, he borrowed huge sums of money to continue his search for another bonanza. Meanwhile Perry remained the same self-confident, arrogant extrovert who had first descended on the Big Bend nearly forty years before. Former associates recall that to dramatize his financial prowess during this period, he carried large sums of money and delighted in confusing waitresses by offering a $1,000 bill in payment for a cup of coffee. When the waitresses failed to have the change, he would reach back in his pocket and remove a large roll of currency and peel off another bill with the same results. Finally, after displaying various denominations of currency, he would ask to borrow a nickel to pay for his coffee.[1]

Whatever delight Perry garnered from these episodes, it failed to compensate for the mounting problems that continued to plague the Chisos Mining Company. Concurrent depressions in the quicksilver market and Chisos recovery were paramount issues during the early 1930s. Between 1929 and 1931, when the market price for quicksilver dropped from $122.15 to $87.35 per flask,[2] Chisos production dwindled from 2,099 flasks to virtually nothing as the com-

---

[1] Interview with Hunter Metcalf, Marfa, Texas, June 6, 1966.
[2] U.S. Department of the Interior, *Minerals Yearbook, 1935*, pp. 713–718.

pany remained closed most of 1931 for furnace modifications.[3] Although the market continued its descent, reaching $57.93 per flask in 1932 and $59.23 the following year,[4] the news from Terlingua was encouraging as the company reported an annual recovery of 1,711 flasks and 1,178 flasks respectively.[5] Most of this came, however, from surface workings at the Mariposa property, which Perry had acquired in 1928.[6]

In 1934, as the quicksilver market rallied to $73.87 per flask,[7] the company failed to report any recovery. During this lull, however, Perry engaged A. R. Fletcher, a geological consultant who had gained a reputation as an expert in tracing metal deposits in the limestone districts of Mexico.[8] This proved to be a wise move; Fletcher laid out a program of work at Mariposa that led to the discovery of a major ore body the following year. The 3,243 flasks the company reported in 1935 mark the company's peak for the decade. From this point forward production records chart the Chisos mine's decline to eventual disaster.

To whatever extent Perry's decisions were directed by his blind faith in the company's future, he must also have realized that economic survival necessitated acquiring production in areas other than the Terlingua district. While his motivation may have been valid, his decisions were less than remunerative. After acquiring the Mariposa property he, in partnership with Thomas V. Skaggs of Lajitas, leased the Bonanza silver mine near Sierra Blanca, which produced nothing and ended as another costly fiasco.

When in 1935 Perry learned of a new French geophysical method being employed in Canada to locate ore bodies, he began a series of negotiations in which he traded his idle Sierra Blanca equipment for stock in the Stanley mine, a Canadian gold mine. Approaching this encounter as a shrewd Yankee trader hoping to fleece some unsuspecting foreign businessmen, he wrote Cartledge:

[3] J. Harlan Johnson, "A History of Mercury Mining in the Terlingua District of Texas," *The Mines Magazine* 37 (July 1947): 21. A producer gas burner was installed in an attempt to increase efficiency.

[4] U.S. Department of Interior, *Minerals Yearbook, 1935*, p. 440.

[5] Johnson, "Mercury Mining," 37 (July 1947): 21.

[6] Perry originally operated the property as the Texas-Mariposa Mining Company, but when in 1936 the property was transferred to the Chisos Mining Company, the corporation was dissolved (transcript of Testimony, *Cause No. 688, Watson-Anderson Grocery Company et al. vs. The Chisos Mining Company*, Federal Records Center, Fort Worth, Texas).

[7] U.S. Department of Interior, *Minerals Yearbook, 1935*, p. 440.

[8] Johnson, "Mercury Mining," 37 (July 1947): 21.

"I have three mining men coming here [Portland] Tuesday night from Canada. They are after the Sierra Blanca mill. I now expect to take them on the boat and sail to New York Wednesday. . . . I hope to be able to throw enough poison into them on the boat so as to be able to make a good trade."[9]

Perry emerged from the negotiations believing he had made "a very fine trade. . . . I should be able to recoup all the losses made at Sierra Blanca." Although for the remainder of the year he waxed enthusiastic about the great potential of his Canadian venture, there appears in his early 1936 correspondence a hint of suspicion that he may have been the victim of his own poison. "I am not so dead sure," he wrote on January 30, 1936, "that this last delay has actually been caused by a small accident or whether it is something else. I will try to find out tomorrow if they are fooling with me." By June, however, Perry was again enthusiastic about his Canadian property. "Everything is all right here," he assured Cartledge, "and the mine is much better than I had supposed and the work is going fine. The mill will be running in two weeks."[10]

Cartledge responded by expressing his pleasure with his employer's apparent good fortune but continued his appeal for Perry to come to Terlingua to make an on-site appraisal of the Chisos operation. Almost pleading, Cartledge wrote: "I can only hope you can make lots of money out of it [the Stanley mine] and can either close this mine down or get that one running to where you can get back down here and dig in and meet true facts and just see where your money is going, that is, just how it is being spent."[11]

Ignoring Cartledge's petition, Perry departed for Mexico with Fletcher to study several Mexican mines working in limestone formations similar to those in the Terlingua district. While in Mexico City Perry received word that new problems had developed at the Canadian mine. Preferring to remain in Mexico, he dispatched a Portland secretary, Allen Stephens, to Montreal to handle the matter. Whatever the problem, it remained a continuing issue, as late in the year Perry finally discovered that instead of being the poisoner, he was indeed the poisoned. "I only learned a short time since," he admitted to Cartledge, "that those Kanucks are mostly snakes in the grass trying to bite me."[12]

Cartledge's response was characteristic. "I can only hope that

9 Perry to Cartledge, August 5, 1935.
10 Perry to Cartledge, August 24, 1935, January 30, June 12, 1936.
11 Cartledge to Perry, November 14, 1937.
12 Perry to Cartledge, November 14, 1937.

you find some way to pull out of that Stanley mine." Referring again to the drain on Perry's time and finances, he continued: "I can't help but feel that if it had not been for Sierra Blanca and the Stanley mine that this place would have never lost the money it has, as you would have been here to have looked after things. As it has been your time for [the] last few years, [that] has been occupied with trouble with these places."[13]

Perry had apparently passed the point of no return at all three sites—Sierra Blanca, the Stanley mine, and the Chisos mine—but, still driven by his inherent optimism, he finally admitted that "I may decide to let Stanley go into bankruptcy in order to clear out the snakes and reorganize it, and get rid of some very bad acting stock holders. The mine itself is OK."[14] His New York attorney disagreed, recalling, "In his [Perry's] later years he tried to recoup his quicksilver losses in unwise Canadian mining adventures about which he knew nothing until he had gotten into trouble."[15]

While still engrossed in the Stanley fiasco, Perry continued to study the Chisos situation from afar, even traveling to Tucson to see a Dr. G. M. Butler at the Arizona School of Mines, "who agreed to come help us find our lost ore at the 600 level. He has a great reputation for finding lost lodes of ore."[16] At Terlingua Butler failed, however, to maintain his reputation, as 1936 production fell to 1,225 flasks of quicksilver.[17] In the face of steadily declining production, and physically, emotionally, and financially drained, Perry decided to close temporarily the Chisos mine. Although the correspondence relating to this event is voluminous, Perry's motives are uncertain. Although closing the mine would reduce payroll expenditures, his decision appears to be directed primarily at disposing of some of the professional mining staff whom he apparently held in low esteem. Explaining his plan to his Alpine attorney, Wigfall W. Van Sickle, Perry wrote:

> Well, here is something which I have to tell you very confidential for the present. . . . We have been struggling for more than a year for ore at the mine. The mine has lost a lot of money and I have sweat blood, but no one at the mine has sweat blood with the exception of Robert [Cartledge] and [E. J.] Dahlgren and [J. E.]

[13] Cartledge to Perry, December 12, 1937.

[14] Perry to Cartledge, December 9, 1937.

[15] James B. Burke, New York, New York, to Kenneth B. Ragsdale, September 28, 1965, in possession of author.

[16] Perry to Cartledge, June 13, 1936.

[17] Johnson, "Mercury Mining," 37 (July 1947): 22.

Anderau . . . but not [Frank E.] Lewis [mining superintendent] or any of the strictly mining crowd. These will all soon be out but I will need Robert, Dahlgren and Anderau and a few others. I now think that I will telegraph orders to close the mine Saturday night.[18]

Apparently this decision weighed heavily on Perry's mind, as he ordered the closing with mixed feelings. While he shows vindictiveness in his pleasure with Lewis' and "the strictly mining crowd's" impending ouster, the closing of the Chisos mine tugged deep on Perry's emotions. His sojourn in the Big Bend had been a thirty-six-year-long love affair with the Chisos property. It was his mine; he had created it, he believed, "building it from the grass roots."[19] But even after making his decision, Perry's overpowering optimism enabled him to envision a return to the days of glory with a new strike of rich ore and a stronger market. With great sincerity and deep personal expression, Perry continues:

I am filled with grief over this thing but I am not thinking of men like Lewis and some of our other mining men. They have all had good pay and pretty easy time, and have indulged themselves in the thought that the mine would go on forever regardless of whether it had ore or whether I lost money. . . . I have walked the floor of my house many a night until the morning light came in the window trying to figure out how to get ore. Do you think any mining man of mine ever walked the floor trying to get ore: . . .

The mine has made a wonderful run in the thirty six years of its life and will run again but it has to rest a little now until I get better acquainted with the mine structure. . . . We just have to wait for a better day. The thing has been wearing me out for quite a long time but it will wear me out no longer. I will have rest. For however much I love my old mine I love myself better. The mine is taking too much out of me, health and strength and money and Mrs. Perry has become somewhat worried. Well, this is a sad tale. It is indeed all too bad. It can't be helped.[20]

A brisk exchange of coded telegrams set up the procedures for terminating production, which by this time was virtually involuntary. The Chisos mine had approached zero yield. Between "January, 1937, and August 31, 1937, the mine produced only 429 flasks of mercury. The August furnace record contains a grim reminder of what was to come. 'Cut tonage ore bins empty' was the simple hand-

[18] Howard E. Perry to Wigfall W. Van Sickle, October 4, 1937, typescript in possession of author.
[19] Transcript of Testimony, *Cause No. 688.*
[20] Perry to Van Sickle, October 4, 1937.

written note."[21] Late in the afternoon of October 19, 1937, a furnace attendant started the fire in the retort and at seven o'clock that evening Dahlgren closed the old Scott furnace and the cleanup operations began.

Van Sickle gave the story to the Alpine *Avalanche*, which reports:

> Production at the famous Chisos Mining Company's mines at Terlingua was temporarily suspended the first of the week awaiting development of new ore bodies, it announced here Wednesday. . . . The mine has been operating at a loss for a number of months . . . [and] the step in suspending production was greatly regretted by the management. Approximately 125 men are employed at the mine, some of whom will be retained in the development work which is to be carried on. . . . The commissary will continue operation during the temporary shut-down.[22]

Although the *Avalanche*—following Van Sickle's news handout—reported the temporary suspension of operations pending the "development of new ore bodies," this was largely Perry's wishful thinking. For all practical purposes, the closing of the furnace on October 19, 1937, pealed the death knell of Perry's once great quicksilver bonanza. A skeleton crew proceeded with cleanup operations, recovering 203 flasks of quicksilver from the furnace condensers while 142 flasks were drawn from the retort. After these 345 flasks were shipped from Terlingua in November, 1937, the remaining Chisos production was negligible.

Blind optimism and frenzied determination apparently denied Perry the vision to realize that he was at the final crossroads of his industrial career. After a decade of failing production—with the exception of 3,243 flasks of quicksilver produced in 1935—common logic, as well as the advice of geologists Butler and Fletcher, should have convinced him that the Chisos resources were exhausted. But emotions apparently conquered reason and logic, and Perry, despite his self-admitted weariness, embarked once again on a frantic search for a new bonanza. The obstacles were staggering.

First, news of the closing attracted the attention of the Chisos' creditors and Perry sent Cartledge $5,800, hoping to maintain a credit flow until new ore could be located. Cartledge responded that he was "going to clean up as much as posible all our small accounts

[21] James M. Day, "The Chisos Quicksilver Bonanza in the Big Bend of Texas," *The Southwestern Historical Quarterly* 64 (April 1961): 445.
[22] Alpine *Avalanche*, October 27, 1937.

with this deposit. Even trying to give all creditors a little on account and tell them that we have closed down to re-organize and that as soon as we get to it, we will start paying on our accounts again."[23] This was, however, a delaying tactic that would be repeated in response to a growing crescendo of creditor complaints.

Perry's greatest problem was, of course, finding productive ore. While diamond drill crews worked around the clock at the Chisos mine, Perry reopened the Mariposa property with some of the staff left idle by the Chisos closing. By November, 1937, the Mariposa prospect shed new rays of hope for the little mine owner, as Fletcher estimated one program of work would yield $80,000 in quicksilver. While pleased to convey this good news, Cartledge also continued to remind Perry of the mounting expenses in the face of no immediate income. On December 14 he reported that "the ore at Mariposa is looking very good. . . . There is little ore on the 300 ft. level of the old workings of Mariposa. . . . [However] we still have too much expense here. . . . I still have on four watchmen but feel that we should cut them down. . . . We have had Alejandro Chacon running the hoist and I find that he really just hanging around up at No. 8."[24]

At this juncture Perry only had eyes for prospecting reports and production records. Rich ore would resolve budgetary problems with the first fire in the old Scott furnace. Still ignoring Cartledge's preoccupation with expenses, Perry turned instead to the Mariposa geology, which now promised renewed activity in the district. His appraisal of the structures reveal the knowledge of a mine owner who had done his homework well. Perry was no dilettante and with almost four decades of tutoring by Phillips, Udden, and Fletcher, he knew well the secrets of the limestone "flags and marls of the Upper Cretaceous." Accordingly, he analyzed the Mariposa prospects: "You probably know that in Mariposa drilling from the bottom of the contrairio that we went through the Edwards 153 [feet] deep, running into a front horizon of ground. I think this soft horizon comes up pretty high in the Edwards in all places on that section where we got a good deal of ore down around 250 or 300 feet deep or even less. This condition may have to do with bringing in Mariposa."[25]

Reports from Terlingua continued to be encouraging and as the Chisos bins began to fill with low-grade Mariposa ore, Perry became impatient for production. He ordered the Chisos furnace restarted

[23] Cartledge to Perry, October 27, 1937.
[24] Cartledge to Perry, December 14, 1937.
[25] Perry to Cartledge, December 27, 1937.

on April 1, 1938, but the results were disappointing. Furnace records show forty-two flasks recovered in June and twenty-eight in July. On July 26 Cartledge reported, "I was ashamed to wire. They only made one flask free run last week." When the August report was equally dismal Perry ordered thirty-five men dropped from the payroll. He had been able to sell only two flasks of quicksilver in six weeks. By September it appeared questionable whether Mariposa was even paying its way, and when Perry saw the October reports, all doubts vanished. Again he was forced to accept the inevitable. "There is no question but what Mariposa is losing money," Cartledge reported, "I am closing it down tomorrow unless you wire to the contrary."[26]

The failure at Mariposa, steadily increasing financial losses, and a virtually closed quicksilver market, all coming in the wake of a sharp recession in the nation's already depressed economy, failed to dampen Perry's zeal for finding a new bonanza. Once again he turned his attention to the adjacent Rainbow property. From Cartledge he learned in January, 1938, that the Rainbow mine was surfacing good ore, and following his failures at Mariposa and at the Chisos, a new venture began to form in Perry's mind. This plan gained momentum when the information reached him that two of the owners of the Rainbow mine would sell their controlling interest in that property. With characteristic enthusiasm he envisioned that with the Rainbow acquisition he could recoup his losses and regain his waning prestige in the district. Appealing again to Eugene Cartledge in Austin to arrange an immediate loan from the Austin National Bank, Perry wrote:

> Seeing that you once got me $75,000.00 from the Austin National Bank to get me out of the claws of those Devils [referring to the Rainbow invasion suit settlement], I think I can now obtain possession of the claws themselves and be rid of them forever. . . . I have been offered the holdings of Starr and Armstrong for $12,500.00 and this will give stock control and the other fellows can be gotten out very cheaply when they learn that control has passed to me.
> Please go to the bank and get me $12,500.00 on demand note, which I am sending you. I will pay it when ever they call and whatever interest they want. . . . We must not miss this great opportunity. The quicksilver market has been getting in better shape for some little time and when the Spanish war is over, the price will soar. I

[26] Cartledge to Perry, July 26, November 4, 1938.

think sooner or later Big Bend [the Study Butte mines] will fall into my hands and then we will control the entire district.[27]

Cartledge responded to Perry's request. Three days later, on May 16, 1938, the Austin National Bank credited Perry's account with $12,500.

In Perry's eagerness to acquire the property he failed to order an inspection of the Rainbow equipment. Therein, he sadly discovered, was the owner's probable reason for selling. While some good ore deposits were blocked out, the furnace needed extensive repairs. This forced Perry to close the furnace during most of October and November, which created another untimely drain on his and the Chisos' funds. Recovery never equalled production quotas and on November 25, Cartledge complained to Perry, "Financing Rainbow out of the deposit you send for Chisos is getting me further and further behind of the Chisos accounts."[28] Cartledge's warning still fell on deaf ears.

Late in November the Rainbow furnace was again fired but the results were disappointing. "Three flasks for the first week," Cartledge calculated, "seemed to me to be too little."[29] Indeed it was. Instead of the Rainbow property helping maintain the Chisos mine, the Chisos and Perry were carrying the Rainbow. Again Cartledge, more than anyone else, saw the warning signs of disaster but still no one paid heed. This time he wrote to Beulah Patterson, Perry's secretary in Portland, to report the sobering facts. "After crediting Rainbow Mines with the $1,250.00, they still owed us [the Chisos] as of December 1, $2,432.00. This balance does not include December purchases from Chisos which are heavy."[30] The Rainbow account exceeded $3,000.00 in December and by February 3, 1939, the Rainbow account owed the Chisos mine $3,828.33.

February production, however, gave Perry renewed hope, as the Rainbow mine produced fifty flasks of quicksilver. This figure was repeated in March, but one month later recovery dropped to twenty-nine flasks. Although some good ore was being surfaced, cracks in the condensing system of the hastily repaired installation prevented complete recovery. After Cartledge inspected the ailing structure and studied the slag tests, his suspicions were recon-

[27] Perry to Cartledge, May 13, 1938.
[28] Cartledge to Perry, November 25, 1938.
[29] Cartledge to Perry, November 29, 1938.
[30] Cartledge to Beulah Patterson, December 6, 1938.

firmed: the company was losing precious metal because of low furnace efficiency. While he recommended closing the furnace and making additional repairs, Perry refused, insisting that the company could not afford to have the plant stopped at such a critical juncture. The economy was false. Though the company continued to encounter sporadic pockets of rich ore, the great bonanza that Perry so frantically sought was not forthcoming. "I am sorry," Cartledge wrote, "but the Rainbow will not make the twenty flasks that I had hoped to make. . . . Also, it appears that Chisos is going to fall down considerable."[31]

With minimal recovery, mounting expenses, and Perry's dwindling finances being slowly siphoned off by the continued search for his lost ore, logic normally would have dictated a tightening of the administrative belt. This never occurred. From the opening of the Chisos mine, Perry never established any clearly defined lines of administrative authority and consequently no one at Terlingua ever exercised total control. The final decisions, which rested with Perry, were made only after negotiating with the different department heads; confusion and misunderstandings resulted. During periods of peak production administrative inefficiency remained hidden in soaring profits, but when recovery dropped, internal confusion became glaringly apparent and hastened the final collapse.

Although organization could never salvage a mine that lacked ore reserves, by delegating authority to a competent administrator —as Cartledge apparently was—the company could have been operated more efficiently and economically. This tightening was never accomplished. Had Perry spent more time at Terlingua, conceivably he would have become aware of the futility of the effort and reacted accordingly. Perry, however, chose neither of the two alternatives, preferring the life of a New England yachtsman to that of a Big Bend miner. All the while his blind confidence in the once great mine drove him forward in search of his lost ore.

For nearly twenty years Cartledge endured this administrative hodgepodge, while warning Perry of its dangers and requesting that he come to Terlingua, examine the facts, and articulate definite operational procedures and objectives. Perry ignored Cartledge's words of caution with tacit admission that all was not right in Terlingua. "Yes, I know a good many matters have been rocking along,"[32] was Perry's typical verbal procrastination. He seemed to feel that if left alone, all problems would eventually correct them-

[31] Cartledge to Perry, January 11, February 29, 1940.
[32] Perry to Cartledge, March 16, 1936.

selves, personal conflicts included. When a power struggle developed between an employee named Fulton and J. E. Anderau, Perry attempted to soothe the ruffled feelings from over 2,000 miles away. Urging Fulton with a childlike rationale to reconcile his differences with Anderau "with a good drink," Perry writes:

> I understand that you and Anderau are not unfriendly to each other notwithstanding all that has happened, and this is fine. I can't run my business with fights going on. . . . Men just have to get along with each other. . . . When you are over to Chisos go down and see Anderau and have a drink with him, for you both dearly love a good drink, and so do I, too. And when I get down there we will all of us have a drink together, many of them.[33]

Perry's pleas for esprit de corps failed to stir anyone, especially Cartledge, for although he was general manager, his was a vacant title. Authority was fragmented between four departments—furnace (Dahlgren), mechanical (Anderau), mining (Lewis), and office and general manager (Cartledge)—and each department head corresponded directly with Perry, who in turn issued orders to each, sometimes with and sometimes without Cartledge's knowledge. The irony of this planned confusion was that Perry probably believed that Cartledge was vested with the authority to administer the operation. Yet while explaining to him his farcical administrative web, Perry cites the pitfalls that foreordained its failure:

> Our hook up at the mine is just exactly as I told you. You are superintendent of all our mining operations at Chisos and Mariposa and really of Rainbow likewise. Merced is superintendent of underground operations at Chisos and above everyone else in the mine, tho Joe has working charge of #16. Joe is assistant to Merced at Chisos, but Joe is superintendent or mine boss at Mariposa and Merced has no connection what ever at Mariposa. . . . Bob Cory really has no authority whatever except *delegated from either you or myself* [emphasis added]. He is . . . rather grasping for authority and you may have to knock him on the head once in a while, but then smile at him and he will be all right. . . . Joe is very grasping for authority also. In fact he tried . . . at Rainbow but I had to knock him on the head. I then smiled at him and he was all right.[34]

Whatever hopes Perry may have entertained for the success of this disarrangement, they were never realized. Personal conflicts increased, efficiency dropped, the deficit mounted, and the hastening

[33] Perry to Fulton [copy to Cartledge], February 12, 1938.
[34] Perry to Cartledge, September 2, 1938.

twilight of the great Chisos quicksilver bonanza drew close.

Throughout the 1930s failure of obsolete equipment also compounded Perry's troubles, much of it the result of neglect and deficit operations. Fletcher described the Chisos facility as being poorly equipped, noting particularly that "all of the electrical equipment was direct machinery, which is no longer used in mining ventures properly equipped of that type."[35] The ancient equipment not only resulted in inefficient operations but frequently caused personal injury to company employees. Cartledge describes one accident as follows:

> Fulton had his air receiver to blow up at Mariposa. . . . The explosion blew Fulton and Tomas Vega a good distance away from the receiver. Fulton is alright and working. However Tomas Vega had several cuts besides his body being plastered with small rocks. Just below his shoulder blade he has a deep whole that looks as it was made [by] a rivet that goes to his ribs. However if no infection sets in he should be over it alright. It wrecked the engine house, which we have had to build over.[36]

While the air receiver blow-up resulted in personal injury, no litigation resulted. Perry was not always so fortunate. The Chisos accident report is a tragedy-laden document of casual disregard for human welfare, and when some of the more horrendous cases attracted outside concern, Perry was forced to resolve the issue in court. The following verbatim excerpts span the twelve-year period from October 27, 1925, to the air receiver explosion on January 26, 1938.

Date	Employee's Number, Name, & Remarks	Days Off
10/27/25	#492, Elijio Ramirez. Rope broke going up high winze in #8. Badly bruised in several places but no bones broken.	48
1/26/26	#128, Jose Rivas. Fell down a steep rise at 650 level in #8. Badly bruised and shaken up but no bones broken.	23
4/28/26	#529, Otanacio Salas. Hit by fall of rock in #9 and badly injured, skull cracked at back of head.	
8/11/26	Sent out to hospital in San Antonio	

---

[35] Transcript of Testimony, *Cause No. 688.*
[36] Cartledge to Perry, January 28, 1938.

8/20/26	Returned, no operation performed	
1/12/27	Died	
1/30/28	#794, Brigido Gallegos. Rope broke in 150 . . . at 1 shaft and he fell about 20 feet—Bruised up but no bones broken. Partially paralized—worked intermittently but was never normal physically—back and legs affected.	4½ mons.
5/8/28	#806, Francisco Hurtado. Coming out of mine at 3 shaft in afternoon—Fell from crosshead at about 50 level—Fell to 750. Dead.	
2/18/30	Abran Avila. Fell 18 to 20 ft. trying to catch companions after his lamp stopped burning.	Off 5 wks.
7/23/30	Inez Perez. Starting Eng. No. 8, foot slipped and caught him in fly wheel, leg broken (both bones) about 3″ above ankle. Leg set short. Returned to work on hoist.	6 mo., 1 da. off
3/14/32	#101, Petronillo Hernandez. Got too cold while sorting and fell down slide cutting head and hand. Not allowed ½ time by Supt.	
6/10/32	#755, Luis Garcia. Stepped on nail while working. Not allowed ½ time because he reported stomach trouble to Dr. instead of the foot.	10
5/10/33	S[C]ipriano Hernandez. Working at Coal Mine. Was barring down when rock loosened, hit him in face, bones of face broken. Nose, jaw, and teeth broken. Hurt bad. Taken to Alpine by R. L. Cartledge same day. Died on way to town. Brought back here.	
2/9/34	#949, Roman Ramirez. Working in 350, #8, fell down winze to 450' level. Back broken, jaw bones, and whole back of head broken. Dead.	
7/12/34	Jack Dawson [Mariposa boss]. Timbering	9 wks.

shaft. Fell off suspended platform. Fell about 25'. Injured head, chest, hand, back, and leg. No bones broken. Backbone dislocated, one rib broken at backbone.

1/3/35	#492, Eligio Ramirez. Moving motor for ventilation fan, Weighed about 500 lbs. Dropped it on his toe. Half-time stopped by Dr. 2/5/35. Herded goats until 2/13/35 when he returned.	1 month
10/16/35	#864, Juan Rodriguez. Tramming on 600. Door standard (pointed rod) pushed through foot at air hoist on 600.	1 month
11/13/35	W. W. Kenyon. Cut on head, two teeth loosened and chipped slightly. Water swivel came loose and fell on his head. Taken to Dr. Miller, who drew cut together with adhesive tape.	7
4/9/36	Carlos Suchi. (Mariposa man) Drunk Sunday night. Pants soaked with gasoline. Burned legs when he started to light a lamp. Not given half-time as the accident had no connection with his work (was stealing gasoline from compressor room).	
3/23/37	Juan Ramos. Working at Coal Mine. Priming engine, back fire, hand badly burned. Chest and front of body from base of neck to belly button burned.	18
7/—/37	#1070, ——— Vasquez. #9 tramming. Cut on foot. Tramming barefoot—stept on sharp rock. Both feet developed boils and became infected—due not washing his feet.	
9/18/37	Eulalio Rodriguez. Hurt right foot—due negligence infection set in later—was working in 740 in hot water apparently causes boil-like infections.	
11/23/37	#658, Mauricio Castro. Night Shift, Mariposa. Running air hoist, 601 tunnel winze about 9 p.m., he evidently started to cross the winze to pull the dumper door up to let the empty bucket down. Apparently	

slipped, fell to bench above bottom of winze. Fell about 25 feet—landed on head—(side near the right eye.) Was apparently killed instantly.

1/26/38   Tomas Vega. Running Sierra Blanca Engine, Mariposa. Air Receiver blew up. Badly bruised, small rock peppering all over his back. Slightly cut head, several slight cuts on body. Worst injury deep cut just under shoulder blade that penetrated to bone. (Paid full time.)[37]

The Hernandez coal mine accident (May 10, 1933) is a blatant example of the company's disregard for employee safety. When the Hernandez family sued the Chisos Mining Company for $30,000 in damages, Perry asked Rainbow mine vice-president C. T. Armstrong to testify to the installation's safety. Armstrong took one look and pronounced the coal mine "a disaster area. I've never seen such sloppy timbering job in my life," he recalls. "It is a miracle that only one man was killed there. The formation he [Hernandez] was working in was soft as clay. I told him [Perry] he wouldn't want to hear my testimony. That ended it."[38]

Armstrong's reluctance to testify failed to dampen Perry's optimism for a favorable court verdict. "In any lawsuit," Perry rationalized, "a dead man is not worth much anyway. . . . When a plaintiff is dead the court loses . . . interest. Judge Van Sickle is always good and he ought to succeed in this work."[39] The judge's success was only marginal, as the Brewster County jury decided in favor of the plaintiff and awarded the Hernandez family $2,500. Perry, however, appealed the case to the Texas Court of Criminal Appeals, which affirmed the decision but reduced the payment.[40] This was a bitter victory for the Hernandez family. After the death of a son and almost four years of litigation they received only $1,400. Perry was elated. "We certainly had a good day in court," he writes; "I hated, of course, to pay out the $1,400."[41]

Perry was financially more successful in the death of employee number 658, Mauricio Castro (November 23, 1937), as Cartledge

[37] Chisos Mining Company Accident Report. Original in possession of Peter Koch, Alpine, Texas; typescript in possession of author.
[38] Interview with C. T. Armstrong, San Antonio, Texas, April 3, 1968.
[39] Perry to Cartledge, September 3, 1937.
[40] Southwest Reporter, 2nd Series, XCVI; 292–297.
[41] Perry to Cartledge, February 18, 1937.

interceded in the company's behalf. Explaining the negotiations to Perry, Cartledge writes:

> I offered to pay the widow $360.00 to be paid in installments of $30.00 per month. I might have to go a little higher or pay them a lump sum in cash. It would be easy enough to have made a settlement with his widow, as young and ignorant as she is but she has her brothers-in-law that are advising her or at least feel that they should have something to say about the settlement. . . . Though I might not be able to settle on my terms, I am enclined to beleive that we will be able to make some kind of settlement with them as two of Mauricio['s] brothers want work from us.[42]

As litigation became a way of life with Perry, the increasing frequency with which suits were filed in the 1930s prompted him to remark, "I am up to my neck in fights—I love them—when I know I am right."[43] But Perry always knew he was right and engaged attorneys in New York, Austin, Del Rio, Alpine, El Paso, and Marfa to prove him so. Their efforts were not always rewarding. While most of the suits involved relatively small amounts of money, they were long and expensive, and settlements were usually negotiated after the case was appealed. According to Robert Cartledge, Perry's stubbornness resulted in court costs that far exceeded the money in question. Recalling the case when a company truck ran over two

[42] Cartledge to Perry, November 26, 1930.

[43] Perry to Cartledge, March 5, 1940. On December 8, 1937, Van Sickle wrote Perry (with a copy to Cartledge), "For your benefit and as a matter of history we have had the following cases but they are not named herein in their regular order."

No. 1141—*Ursulo Llanez* vs. *Chisos Mining Company*. Ask damages for $5,000 and got judgement for $1,400.

No. 1252—*M. B. Whitlock, et al.* vs. *Chisos Mining Company*.

No. 1519—*Guadalupe Hernandez, et al.* vs. *Chisos Mining Company*, July 25, 1931. This was a hard fought case and the defendant received an instructed verdict.

No. 1520—*R. C. Garnett, Temp. Admr., Estate of Cipriano Hernandez* vs. *Chisos Mining Company*, July 25, 1933.

No. 1670—*Alpine Mining Company* vs. *Chisos Mining Company, et al.*, June 13, 1936.

No. 1677—*J. J. Dawson* vs. *Chisos Mining Company, et al.*, July 7, 1936.

No. 1577—*Guadalupe Hernandez, et al.* vs. *Chisos Mining Company*, November 28, 1934.

No. 1668—*Roy A. Downey, Admr. of Cipriano Hernandez Estate* vs. *Chisos Mining Company*, April 11, 1936.

No. 1698—*Fernando Valenzuela* vs. *Chisos Mining Company*, November 18, 1936.

children at Terlingua, killing one and seriously injuring the other, Cartledge explained that "Perry spent more money defending the case than it would have cost to settle out of court."[44] Mae Ament, who represented Perry in the Big Bend after Van Sickle's death on September 14, 1941, recalls: "I couldn't get him to pay his debts. He was just that stubborn. He wouldn't do anything he didn't want to. Just take the Chicago Pneumatic Case, it only involved $635.25 and we appealed it to the Supreme Court of Texas. Just goes to show he would go to any length if he believed he was right."[45]

The way he played the legal game, according to Perry, was more important than the points he posted on the board. When he lost the first case in the Chicago Pneumatic suit, he reassured Mrs. Ament: "It is really just too bad that we lost—but cheer up and we'll carry it up higher—even to the Supreme Court of the land. We do not owe that money—they will have to work *for it*—if they get it. You did a wonderful job in that brief and deserve success."[46] Then expressing his disgust with the plaintiff, he wrote Cartledge, "They are a lot of God Dam Mics, anyway."[47]

These peripheral matters, while meaningful to Perry, were expensive and time consuming and contributed nothing to the salvation of the Chisos Mining Company. As production dwindled, so did the labor force, and as life ebbed from the once great installation, periodic payroll records chart a steady course to bankruptcy. From September 7, 1934, to October 27, 1938, the number of employees steadily dropped from 150 to 20.[48]

In the face of mounting production problems and Cartledge's constant bombardment of what should have been sobering facts, Perry still found time to enjoy the pleasures of the international yacht racing set. But his persistent neglect of the business failed to dampen either Cartledge's dedication to duty or his devotion to his heedless employer. This is one of the tragic ironies of the Terlingua experience. Even as the self-sacrificing Cartledge neglected his family, alienated friends, offended colleagues, and endangered his life to serve the master of the Chisos mine, he continued to express concern for Perry's personal pleasures. Concluding one burden-filled letter, Cartledge wrote: "I heard last night . . . that you were in Newport [Rhode Island]. I only hope that you have a good time

---

[44] Interview with Robert L. Cartledge, Austin, Texas, October 17, 1965.
[45] Interview with Mae Ament, Alpine, Texas, June 6, 1966.
[46] Perry to Mae Ament, October 8, 1941, in Ament estate, Alpine, Texas.
[47] Perry to Cartledge, September 22, 1939.
[48] Payroll records, Chisos Mining Papers.

at the races."[49] Responding on his New York Yacht Club stationery, Perry reassured his votary employee, "The cup races are over—English fast and her skipper no good—not good enough."[50]

At times, however, when Perry appears to have grasped the severity of the company's financial condition, he overreacted. In dismissing a member of his dwindling staff without notice, he instructed Cartledge: "In paying off Campbell do not allow him any time for notice. It is just as well to let every man who works for us know that no advance notice will be given if we have to dispense with their services. . . . I like Campbell first rate, but being sick . . . we can't keep him. . . . I would not keep Jesus H. Christ there . . . [if] he couldn't or wouldn't make good."[51]

Conditions grew steadily worse. On August 26, 1938, Perry telegraphed Cartledge that "we have to remake our plans immediately about Chisos in keeping with our finances, therefore lay off 35 Mexicans tonight. . . . [Also] cut down on store purchases what you can . . . somewhat like last winter." Communications at the mine apparently broke down. For instance, when Perry saw the August payroll record, he was "surprised that we had on about 65 men whereas you [Cartledge] know the limit was 57 men."[52] Cartledge's explanation was a gross understatement: "Things at the mine are running more loosely now than before." Attempting again to convey the hopeless financial debacle to Perry, he explained: "The Chisos payroll alone for August was over $5,000. . . . You will see that . . . the $5,000 . . . amount deposited each month does not cover our expenses and that our accounts payable are increasing considerable."[53]

While Perry appears to have held Cartledge responsible for the company's financial imbalance, this condition was not of Cartledge's making. He continued to insist that Perry come to Terlingua "and go over the situation and lay out your plans and give someone of us definite orders."[54] Perry instead responded to his doctor's orders for a rest and embarked on his yacht for a winter vacation in Florida. The problems of saving the mine vanished temporarily from the Terlingua correspondence as the pleasures of the cruise captured Perry's interest. On January 39, 1939, he reported that "the voyage

[49] Cartledge to Perry, August 4, 1937.
[50] Perry to Cartledge, August 6, 1937.
[51] Perry to Cartledge, January 18, 1938.
[52] Perry to Cartledge, August 26, September 14, 1938.
[53] Cartledge to Perry, September 20, 1938.
[54] Cartledge to Perry, February (undated), 1938.

through the various passages and Lake Okeechobee and through the several bays was interesting, but not exciting."[55]

Although Perry remained in Florida most of the winter, he corresponded almost daily with Cartledge, with mounting expenses the priority topic. But two thousand miles away from Terlingua and basking in the warm winter sun, Perry conceived a new plan, spawned through necessity and nourished by a determined creativeness; the abandoned Chisos dumps suddenly assumed an unrealized production potential in Perry's fertile imagination. As Perry's interest mounted, he envisioned an unrealized recovery potential in those mountains of pinkish slag. Cartledge recalled later that Perry's enthusiasm stemmed from earlier conversations with Big Bend miners William D. Burcham and E. A. Waldron, who told him that they believed that the dumps contained ore that should be furnaced. At this point, grasping at straws for survival, Perry multiplied casual conversation into reality and started outlining plans for handling the rejected material. While both Cartledge and Dahlgren, the furnace operator, regarded the venture as both unwise and speculative, Perry nevertheless purchased earth-moving equipment to expedite this operation. "After damnable poor luck we got a break yesterday," Perry writes, "and got a wonderful trade on a Power Shovel and very cheap—$2,600 and fine condition. . . . With this wonderful shovel, Dahl[gren] is going to get his production up to 50 fl. per month."[56]

When Cartledge explained that the dumps contained extremely low grade ore that most operators would have left underground in abandoned workings, Perry insisted that it be surfaced for fear it might prevent the discovery of some rich deposit. "But by this time," Cartledge reminisced, "he wouldn't listen to nobody." The venture proved to be another costly fiasco at a time when Perry could scarcely afford failure.[57]

While Perry's Florida sojourn may have refreshed him physically and mentally, it did nothing to alter conditions at Terlingua, and it fell to Cartledge to face the issues alone. Throughout most of 1939, with work now confined largely to the heretofore abandoned dumps, the company's financial posture remained unchanged. Delinquent accounts, however, had been a long-festering sore at Terlingua, which over the years Cartledge had handled with skill and

[55] Perry to Cartledge, January 30, 1939.
[56] Perry to Cartledge, April 20, 1939.
[57] Cartledge interview, September 28, 1965.

diplomacy when he had money and sometimes when he did not. When pressure increased to settle past-due accounts, Cartledge's first thought always was to protect Perry. "Since most likely you were wanting to know where he [Perry] was so as to try and get some money out of him," Cartledge wrote Albert Mathis and Company in El Paso, "I am going to be good to him myself and am sending a check for $1,400.00 to apply on our account so as you will not have to bother him for a while yet."[58]

As more accounts became delinquent, Perry remained optimistic. He believed that he recognized in the news from Europe his hope for salvation. German aggression would prove as a boon to the quicksilver industry, and, apparently recalling the bonanza years of World War I, he wrote Cartledge enthusiastically, "I think, beyond a doubt, within a year anyway, there will be a great war in Europe and then quicksilver will sell easily at $100.00 higher [than the present market, $79.92]."[59] But his optimism was premature, as a New York buyers' strike stymied quicksilver sales and news from Terlingua remained grim throughout the year.

On January 17, 1940, Cartledge wrote Perry at the Hotel Alcazar in Miami asking for "a little more money this month. . . . I can't pay on some accounts that I need to pay on." Needing fuel supplies to maintain operations, Cartledge explained that he could not place an order "until [I] can send check on a/c [account]."[60] The buyers' strike continued into 1940, forcing Perry to admit temporary insolvency: "I had to borrow money for my last check to Austin [National] Bank."[61] While Perry sent Cartledge $2,600 in February, that amount failed to ease the company's financial bind. "I certainly would appreciate your sending in another deposit," Cartledge explained, "as this was really gone before I got it and I still have Radford [wholesale groceries], Darbyshire [machine company], Standard Oil Co. and a few others hollowing for money." Perry responded that he would send "all we have at this time—something under $3,000."[62]

As the unsatisfactory dialogue continued, the market reacted to the United States' move toward military preparedness; prices for quicksilver rose to $103.94 per flask in 1939 and $176.87 a year

58 Robert L. Cartledge to E. H. Krohn, April 4, 1935.
59 Perry to Cartledge, January 28, 1938.
60 Cartledge to Perry, January 17, 1940.
61 Perry to Cartledge, March 1, 1940.
62 Cartledge to Perry, March 4, March 15, 1940.

later.⁶³ But Perry was unable to share the rewards, as he had nothing to sell. In the summer of 1940 less than a dozen men worked the Chisos No. 8 shaft, the once fabulous "glory hole," searching for veins or stringers that might lead to another strike, but none proved worthy of their effort. Late that summer, with pressure from irate creditors mounting and the gap between company expenses and Perry's bank deposits widening, Cartledge, steadfast as ever "paid Radford Grocery Company $1,000 of my own money just to keep them from suing Mr. Perry."⁶⁴ But his devotion went unrewarded. On August 24, 1940, Perry fired his vigilant observer, reliable confidant, and most trusted and dedicated employee.

In the six barely legible handwritten pages of the dismissal letter, Perry, in an apparent rage of frustration, assesses to Cartledge total responsibility for the mine's decline and concludes that the company's salvation is contingent upon his leaving. In assembling a potpourri of irrelevant, unrelated, and some almost ludicrous events to justify his decision, Perry destroys the credibility of his explanation. He begins as follows (Cartledge's marginal comments appear in brackets with an asterisk):

"For a considerable time I have been much concerned over your remaining here. . . . It is much like the case of a man's wife who has 'Strayed'—Does the husband know anything about it— No—but other men know." If Cartledge was the supposed errant one, Perry fails to clarify that point, instead moving ahead to the death of Feliz Valenzuela, which had occurred more than two years prior to the dismissal. Perry nevertheless accuses Cartledge of being the "master mind" of the episode, noting that while Cartledge was lucky enough to save himself "behind the Mexican Shield . . . the animus remains—Our Mexicans and all Mexicans down here are against you. . . . You cannot safely remain." Perry dwells next on Cartledge's personal welfare, recommending:

First of all you have to somewhat remake yourself—your attitude of mind—you will have to see many things differently—and you will, little by little, as the time goes by, in your home with your family, and indeed you need to get acquainted with your family anew and this will be worth much to you. [*Lost my family to take care of you.] I trust that you will finally come back to us reconstructed in many ways like you used to be—and when your only thought was to do as suited me—[*Allways tried to do your way.] Now and for some time, your thought has been quite different—you have wanted

⁶³ U.S. Department of Interior, *Minerals Yearbook, Review of 1940*, p. 655.
⁶⁴ Cartledge interview, October 12, 1965.

to Control me—you and Ed. Babb together. He used to belong to me—I brought him here—but now you have tied him to you, when I want him to do something—his first thought is whether you will be pleased—if not to demur. [*Always told men to do what ever you wanted.] I have to have instant adherence to my orders, for one thing I run my business at long range—2,000 miles away . . . I can't run my business unless orders are duly executed good or bad. [*Always tried to execute orders.]

Referring next to the size of the Chisos staff, Perry states, "I had distinctly limited you to 34 or 35 and no more *positively*. [*I was against it but Fovargue (superintendent of mining) told me you said to put them on.]" Perry notes also the Mexicans' failure to find ore, Cartledge's failure to collect the Billalba store account while serving as executor of that estate, and "your unrelenting fight on Mr. Anderau [Chisos master mechanic and a Perry favorite] which you have waged for years and years is without a single redeeming feature."

Cartledge's loss of prestige among the Anglo staff, according to Perry, limited his effectiveness as general manager. "The white men in my employ are against you (excepting Babb) almost to a man," Perry claims. "I could write you for hours about the things you have done, which you should not have done." But Perry places the greatest emphasis on Babb's refusal to accept Cartledge's vacant position as general manager. Perry continues: "I had the idea to let Babb take your place . . . [and I told him I] would raise his salary—He flatly declined. Said he would not take your place under any circumstances and he would pack his things up and vacate his house and go away—which I accepted and told him to pack up and leave promptly, all which showed that he is tied to you. [*Babb tells a different story.]"

Interpreting Babb's actions as an attempt to "bluff me into not having you go away," Perry displays again the vindictiveness in his psychological makeup which he reiterates when he claims proudly that "many years ago I discharged every white man on the property at one time in one day—they sure deserved what they got."

Demanding that Cartledge resign "at once as J. P. [justice of the peace] & Notary and turn over to me all diamonds for drilling in your possession," Perry emphasizes that his former general manager is "not to come on this property or any of my properties nor stay at the Hotel—you can send for your mail [and] you have in Deposit here enough money to protect the company. . . . Your salary will go on until tomorrow night." Then in conclusion he explains that "I have to have complete harmony and good will all along the line and going to have it—I can't get it with you and Babb here—at least not at the present—but in the end I expect to save you Mr. Robt. L. Cartledge."[65]

Whatever reasons Perry had for firing Cartledge, they probably do not appear in his dismissal letter but are found rather in his psychological makeup. Of all the truths, half-truths, and fabrications in Perry's letter, none of the accusations can logically justify dismissing an employee of twenty-nine years. If some of the Mexicans, for example, feared and distrusted Cartledge, his almost three decades of service in an administrative position tends to negate the validity of the accusation. And if Cartledge's life was endangered, this apparently had not bothered Perry previously. Neither should it have been a surprise for Perry to learn that the staff size exceeded his instructions, as Cartledge's almost daily letters explained that some of the unit managers (at the Rainbow, Chisos, and Mariposa properties) continued to hire in the face of orders to limit the staff. Also, Perry's claim that "the white men in my employ are against you" obviously stemmed from the confused administrative scramble that Perry stated he created and that Cartledge asked him repeatedly to correct—go "over the situation and lay out your plans and give someone of us definite orders."[66] If Cartledge attempted to "control" Perry, it was through urging the mine owner either to assume or delegate administrative control of the Terlingua properties. Perry's reference to Cartledge's having "enough money to protect the Company" is totally ambiguous. In all probability the company did not have enough money to protect Cartledge, as he stated later that the company owed him $4,358.12 and he filed an approved claim in the bankruptcy proceeding for $1,506.96.[67]

The most revealing episode Perry cites in the letter, however, is Babb's refusal to accept Cartledge's vacated position. This independent action by an employee who "used to belong to me" undoubtedly wounded Perry in his most sensitive part: his ego.

As sole master of the Chisos mine, Perry gave orders, demanded responses, and remained aloof and above reproach. In his own lofty self-appraisal he made no mistakes himself nor tolerated shortcomings in others. "There has not been many mistakes made at the Chisos Mine," he once wrote Cartledge, "no other in the world

[65] Perry to Cartledge, August 24, 1940, typescript in possession of author.
[66] Cartledge to Perry, February (undated), 1938.
[67] Day, "Chisos Quicksilver Bonanza," p. 449; Bankruptcy proceeding file, Cause No. 688.

could have created the Chisos like I had done. And scarcely any other man in the world would have lost it many years ago."[68] But now the Chisos mine was lost—or if not lost, certainly in dire straits —and Perry, through his distorted rationale, could absolve himself of all guilt and responsibility. On the other hand, if mistakes had been made, someone must be made to accept the blame. Vindictiveness, which he expresses with apparent pleasure in the dismissal letter, was an integral part of Perry's thought process: ". . . many years ago I discharged every white man on the property at one time in one day—*they sure deserved what they got*" [my emphasis]. And now Cartledge, the apparent top administrative official at the mine, again standing alone and unprotected, was selected to get what no one apparently deserved—except possibly Perry.

That Cartledge was never given an opportunity to discuss his dismissal casts a further shadow of doubt on Perry's integrity. "He handed me a letter at the mine office," Cartledge later recalled, "and told me not to read it until I got home." That night, after studying the contents of the letter, he walked up to Perry's house to discuss the matter, "but he had locked himself up in the house and wouldn't see me."[69] Although Cartledge remained in Terlingua two or three days, Perry still refused to see him and discuss his dismissal. Finally, on August 27, 1940, after serving the Chisos Mining Company for over twenty-nine years, Cartledge got in his Buick sedan and drove east toward Study Butte. He never saw Howard E. Perry again.

If Perry believed that Cartledge's departure would alter the company's fortunes, he was misled. Only productive ore could save the company, and Fletcher had already given Perry his negative verdict. "The mine," according to the geologist, "had been worked out. . . . In 1939 most of the program I had laid out had been concluded, and none resulted in ore. . . . I told Mr. Perry that before I laid out that development that I considered these chances remote."[70] But a remote chance was all that Perry required. Too much of a gambler to hesitate in the face of poor odds, he forged ahead, trying to keep the company operative until that remote chance materialized in the elusive red cinnabar.

Money and creditors continued to plague him, but with no apparent effect. "I am used to trouble," he once wrote his former general manager. "Mrs. Perry says that I am never quite normal

68 Perry to Cartledge, February 15, 1938.
69 Cartledge interview, October 12, 1965.
70 Transcript of Testimony, *Cause No. 688*.

unless I have at least three serious troubles. I work like a tiger every day and I try not to let things bother me."[71] He apparently was not bothered too much when the Austin National Bank obtained judgment for nonpayment of the Rainbow loan, as he subsequently negotiated a $25,000 unsecured loan from an old San Antonio friend, Charles Baumberger.[72] Feeling these funds were not adequate to meet his immediate expenses, he turned next to his wife, from whom he received an additional $10,000 on January 28, 1942. As security for this loan, he executed a mortgage on his home, which Mrs. Perry accepted in lieu of payment.[73]

Perry's funding yielded little, and the Chisos furnace records tell a tragic story of the once great mine gasping its final breaths: June, 1940, fifty-six flasks; July, 1940, fourteen flasks; and January, 1942, forty-four flasks.[74] By June 22, 1942, when it appeared productive ore would not be forthcoming and Perry would not be able to repay the loan, Baumberger recovered judgment in Bexar County against the Chisos Mining Company for $27,775.[75]

Perry had still other problems to face. In Cartledge's absence, it now fell to Perry and bookkeeper Arthur Ekdahl to answer irate creditors. Hal W. Harman, one of the more vociferous creditors, writes: "A short time back I asked you like a gentleman to pay me the past due account of $160.00 owed me by your firm. . . . I have had a mighty good opinion of you folks, but it is sure slipping now. Of course you understand that I could get hard-boiled and say a lot of things, but I can still be a gentleman." Then threatening to instruct his attorney "to spare no expense to see that you do pay the bill," Harman concludes, "I am not fooling and if you still think that I am you are going to be badly mistaken."[76]

Ekdahl replied that since he "was out of funds at the time," he had forwarded the previous correspondence to Perry in New York. Since litigation was a language Perry understood, Ekdahl soon had temporary cash reserves to settle the Harman account. "We are really sorry," Ekdahl explains, "that you had to wait so long a time for this remittance. With very kindest regards, and in all friendliness, Arthur Ekdahl."[77]

[71] Perry to Cartledge, October 3, 1936.
[72] Bankruptcy proceeding file, *Cause No. 688.*
[73] Transcript of Testimony, *Cause No. 688.*
[74] Furnace Records, Chisos Mining Papers.
[75] Transcript of Testimony, *Cause No. 688.*
[76] Hal W. Harman to Arthur Ekdahl (undated), 1941, Chisos Mining Papers.
[77] Ekdahl to Harmon (undated), 1941, Chisos Mining Papers.

Perry, while more verbose and boastful, lacked Ekdahl's friendly style in his communication with the unhappy vendors. Both tact-
ful and belligerent, he is nevertheless steadfast in his refusal to pay
the Goodyear Rubber Company an overdue balance of $1,048.30.
Excerpts from his voluminous reply, addressed to R. V. Rinehart,
follow (my paragraphing):

> Your good letter of the 20th instant has just been handed me. I
> don't think you are any more disappointed over the slowness of our
> payments to your company than we are. However, our finishing up
> of our expansion program here has been somewhat delayed, partially
> on account of the difficulty in getting many things which we have
> needed. . . . I think I have written you that we have received a man-
> date to make quicksilver and we have been for quite a long time
> nearly breaking our backs to comply. . . .
>
> It is quite impossible to clean up your account just at this time
> and furthermore your good company is not suffering in the least for
> this money. Small accounts are cleaned up quickly all the time and
> many large accounts are cleaned up promptly where some poor con-
> cern is liable to go broke for want of money. Your company has had
> our business for well over a quarter of a century and are a long way
> ahead of the game. . . .
>
> I think I have already written you that every shot and every shell
> requires for its detonation a particle of fulminate of mercury—war
> cannot be made without it. During the World War we furnished our
> country more than twice as much quicksilver as any other mine in
> America. . . . I stabilized it for the government during the World
> War and when in Washington a short time ago they grabbed me ask-
> ing for my help . . . but there is a strong trend for higher prices and
> as soon as we are in the war with both feet the price will rise and
> may sell as high as $325 a flask like it sold for a long time during the
> World War. . . .
>
> Please do not get the idea that this business is poor or that I am
> personally poor. Another thing worth thinking about is that this con-
> cern during the thirty-nine years of its existence has never but once
> given its note for merchandise, to any bank, trust company, or finan-
> cial concern. You will have to look far afield for any concern doing a
> considerable business which can make a statement of this kind. I ask
> you do you not think this very unique? . . .
>
> I own personally about a million dollars worth of property in
> Chicago where there is quite a boom going on so therefore I am not
> selling any of my Chicago property. I own property in Cleveland,
> Ohio, my native town, and my home, a large country estate seven
> miles east of Portland [Maine] on Casco Bay that cost well over
> $500,000 and has been called the finest in the state between Beverly,

Mass. and Bar Harbor. Also, I have property in New York City and Mrs. Perry is easily worth over a million dollars and all of her properties and my own are free from incumbrance. One would think from your letter that there might be danger of this company or myself going fluey. . . .

I am not in the least afraid of you or anybody and no one has better legal representatives. You could not possibly collect for many years and in the meantime you would absolutely lose the good business which we have favored your company with for many years. In view of our friendly relations and the good business which we have favored you with for such a length of time I ask you not to bother us but rather remain friendly. . . . I have used Goodyear tires in my seven or eight cars in my garage at home; they are good but they are not the only good tires to be had in this country.[78]

Perry must have reassessed his financial posture and eventually paid the Goodyear account, as no further communications with Rinehart appear in the company records. But Rinehart was only one of many who were challenging Perry in his final charge to salvage the Chisos property. Time was of the essence and Perry, apparently still thriving on trouble, quickened his pace during June and July of 1942. He visited the mine twice, consulted with attorneys in Alpine and Del Rio, and kept business appointments in San Antonio and Houston. The news from Terlingua, however, remained unchanged, creditors continued to agitate, and the advice from his various legal staffs gave him scant hope of rescuing the dying Chisos mine.

Undeterred, Perry still believed that all was not lost and if given a little more time, he would soon be riding the crest of the new quicksilver boom. The United States' entry into World War II did indeed precipitate a boom climate; the nation's 1942 production of quicksilver, 50,846 flasks, was the highest since 1882, and the price climbed to $196.35 per flask. Even this record recovery failed to meet wartime needs and 38,911 flasks were imported from Canada and Mexico.[79] But Perry's hopes for a role in this new prosperity went unfulfilled and the news from Alpine indicated that inevitably the tide was turning. On July 28, 1942, Brewster County Sheriff Clarence Hord received an execution issued pursuant to the Baumberger judgment, empowering him to post a "notice of sale of the said property of the Chisos Mining Company . . . for the first

[78] Day, "Chisos Quicksilver Bonanza," pp. 446–448.
[79] U.S. Department of Interior, *Minerals Yearbook, 1942*, pp. 713–718.

Tuesday in October [6], 1942, at the Courthouse door in Alpine, Brewster County, State of Texas."[80]

The clock was now a factor. With just two months remaining, Perry, accompanied by his wife and Portland secretary, Beulah Patterson, hastened back to Alpine, where they met with attorney Mae Ament on August 1 "to discuss the matter of reorganizing the Chisos Mining Company under the applicable chapter of the Bankruptcy Act." Five days later Perry filed a petition for corporate reorganization in the United States District Court in El Paso. Listed as *Cause No. 687*, this flagrant document states that the company "has rich mineral properties capable of giving up great amounts of such minerals [quicksilver] but certain of its creditors are harassing and interfering with its operations and particularly its credit by instituting suits on their claims."[81] This action placed the company in voluntary bankruptcy, thereby allowing Perry time to seek additional financing while halting temporarily further action by the creditors. This maneuver, however, had an ominous ring and when the news of bankruptcy reached Terlingua, the last thirty-two miners walked out of the Chisos mine and sought other jobs.[82] Apparently an era was at an end.

Perry did not think so, as on August 13 he filed a Debtor's Motion for Dismissal of the Original Petition stating that he "believes that he will be able to raise some funds in the course of the next few days, by making a trip out of town for this purpose, and that he believes he will get sufficient funds to pay all or satisfy all of the creditors of the debtor corporation."[83] The court concurred and Perry returned to New York to complete negotiations with a suspicion-shrouded firm, Natural Resources, Inc.

The creditors, whose patience was growing short, found little comfort in the legal flotsam and jetsam that continued to be tossed in Perry's wake. Meeting in Alpine they agreed to take action and on September 5, 1942, filed a petition of involuntary bankruptcy in the name of *Watson-Anderson Grocery Company et al.* vs. *The Chisos Mining Company*.[84] This news reached Perry in New York

---

80 Referee's Record and Index, *Cause No. 688*.

81 Bankruptcy proceeding file, *Cause No. 688*. Perry's voluntary bankruptcy plea, *Cause No. 687*, was changed to *Cause No. 688* when the creditors filed an involuntary action against the Chisos. Records of both actions are filed as *Cause No. 688*.

82 Day, "Chisos Quicksilver Bonanza," p. 448.

83 Debtor's Motion for Dismissal of the Original Petition, *Cause No. 688*.

84 Bankruptcy proceeding file, *Cause No. 688*.

and on September 24, his attorneys there, Burke and Burke, telegraphed Maxey Hart, district court clerk, Western District of Texas, in El Paso: "Chisos Mining Company and Howard E. Perry, president, are in negotiation here for an agreement which would furnish new management mine and provide plan for payment all creditors and immediate opening of mine at Terlingua. Kindly submit this telegram to the judge and copies to any dispositions by court be delayed for at least ten days."[85]

On the following day Natural Resources, Inc., a firm based in Covington, Kentucky, and Howard E. Perry completed an operational agreement that allowed Perry to maintain an active interest in the property and be paid a salary in addition to a certain percentage of the profits. Although Natural Resources, Inc., would provide no immediate cash, it agreed to pay the creditors as profits accrued. The contract, signed on September 25, 1942, was a vacant triumph for Perry, as it was invalidated by the court as an illegal transfer of property that now belonged to the creditors.[86] Finally on October 1, 1942, unable to reorganize his property, Howard E. Perry and the Chisos Mining Company, Inc., filed its Amended Original Answer admitting insolvency.[87] The final entry in the Monthly Furnace Record, however, states even more graphically the last death struggle of the once great mine: "Furnace closed down and Rotary closed down all month. 10 fl quicksilver produced from soot made in August."[88]

Perry had finally accepted the inevitable. This Chisos mine was producing no more.

[85] Telegram from Burke and Burke to Maxey Hart, September 24, 1942, Bankruptcy proceeding file, *Cause No. 688.*

[86] Bankrupt's Petition for Amendment of Present Bankruptcy Proceedings to Chapter XI of Said Act, *Cause No. 688.*

[87] Bankrupt's Amended Original Answer, *Cause No. 688.*

[88] Monthly Furnace Record, Chisos Mining Papers.

CHAPTER 12

Bankruptcy

THE legal maneuvers that followed the bankruptcy mark the beginning of a new chapter in the Chisos drama. *Cause No. 688* would prolong the company's legal life for more than two years and, in the end, would reveal publicly the confused and ill-administered inner corporate workings of the once great quicksilver producer, as well as sculpture permanently in legal vernacular the strange and complex image of its president, Howard E. Perry.

Even before the bankruptcy proceeding was initiated, a legal struggle between the creditors had already begun. On the same day (September 5, 1942) that Fred C. Knollenberg, attorney for the petitioning Watson-Anderson group,[1] filed the bankruptcy suit against the Chisos Mining Company, he wrote the San Antonio law firm representing Baumberger—Moursund, Ball, Moursund, and Bergstrom—requesting that the execution of the court levy authorizing the sale of the Chisos property be withheld. When the San Antonio counsel refused to concur, Knollenberg filed his Application for Restraining Order on October 5, the day prior to the scheduled sale of the bankrupt mine. The application, focusing on three key themes, states:

> VI. If the sale is made in accordance with the levy . . . substantially all of the assets of the corporation will be gone, and plaintiffs . . . believe that the property is worth many times the debt and sufficient to pay all of the indebtedness of the Chisos Mining Company, so all of the creditors will suffer irreparable injury unless a restraining order . . . is issued by this court.
>
> X. That [income] . . . from the sale of quicksilver which was produced by the said mining company was used by the said Howard E. Perry to purchase the other property standing in his name [the Rainbow and Mariposa property], and a legal fraud was perpetrated upon the creditors of Chisos Mining Company by said transfer of title in the personal name of Howard E. Perry. . . .

[1] In addition to the Watson-Anderson Grocery Company, other petitioning creditors are J. B. Keefer, Jr., Alpine Lumber Company; S. R. Sharp, doing business as Surebest Bakers; and N. S. Starr, doing business as V. B. Oil Company.

XI. That said judgment in favor of C. Baumberger . . . [be denied because] practically the entire assets of the bankrupt estate, which plaintiffs are informed and believe amount to $50,000 or $600,000, would be lost, and the creditors . . . would recover nothing on their claims.[2]

When later that day Judge Charles A. Boynton issued the restraining order, Baumberger's counsel replied, "if said allegation be true [Chisos valued at $500,000] they are unable to see how the Chisos Mining Company would be insolvent as claimed by Petitioners."[3] That question would be raised many times in the coming months. On the same day (October 14, 1942) that Baumberger's attorney filed their countermotion, the Chisos attorneys filed with Judge Joseph G. Bennis, referee in bankruptcy, a Petition and Schedules listing a total indebtedness of $377,026.49 and assets of $1,236,285.59.[4] Those figures would also be subject to scrutiny by all parties to the proceeding.

If those attending the first meeting of the creditors in El Paso on November 5, 1942, anticipated confronting Perry in person, they were disappointed. Attorney H. P. Talley, representing the Chisos mine, presented the referee a written statement by Dr. W. W. Herrick of New York City, certifying Perry's inability to attend the meeting. Herrick claimed that "because of disabilities incident to his [Perry's] age . . . to take a trip to El Paso would jeopardize his health seriously and should not be considered at this time."[5] When Perry failed to appear on November 5, the judge recessed the meeting until November 20 and ordered Perry via certified mail to present himself in person for examination by the creditors. Again the referee received what would become an all-too-familiar response from Dr. Herrick.[6]

Even in Perry's absence the creditors proceeded with plans for disposing of the bankrupt property and distributing the receipts to the approved claimants. They first elected Jim L. Vance, trustee in

[2] Application for Restraining Order, October 5, 1942, *Cause No. 688, Watson-Anderson Grocery Company et al.* vs. *The Chisos Mining Company*, Federal Records Center, Fort Worth, Texas.
[3] Motion to Dismiss and Answer of C. Baumberger and his Attorneys, October 14, 1942, *Cause No. 688.*
[4] Petition and Schedules, *Cause No. 688.*
[5] Dr. W. W. Herrick, New York, New York, to Whom It May Concern, November 5, 1942, Bankruptcy file, *Cause No. 688.*
[6] Other letters in the bankruptcy proceeding file from Herrick are dated January 4, 1943, September 4, 1943, and January 14, 1944.

bankruptcy, as their court representative in the forthcoming litigation. Jack Ball, representing the Baumberger interests, gave Vance his first opportunity to fulfill his new assignment. When Ball filed his proof of secured claim, $27,775 inclusive of $2,500 allowed as attorney's fees, Vance objected on the ground "that judgment was obtained within four months preceding the filing of the petition in Bankruptcy, at which time the Bankrupt was insolvent, the lien asserted being rendered void under applicable Bankrupt statutes."[7] Vance thereby raised the prime question that would be debated during the ensuing months: was the Chisos Mining Company insolvent when Perry filed bankruptcy? The search for the answer would provide some of this episode's more dramatic moments.

Following the El Paso meeting Vance conducted an on-site inventory of the property, contacted prospective purchasers, and accompanied them to the mine to examine the company's physical assets. Vance also made arrangements to keep the mine pumped free of water, provided around-the-clock security, and employed a small staff to operate the company store, hotel, and tourist courts, and to maintain the electric light plant. Arthur Ekdahl, the former Chisos bookkeeper, was also engaged to bring the company books up to date and execute current tax reports.

The creditors' key decision was what disposition should be made of the bankrupt property. They had two alternatives: one, keep the property intact for sale as an operating mine; or two, sell the equipment and surface facilities for salvage. Vance engaged L. A. Nelson, associate professor of geology at the Texas College of Mines (now University of Texas at El Paso), to examine the surface and underground geology and report his findings to the referee. His appraisal, no source of encouragement for the creditors, follows: "No definite valuation in dollars can be placed on the mining property as I view the condition. In the first place, the value of a mining property naturally depends upon many things. Primarily ore [and] . . . since the ore supply on both the section 295 [Chisos] and 59 [Mariposa] has been apparently depleted, the value of these properties from the mining standpoint has decreased to practically nothing."[8]

Nelson recommended further that both underground and surface equipment "should not be removed from the mine until a definite decision to permanently discontinue mining has been

[7] Referee's Record and Index, *Cause No. 688.*
[8] Appraiser's Report, *Cause No. 688.*

reached." He also stated that the Scott furnace "is obsolete for future use at its present location," as this type of furnace was "designed primarily for a rather high grade ore." Considering the quicksilver embedded in the masonry furnace structure, Nelson advised that this facility "be carefully dismantled and the brick and mortar treated metallurgically to recover the Quicksilver that is contained therein. The estimate of recovery of the mercury in the furnace has been from $10,000 to $20,000."[9]

According to Nelson's appraisal, the rotary furnace also had little value and should be sold. "Other machinery," the geology professor added, "mostly obsolete, housed in the old Power Plant has a value principally for what they could be sold for scrap material. . . . The buildings . . . have only a potential value. The value would be higher if the camp should be kept open." If the facility was to be kept operative, Nelson suggested that mining operations be confined to the Mariposa property.[10]

Vance also solicited former superintendent Robert Cartledge's opinion on the future disposition of the bankrupt property. Recommending that the facility be left intact, Cartledge wrote: "It would be a big mistake to attempt to tear down at this time the Chisos Furnace. It would be purely speculative as to the number of flasks of quicksilver that would be recovered. A propective buyer of the Chisos property would have nothing to speculate on if the furnace had been torn down. On the other hand, the tearing down of the furnace would certainly hurt or ruin a sale to any one expecting to by the property for the purpose of mining."[11]

The creditors elected to leave the Chisos facility intact and on February 16, 1943, Judge Bennis ordered the bankrupt mine advertised for sale to the highest bidder. Eight sealed bids were opened at two o'clock on the afternoon of March 15, 1943, in Room 208 of the O. T. Bassett Tower in El Paso, which, pursuant to his authority "set forth and described," Vance rejected. He then offered "the properties, assets, and effects" for sale in open competitive bidding, which the Texas Railway Equipment Company of Houston purchased for $81,000.[12]

[9] Ibid.
[10] Ibid.
[11] Cartledge to Vance, January 14, 1943, *Cause No. 688.*
[12] Report of Sale, March 19, 1943, *Cause No. 688.* The Texas Railway Equipment Company is a salvage firm, at that time owned jointly by George and Herman Brown, founders of the Brown and Root construction firm, and F. L. Dahlstrom. Dahlstrom, who investigated the property and appeared at the

With the funds from this sale, plus prior receipts from the sale of twenty-five flasks of quicksilver on hand, as well as the store and hotel income, Vance began disbursing funds to the creditors. On April 2, 1943, checks totaling $4,297.85 were sent to thirteen prior labor claimants, and on May 29, the Austin National Bank received $9,833.05 as payment of its lien.[13]

At their May 27 meeting the creditors declared a 5 percent dividend for seventy-one general and unsecured claims from the 100 applicants. While this group received $14,186.70, those claimants omitted from these benefits are especially significant. They are: Claim No. 76, Grace Perry, $4,250.00; Claim No. 88, Grace Perry, $14,975.00; and Claim No. 92, Howard E. Perry, $183,207.10. Judge Bennis ordered that since these claims "have not been finally determined or allowed, opposition to certain of said claims having been heretofore filed . . . it is held that said claims shall not, at this time, participate in the payment of said dividend." Funds were allocated, however, pending determination of the validity of these claims. Bennis ruled further that while the same principle applied in the matter of the Baumberger claim, money sufficient to pay that claim fully should be set aside and the claimant receive also the 5 percent first dividend.[14]

In the Trustee's Second Report Vance reported receipts totaling $87,718.88, which included the $81,000 from the sale of the property.[15] Judge Bennis then ordered payment of a second 5 percent dividend on November 26, noting specifically that the status of the Baumberger and Perry claims should remain unchanged.[16]

----

sale, based his bid price on the salvage potential of the property. Herman Brown, who made the decision to continue the quicksilver operation, according to Dahlstrom, unwisely acted on the advice of E. A. Waldron, a Terlingua district quicksilver miner. Operated as the Esperado Mining Company, the mine lost over $1 million, according to Dahlstrom. Other estimates are higher.

[13] Disbursements to Creditors, *Cause No. 688.* Ten quarterly interest payments of $187.50 were paid on the $12,500 Rainbow mine purchase loan between August 14, 1938, and November 18, 1940. In October, 1942, the bank charged off $6,500 as delinquent, and, according to Austin National Bank assistant vice-president Bernard Goodstein, this was probably paid by bank vice-president Morris Hirshfeld, who had approved the loan.

[14] Order Allowing First Dividend of Five Percent, July 9, 1943, *Cause No. 688.*

[15] Trustee's Second Report, September 22, 1943, *Cause No. 688.*

[16] Order Declaring Second Dividend and Ordering Payment, November 26, 1943, *Cause No. 688.*

With Perry's chances of receiving cash dividends from the bankruptcy settlement growing steadily dimmer, he instructed his El Paso law firm, Kemp, Smith, Goggin, and White, to propose a compromise settlement. On March 16, 1944, J. M. Goggin wrote Judge Bennis that "we wish to make the following offer of settlement to you and the various creditors in the bankruptcy proceeding. . . . This offer is being made on the part of Mr. and Mrs. Perry in order to try to eliminate . . . extensive and lengthy court proceedings." Goggin added that Mrs. Perry is willing that her claims be allowed in the amount of $11,725 [reduced from $19,225] and that Perry agrees to compromise his claim at $20,000 [reduced from $184,207.10]. "In consideration for this," Goggin concluded, "the creditors who have personal judgments as against Howard E. Perry, are to release him from all further liability in connection with such personal judgments."[17]

With two dividends paid and both Perry's and Baumberger's claims still being challenged by Vance and the creditors, Judge Bennis scheduled public hearings on these claims before ruling on their validity, as well as their priority. The purpose was to gain additional evidence to resolve several key issues: one, was the Chisos Mining Company insolvent when Baumberger obtained the judgment; two, did the Terlingua books provide a complete record of the Chisos' true financial condition; three, were the property values claimed by Perry excessive; and four, had the statute of limitations rendered Perry's claims invalid? Although the ensuing hearings did nothing to strengthen either party's legal posture, the questioning did bring into sharp focus many aspects of the Chisos operation that, had Perry been concerned with posterity's view, would have been better left obscured.

Testimony began on April 27, 1944, in the Grand Jury Room of the United States District Court in El Paso.[18] A. R. Fletcher, Perry's former consulting geologist and the first witness in the two-day hearing, focused his testimony on both the current value of the mine and the method of past operation. Evaluating the mine's present worth, Fletcher stated: "That mine is worked out, period. . . . When he [Perry] first put me on the job, I said, 'Mr. Perry, I should not undertake this, the fact the mine had been very, very thoroughly studied by Mr. Lewis [former mine superintendent],

[17] Goggin to Bennis, March 16, 1944, *Cause No. 688.*
[18] The hearings covered four days: April 27–28 and May 26–27, 1944.

apparently all bets had been taken,' I said, 'I want you to understand I am doing this despite the fact that I cannot give you very much hope.' That was the basis under which I took that job."[19]

The litigants pursued next the subject of Perry's administration of the mine. The questioning of Fletcher continued:

Q. Isn't it a fact that Mr. Perry had at that time been operating forty years and has become a little old in years and somewhat contrary?

A. Mr. Perry was contrary?

Q. Yes, Sir, I don't know any better word. We will say self-opinionated?

A. I will say Mr. Perry was eccentric.

Q. The community at large knew, and you realized and discovered the reason that he was not making money with that mine was that he didn't have the proper machinery, and in your opinion, the proper management?

A. No. The real reason was he did not have the ore.

Fletcher testified further that the underground mechanical operation "was very poor. . . . The mine is extremely hot, horribly hot, and there wasn't any provision made for ventilation. Air pressure was low, the machinery broke down frequently, sometimes . . . the breakdown in that machinery would run twenty per cent of a shift. . . . On the other hand, Mr. Lewis' work in hunting for ore and his mapping had been very good, he had done good work but under great handicaps due to equipment."

Arthur Ekdahl, former Chisos bookkeeper, followed Fletcher on the witness stand as the questioning focused on the company's accounting system. W. C. Roche, attorney for the trustee, asked Ekdahl about a duplicate set of company books kept at Portland. Ekdahl's response, however, was interrupted by J. M. Goggin's objection on grounds that Ekdahl's answer would be speculation, since he worked at Terlingua. The former bookkeeper did state that while duplicate sets of ledgers, cash books, and journals were purchased by him, only one set was kept at the mine.

The success of the Baumberger claim rested upon Ball's establishing that the Chisos mine was solvent at the time Perry claimed bankruptcy. Ball, no doubt, was encouraged by Ekdahl's following responses:

[19] Transcript of Testimony, *Cause No. 688.* All material pertinent to the proceedings is from the Transcript of Testimony unless otherwise noted.

Q. You do know that those books do not show all of the accounts of the company?

A. Well, no, I can't say that because I know there was a system of bookkeeping up in Portland, and I don't know what transpired there. All I know is what transpired on these books.

Q. And your books there were in balance and showed the company as being in balance?

A. Yes, sir.

Q. Showed it as being solvent?

A. Yes, sir.

Knowing that the premise of solvency would be challenged by Perry's counsel, Ball moved next to show that if statistical inaccuracies in company records were revealed, they were accountable to Perry's arbitrary business methods and would, if accepted as admissible evidence, strengthen further Ball's argument. He had learned, for example, that the credit column of the Chisos ledger did not show the full value of quicksilver income. As the price of quicksilver fluctuated, Perry consistently allowed the Terlingua office to take credit for only $40 per flask, always less than the actual sale price. Ball questioned Ekdahl as follows:

Q. The only credit you had on your produce, quicksilver, was just arbitrary amounts furnished you either by Mr. Perry or the Portland office?

A. That is right.

Q. And in figuring up profit and loss, or trial balance—get back to this loss and expense—that was based on arbitrary figures you received on quicksilver?

A. That is right.

The Baumberger attorney questioned Ekdahl next about Perry's difficulties with company superintendents, the failure of obsolete equipment, and finally, attempting to assess to Perry personally the responsibility for the company's final collapse, Ball asked, "You do know it was common knowledge, if operating at a loss or failed to make a profit, it was due largely to his [Perry's] arbitrary methods, obsolete methods of operations down there, wasn't it?" Ekdahl responded, "Yes, sir, I think it was common knowledge."

Ball continued to dwell on the incomplete records kept at Terlingua. In response to his questioning Ekdahl admitted that the Mariposa property did not appear on the company's list of assets and that one $305,000 depreciation item was carried on the books the entire time he was employed by Perry—or, since 1918! Ekdahl also stated that even as new facilities and equipment were added—the Oasis, the motion picture theater, new trucks, and automobiles—the $305,000 depreciation item remained unchanged. In addition, some of the accounts receivable, such as bad debts, were as old as twenty years and, according to Perry's former bookkeeper, none were ever charged off.

As each facet of the defunct Chisos Mining Company was examined, the reasons for Perry's continued absence from the courtroom became more apparent. His El Paso legal staff was finding much to object to. And as his personal actions in the company's downfall were revealed to the irate creditors, the recurrent suspicion of fraud and intrigue continued to follow in Perry's wake. He would long regret the claims and declarations made to Earl F. Metcalf when they prepared the court-rejected Natural Resources, Inc., contract.

When asked if at any time subsequent to 1940 he had seen Howard Perry remove any books, records, or papers from the bankrupt properties at Terlingua, Metcalf answered in the affirmative. He stated that about two o'clock on a July night in 1942 he discovered Perry in a parked truck near a place known as Jackknife Bridge about forty miles north of Terlingua. The driver of the stalled vehicle was Thomas V. Skaggs, Perry's former business partner from Lajitas. Metcalf continues:

> I stopped my car to find out what was going on [and] . . . it was at that time that I discovered Mr. Perry in the front seat, very much in a state of excitement which was explained by him as [not] knowing whether or not it might be some robber approaching him. When I asked him why he should fear a robber in a country . . . where he was so well known . . . he stated that he was—as nearly as I can recall his words—that he had things with him which he desired to protect, even to the extent of his life.

Realizing that Perry wished to continue on to Alpine immediately, Metcalf offered to drive him there in his automobile. As he assisted Perry in transferring his bags and personal belongings from the truck, Metcalf discovered the source of Perry's concern. There were two valises, "both of which," according to Metcalf, "were bulging so much that [only] the straps on one and some ropes and string around the other would permit them to be closed. It was quite obvious that they were full of papers, envelopes."

When asked if he had seen the two valises subsequent to that roadside encounter, Metcalf testified that when he met Perry in New York to prepare the Natural Resources contract, "he dug deeply into both of them picking out various papers which he used to quote information to me, several of which he gave me to read and audit and a good number . . . were selected by him to take to his attorney's office the next morning. . . . I asked him why he used that method of carrying his books and records, and he said that was his way of keeping track of his business affairs." Under further questioning by Roche, Metcalf recalled that he had first seen "those cases in Mr. Perry's house at the mine . . . laying on the floor near a bunch of cigar boxes."

Metcalf's testimony about claims Perry made during the Natural Resources contract negotiations further weakened Perry's case in the bankruptcy settlement. His claim that the company owed him $200,000 [cited as $183,207.10 in his claim against the bankrupt mine] had also been previously challenged by Metcalf and his associates. Perry claimed, according to Metcalf, that was the amount he had advanced the company over the years for operating costs when the $40 per flask allocation failed to keep the company solvent. When the Natural Resources, Inc., staff told Perry that they were not interested in taking over a company with that large indebtedness pending, Perry agreed that the items should be withdrawn. Metcalf also testified that prior to drawing the contract, he had learned that most of the money Perry claimed the company owed him was beyond the statute of limitations in Texas.

The inflated values Perry placed on his mining properties were also challenged by Metcalf and his associates, as well as by the attorney for the referee. All agreed, including Perry, that the $100,000 evaluation placed on Sections 41 and 59, the Mariposa property, was unrealistic. (The Petition and Schedules value the combined sections as $137,500.) The questioning of Metcalf continued:

> Q. In any of these conversations that you had with Mr. Perry did he, at any time, agree or admit that this $100,000 valuation was worth only $10,000?
> A. Yes, sir.
> Q. What was the essence of the exact words?
> A. In [an] argument with his attorney, Mr. Daniel Burke, on property values, I stated that I would not under any circumstances . . . bid more than $10,000 for the Mariposa properties. And Mr. Perry agreed that if we were able to arrive at what he considered proper values on the Chisos Mining Company properties . . . he would accept $10,000 for those sections of land, 41 and 59. . . .

Q. Although he had previously stated to you that they were on the books at $100,000?

A. Yes, we were trying to settle down the values shown on the books to what we considered to be their real worth.

Q. And you agreed on the sum of $10,000?

A. Yes, sir.

When it became apparent that the company journals and ledgers provided the court did not contain a complete record of the Chisos transactions, Ekdahl was called again to the stand to explain the unorthodox Chisos accounting procedures. He stated that some functions were handled jointly by both the Terlingua and Portland offices. For example, when large expenditures were made for machinery, "those accounts would be paid by the Portland office and . . . we would just get a yellow memorandum and we would charge a particular account that he [Perry] would pay with the Portland office."

Sources of cash entries on the Terlingua books were sometimes vague and the money frequently found its way eventually into Perry's pocket. Attorney Goggin asked Ekdahl to identify a $3,024.68 item.

Q. And this balance of $3,024.68, you say that came from, in any part anyway, from sales?

A. I think it was the sale of property . . . the sale of some land down there and the company . . . received the check. It was charged on the cash book, and then Mr. Perry wanted it, so we turned around and charged the cash and sent him the check and charged it to his account.

Ekdahl admitted that because of the incomplete records maintained at Terlingua, he did not know whether the property belonged to the Chisos Mining Company or was Perry's personal property. He admitted under further questioning that the company books contained no entry crediting $100,000 to Perry for the transfer of the Texas-Mariposa property to the Chisos Mining Company, nor was there a record of the $25,000 loan from Baumberger or the $12,500 Austin National Bank loan.

If Perry hoped to base his claim on the Terlingua mine records, his attorneys must have now realized that they lacked sufficient evidence with which to represent their client. And even if they believed they had a case when the proceeding began, Judge Bennis destroyed those hopes as the second day of hearings drew to a close. Referring to the incomplete records exhibited by the bankrupt, Bennis asked Ekdahl:

> Q. From the manner in which the records were kept, particularly in view of the fact that part of the records reflecting the sale and disposition of your quicksilver production was maintained at Portland, an audit of the transactions of the Chisos Mining Company, to be true and correct and full, would necessitate an audit of both sets of books, would it not?
>
> A. I would think so, it would be necessary.
>
> Q. Your books at Terlingua would not suffice for the purpose of a complete and accurate audit?
>
> A. No, Sir. They would not suffice, they wouldn't reflect a true value.

As Bennis began his summation of the first two days' testimony, Perry's counsel no doubt realized that for the remainder of the proceeding they would be providing their client only a perfunctory service. Their argument now lacked substance. Bennis stated to the court:

> It is stipulated by the trustee through his attorney of record, and Howard E. Perry and Grace H. Perry, by their attorney of record, and Mr. Goggin, that the books and records maintained at the mine at Terlingua did not disclose the indebtedness owing to Mr. Perry, which has been made the basis of his claim against the bankrupt corporation. . . . I am not fully persuaded that Mr. Perry was physically incapacitated from attending the hearing of November 5, 1942, and the later hearings and submitting to examination at the request of the creditors. In view, however, of the advanced age . . . I am prepared to give him the benefit of the doubt in this respect. . . . In view of the close and intimate relationship of Mr. Perry to the Chisos Mining Company, and in view of the disclosures made by this record of an apparent failure to deliver to the trustee full and complete records of accountings between Mr. Perry and his corporation, I shall require very clear and convincing proof on his part of the justness and validity of the claim he has presented, together with the claim of his wife, Grace H. Perry.

Bennis then recessed the hearings until ten o'clock on the morning of May 26, 1944.

When the testimony resumed the referee announced, "Let the record disclose that neither Mr. Perry or Mrs. Perry is in attendance." After Perry's vigorous travel schedule just prior to the bankruptcy was read into the court record, the opposing litigants viewed his continued absence with increased suspicion. Although the Perrys' testimony was taken in New York and admitted as evidence in the El Paso hearing, the former president of the Chisos mine again justified his failure to appear as the orders of Dr. W. W. Herrick, "one of New York's greatest physicians."

While the first two days of hearings dealt with the priority of the Baumberger judgment, the final two days were allocated to the Perrys' claims. The opening objections of the other claimants sounded like the recapitulation of a familiar theme. The attorney for the trustee, W. C. Roche, objected because "it is impossible to determine the nature and scope and validity of these claims until we have had an opportunity to examine Mr. Perry." As attorney for Baumberger, Ball objected on the grounds "that the claims of Howard Perry are not shown on the books of the corporation kept at Terlingua" and "there is no competent evidence to show any amount claimed was at the time advanced." Jack Rasberry, also representing Baumberger, objected further "that the proof is or will show that Perry owned ninety percent of the company, managed it with alter ego, and that in addition to the fact the books did not show any of the advancement from Perry, such information was actually withheld from the creditors." Roche objected on the further ground that "there has been no proof whatsoever offered for what uses the various sums of money was expended or were actually received by the bankrupt corporation, and that there was no proper book entries on the corporation books proving such claims."

While the Perrys' responses to interrogatories previously submitted by the opposing attorneys were read into the court record, they were seldom enlightening. Perry categorically defended his company policies on the basis of ownership. "I own the business," he testified, "I 90% and Mrs. Perry 10%. I would just draw some money for traveling expenses and spend it and when I had spent it I would just draw some more. I own the business 90% and it didn't make any difference." When asked if it was a fact that he was indebted to the bankrupt, Perry replied, "No, certainly not. There is no reason to explain why I am not indebted to the Bankrupt because the Bankrupt owes me a large amount. The Bankrupt needed money for operation and obtained it from me and Mrs. Perry."

Perry also denied both the removal of records from Terlingua as well as the "juggling of funds" between his and his wife's personal accounts during the last five years. "Anyone that knows Mrs. Perry," the former mine president testified, "knows that there couldn't be any 'juggling' of funds with her. Mrs. Perry at or about

January 28, 1942, loaned me personally $10,000 for which I gave her a mortgage on my home property and she subsequently accepted the property in payment of the debt."

As the trial continued, it became apparent that Perry failed to grasp the simple basics of business management. When asked if any records were kept showing the cost of producing the quicksilver, he replied simply, "No cost accounting was maintained by the company, [it is] too difficult in quicksilver mining."

Mrs. Perry's answers, though more cordial in tone, were no more helpful in unravelling the twisted mass of Perry's business activities. When asked about various specific amounts loaned to her husband, she would reply, "I don't know what was done with it. . . . but in just what way it was used, I can not say. . . . Mr. Perry was out there and he needed money for the company and I wired the $200."

Following the submission of Mrs. Perry's testimony, Goggin offered as evidence a book which he claimed was the Portland ledger. Attorney Ball objected to its presentation as evidence "on the grounds it is not shown to be a part of the records of the Chisos Mining Company, not a book of original entry, and none of the entries therein have been proved up by competent evidence." Roche, representing the trustee, objected on the further ground that "being the Portland ledger, [it] was kept merely as a matter of personal convenience by Howard E. Perry, is in the nature of a private memorandum, [and] has not been properly proved up; said ledger should therefore be excluded."

The hearings that opened with objections ended with objections, and the "convincing proof" of Perry's claim that Judge Bennis stated he was seeking was never forthcoming. The procedural steps of the settlement continued; on December 29, 1944, the third dividend of 50 percent was paid, and on February 12, 1945, Judge Bennis declared a final dividend of .0896 percent, making 68 percent of the indebtedness recovered by the petitioning creditors. Claims totaling $216,837.71 were disallowed, while claims aggregating $81,125.89 were allowed and paid from total receipts of $88,016.06.[20] Of this amount, Baumberger recovered $19,504.59 as a general and unsecured claimant,[21] but Perry's name is noticeably missing from the list of recipients.

Perry never knew the results of the settlement as time finally ran

[20] Referee's Record and Index, *Cause No. 688.*

[21] Attorney Jack Ball states that following the sale of the La Harmonia ranch, which Perry owned jointly with Wayne Cartledge, the Baumberger estate eventually recovered the full amount of the original loan.

out for the former mine owner. There seemed to be a premonition of death contained in a letter to Maude Cartledge acknowledging the death of her father, Eugene Cartledge. "Thank you so very much for your recent letter," he wrote. "I am so glad you father's going was so peaceful—When my time comes I hope my going will be like his."[22] On December 6, 1944, Howard E. Perry died in his sleep in Boston while en route to a Florida vacation. His wish was apparently granted.

Following the bankruptcy sale, Terlingua was never the same. The Chisos mine was reopened as the Esperado Mining Company, which operated the property unsuccessfully until the end of World War II. At that time the mine was closed, the village abandoned, and the surface installations were converted to scrap.

Though still listed on the maps of Texas, no highway signs guide the traveler to Terlingua. The site can no longer be called a town. From a distance the casual passerby can still see the vacant buildings and the rotting head-frames standing like weary sentinels over the gaping shafts. And those who venture close to the darkened excavations might even imagine they hear the sounds of laboring men as they helped Howard E. Perry build his empire in what was one of the most remote and isolated areas of the southwestern frontier.

Much about Perry and his company, by necessity, must be left to conjecture. Why he so seldom inspected his Big Bend properties; why he placed such high trust in Robert Cartledge and then dismissed him without notice; why, despite his attitude toward other people, he inspired enduring loyalty among members of his staff; and why, in the face of imminent failure, Perry struggled so desperately to salvage his mine? These are questions that only Howard Perry might have answered. These gaps in the data, while essentially unimportant, nevertheless add mystery and intriguing appeal to the story of the man, his company, and his town.

Some of those who knew the Chisos mine as Perry did, believe that the mine, like Perry, was just tired and needed a rest—that someday it will produce again. One of those optimists, for example, was Harry Fovargue, the Chisos' last mining superintendent, who stated: "It is difficult to say why he did that [plead bankruptcy]. My opinion is that he was tired. He was 80 years old. He had spent too many years in that burning desert of treeless, yellow hills and

[22] Letter from Howard E. Perry to Maude Cartledge, March 22, 1942, in possession of Maude Cartledge, Austin, Texas.

gray mountains. The Big House on the hill no longer appealed to him. . . . It has been said often that the Chisos was worked out but I do not think so. I believe there still is cinnabar waiting to be mined."[23]

Since the dawn of civilization hidden mineral deposits have intrigued man. They will continue so long as material gain remains a prime motivating force. But whether there is still cinnabar in the abandoned Chisos workings—and it is doubtful—is also essentially unimportant as this chapter in the Terlingua story is ended. Yet the latent possibility still lends romantic appeal to the memories of Howard E. Perry, his Chisos Mining Company, and those who enjoyed the good life at Terlingua.

[23] San Angelo *Standard-Times*, June 8, 1949.

Epilogue

THE historian writing local history, in contrast to works of broad national and international scope, is faced with a two-fold problem: one, while he establishes an interpretive premise on a highly circumscribed topic, he must at the same time avoid being provincial as he projects the topic in a more expansive historical context; and two, on achieving this interpretive ideal, he must also seek an appropriate balance between that which is abstract, universal, and general on the one hand and that which is highly particular to a time and place and person on the other. This study of the Chisos Mining Company at Terlingua symbolizes this challenge.

Excepting various peripheral episodes, the entire Chisos drama was played out in a single village, probably encompassing no more than one square mile of desert terrain in southern Brewster County, Texas. Likewise, Howard E. Perry is the single outstanding personage in the Chisos story, and while his directives emanated from various points across the United States, as well as from Canada and Mexico, only his four-decade association with this company and this region were examined. Perry, his company, and his town, therefore, emerge as the focal points of this study—subjects "highly particular to a time and place and person."

But to fail to cast the man, his company, his policies, and his product in the broader context of a state, national, and international experience is to fail to grasp the total ramifications of the Terlingua experience. This more comprehensive examination, therefore, reveals the social, cultural, and economic tentacles of Perry's Big Bend experience extending far beyond the perimeters of his village. Only by viewing "local history" through the "small end of the telescope" can the historian relate properly a local experience to the broader context of life on this shrinking planet. When this premise is achieved, "local history," as a literal concept, becomes an illusion.

When viewed from a local perspective, Perry's venture at Terlingua appears as an experience unique to that remote area of Texas. That Terlingua's economic base was quicksilver mining in a state where mining is comparatively rare reemphasizes this assumption.

Yet when this experience is projected in a national and international context, an entirely different picture emerges. Instead of being unique, the four-decade history of Perry's Chisos Mining Company falls into three historical categories that give the Chisos story larger social, cultural, political, and economic overtones.

First, Perry's investment in the Big Bend quicksilver properties is analogous to the total late nineteenth-century foreign investment pattern established by United States and European entrepreneurs (British investment in Western cattle, land, and mining, and United States investment in Mexican and South American mining and petroleum ventures are comparable studies); second, an analogy exists between the Terlingua experience and Eastern investments in the American West during this same period.

The analogy is easily discernible in Perry's Terlingua venture, as in both United States and European foreign investment patterns, the undertakings represent attempts of conservative business to enter unfamiliar environments seeking high-yield investments, many times with government sanction and protection. Although Chicago and Terlingua are within the continental United States, in 1902 the geographical, social, political, and ethnic differences between the two areas were analogous to those between the United States and many areas abroad where international business investments were consummated. And although the United States military establishment flexed its muscles in Central America and the Pacific Islands in response to private economic interests, the Big Bend of Texas provided no exception when Perry claimed the security of his interests was threatened by Mexican revolutionaries.

A parallel is also definable in the matter of the basic economic operating premise. Profits—when there were profits—frequently resulted from resource abundance, rather than sound business practices, and in many cases depended on a large domestic labor force— cheap, primitive, and largely untrained—supervised by a skilled management that represented a different cultural and ethnic society. When conflicts arose between the immigrants and the domestic society, local claims of exploitation were countered by claims that social and economic rewards accrued to those claiming exploitation. No doubt these were familiar sounds to Perry.

A third category of historical experience in which the local Perry-Terlingua experience may be projected deals with Perry the man and the source of his political and economic philosophy. In studying the president of the Chisos Mining Company, the student must keep Perry within the social and economic climate from which he emerged. Within this context Perry looms large as the traditional nineteenth-century entrepreneur whose ideological roots were nourished in the speculative environment of post-Appomattox America. Since the economic harvest was still abundant, vast opportunities awaited for men commensurate to the challenge. Perry entered the economic arena and won.

To call the Chisos president controversial is understating the case; complex and provocative are more appropriate adjectives. And while the robber baron nomenclature may be applicable, it also raises an outdated question that has long since become frayed from overuse. The perceptive reader can decide whether Perry the businessman was good or bad; for this study that question is for the most part peripheral. His central role in the Chisos drama runs far deeper.

While admittedly Perry was not above cheating when it appeared he might get away with a bundle of cash, he, like many of his contemporaries, possessed other more substantive traits of character. To achieve success in the industrial melee of post–Civil War America required men of breadth and capacity. And while they were sometimes ruthless in their methods and pretentious in displaying the rewards of victory, their successes were neither frequent nor accidental. Although Perry displayed his trophies of the economic chase—two mansions, a yacht, and a fleet of automobiles—he also established himself in the world of business and finance and brought a measure of civilization to a frontier community. And if he was vain, determined, opportunistic, vindictive, and rough, he could also inspire unbelievable loyalty in the men who worked for him. It is difficult to see as a pirate a businessman who said: "If we dissect men we destroy them. We have to strike a balance and see if on the whole they are not good."[1]

When Perry is projected into his proper time slot, the emerging image gives the subject greater definition and relevance. Examining the mine owner in this frame of reference may not provide justification for his every action, but it does nevertheless help explain his motivations. During Perry's forty years in the Big Bend, America's industrial economy lacked uniformity and was changing unevenly. While United States Steel shipped ore to its Pittsburgh foundry in steam-powered gondolas, the Chisos Mining Company moved its material through a primitive wilderness in wagon trains. While Eastern skilled laborers were thinking about the benefits of union-

[1] Perry to Cartledge, August 4, 1936.

ization, the Mexican muckers at Terlingua were light years away from the trade union movement, for reasons over which Perry exercised no control. Yet as he functioned within this changing economic environment, he based his decisions on managerial precedent, the dictates of the market place, the relative freedom granted by remoteness, and the permissive mores of contemporary society. This was Perry's world in the Big Bend of Texas, far removed by time and distance from the nerve centers of business and government. Within the security of his desert fastness Perry experienced a freedom of choice unknown in today's sophisticated society.

Early in this century the successful entrepreneur also stood tall in American esteem. Movie stars, golf professionals, television personalities, and Indy 500 winners had not yet achieved mass identity; instead the national heroes, as well as the philosophic ideals by which the nation's masses measured success, were its business leaders. America was above all a business-oriented society in which business activity and the business point of view were major forces in shaping our fundamental institutions, as well as our attitudes toward business in particular and society in general.

Society therefore gave Perry and men of his ilk a voice that spoke with resounding authority, and if the decisions Perry articulated at Terlingua in the 1930s offend a more socially conscious America in the 1970s, then the finger of guilt should be pointed at society, not Perry. If Perry practiced racism in the Big Bend, it was because a racist society placed no restrictions on his actions. If Perry was guilty of exploitation, then it should also be remembered that the exploited came begging for the opportunity. The Mexican to whom Perry offered a dollar a day and a place to live considered himself fortunate. If he could have found work in the Mexican mines of Sierra Mojada, he would have received less than half the amount Perry paid. Thus, in the Mexicans' appraisal, Chisos employment represented a step up the social and economic ladder. And if, from a contemporary perspective, this projects Perry in an unfavorable light, the Mexican who solicited employment at Terlingua probably would not agree. He saw no basis to complain, because Terlingua appeared strangely familiar: the same 5 percent of that society who made the decisions, received the good salaries, and lived in the better homes differed little from a comparable segment of society he had known in Mexico. To the average Mexican immigrant who crossed the Rio Grande to work for Perry, little had changed except he had progressed toward a better life in America.

If America was a business-oriented society dominated by the

business point of view when the twentieth century emerged, so it remains today. While change has come with the passage of time—new products, new methods, new markets, and new leadership—the basic philosophy of the business community continues virtually unaltered. Profit is still the businessman's guiding principle. But as the shadows of Perry's Terlingua twilight grew longer, social change was in the offing. The economic upheaval of the 1930s cast in sharp relief social inequities that had too long gone unnoticed. What emergency legislation began in the mid-1930s, additional legislation and expanded unionization continued in the 1940s and 1950s.

An opposing struggle between the market place and the union hall, now watched over by a more concerned society and a sometimes responsive legislature, partially restrains the business community's freedom of choice. But the balance is tenuous and when the scales are tipped in business' favor, an instant replay of a familiar drama begins. While the episode and the performers may place an occurrence in its proper time slot in history, the basic premise is universal: Teapot Dome, Watergate, IBM and Motorola, Gulf Oil Corporation and its independent distributors, ad infinitum.

Thus the drama continues. And the performers, unknowingly and without pen and mantle, are eventually accorded membership in the same philosophic fraternity that still lists Howard E. Perry as a charter member.

While the business point of view remains, some citizens are beginning to question its yields in terms of enduring values. A half-century ago the perceptive C. A. Hawley, recognizing industry's ravages of the nation's natural resources, expressed the feelings of many Americans of the 1970s, when he wrote:

> By comparison [with the Spanish system of prorated quicksilver mining], our American system seems foolish and wasteful. We rush in, in a mad scramble to get rich . . . produce all we can as fast as we can, with little regard for market demand and none at all for future generations, make a millionaire or two, exhaust the deposit, close down the works, and call it done. Ghost mining towns found throughout the West are tragic proof of the lack of national foresight in the management of our mineral resources. Future generations will someday have to pay for all this wanton waste and destruction.[2]

These were indeed prophetic words.

While the names of Perry and the Chisos Mining Company do

---

[2] C. A. Hawley, "Life Along the Border," *Sul Ross State College Bulletin* 44 (September 1964): 32.

not appear in this statement, Hawley's beliefs were, no doubt, inspired by the Terlingua operation, as apparently this was his only contact with the mining industry. His narrative also touches the high points of the Chisos operation: ". . . produce all we can, as fast as we can . . . make a millionaire or two, exhaust the deposit, close down the works, and call it done."

October 1, 1942, when Perry finally "closed it down," marks the point of demarcation in the history of the Terlingua Quicksilver District. During the almost half-century preceding that date, the industrialization of the quicksilver mines opened that remote region of Texas to settlement and pushed back by a few miles one of America's last frontiers. In its wake came settlement and the "facilities of civilization," enjoyed by some who had experienced them before and by some to whom they meant an introduction to a new way of life.

And with settlement and civilization came discovery. The influx of people drawn by industry to this primeval land were astounded by its rugged beauty: the mile-high Chisos Mountains, the imposing canyons, colorful arroyos, the flowering desert, and the expansive plains. All seemed to agree that this area possessed a geographic distinctiveness found nowhere else in the nation. Their exalted praise, however, lacks the humbling impact of a statement made by former Brewster County sheriff and Texas Ranger Captain E. E. Townsend: "It made me see God as I had never seen Him before."[3] As a member of the Texas legislature, Townsend supported the establishment of Big Bend Park, and on November 21, 1955, the 788,682-acre reservation was dedicated. To date more than a million people have enjoyed its unique recreational facilities.

If Perry and the other members of his professional fraternity are guilty of "wanton waste and destruction" on the one hand, their efforts yielded social and cultural benefits on the other. In this projection the Terlingua Quicksilver District experience emerges as a microcosm of the history of mining in the United States. Just as the discovery of minerals in the trans-Mississippi West was the greatest single factor in the settlement of the American West and the final filling of the land, the opening of the Terlingua Quicksilver District brought settlement and civilization to the Big Bend of Texas. If in this process inequities occurred, so also did rewards. Of necessity, they go hand-in-hand.

[3] Virginia Madison and Hallie Stillwell, *How Come It's Called That?*, p. 38.

In evaluating America's industrial past in the light of present and future civilization, each facet of the question must be equated with the other. No one can be entirely positive or entirely negative. Likewise, the business leaders themselves—Perry and all men of his kind—should also be accorded the same measure of adjudication in determining their posture in the nation's progress to the present. Probably unknowingly and without foresight, Perry himself articulated the premise upon which the industrialist's role in the scheme of progress ultimately will be determined: "We have to strike a balance and see if on the whole they are not good."

A

History of Quicksilver

"IN recorded history quicksilver is first mentioned in documents by Aristotle (384–322 B.C.) and by Theophrastus about 315 B.C. Theophrastus credited the Athenian Callias with the invention of methods to beneficiate cinnabar in 415 B.C. . . . Plinius and Claudius Galenus wrote of quicksilver as a poison. Pliny (23–79 A.D.) recorded that some 10,000 pounds of cinnabar were brought to Rome each year from what is now the Almaden Mine in Spain. Although the production was no doubt small in these early times, it seems probable that a primitive form of quicksilver mining, with concentration of cinnabar by panning, and a method of extracting quicksilver from cinnabar by heap roasting or retorting was practiced at least 2,500 years ago.

"The early consumption was probably mainly in employing finely ground cinnabar for rouge; perhaps there were also some medicinal uses. Amalgamation seems to have been known to Pliny, who recorded in his seventh book that quicksilver was used in recovering noble metal from earth and also for gilding. . . . Paracelsus (1493–1541) introduced the use of quicksilver as a specific in the treatment of syphilis. From that time on its special properties were more fully appreciated, and the metal and its compounds have been held in high, though fluctuating, regard for medical purposes. Torricelli, a pupil of Galileo, the father of experimental investigation, used quicksilver to determine the pressure of the atmosphere in 1634. This feat marks the introduction of quicksilver into scientific research by the invention of the barometer. . . .

"In 1720 Fahrenheit invented the mercury thermometer, an invaluable aid to industry, which forms the basis of our civilization. Professor Braune, of St. Petersburg, first succeeded in freezing quicksilver during the winter of 1759–60, and this date also marks the common acceptance of the proper classification of quicksilver among the true metals. . . .

"Still later the gold mining of California and the silver mining

Reprinted by permission from C. W. Schuette, *Quicksilver: Bureau of Mines Bulletin Number 335*, pp. 3–5.

of the Comstock [Lode] depended upon the supply of this metal. The fortunes of the Republic were decided by the wealth produced by the aid of quicksilver from California. Since 1799 quicksilver has played another part of prime importance in the history of nations, due to the invention of fulminate by Howard in that year. No modern war can be fought without its aid, and the blasting cap is a necessity in mining."

B

The Controversial Survey 295

THE original survey of Survey 295 was made by John T. Gano, deputy surveyor of Presidio County, on March 7, 1882, by virtue of Land Script Certificate No. 3165, issued to G. C. and S. F. Railroad, March 13, 1881 by W. C. Walsh, commissioner of the Texas General Land Office. This survey was made on request of R. M. Gano and Sons, to the County Survey of Presidio County, Fort Davis, Texas, on February 28, 1882. The plat of this survey was filed in the General Land Office on May 28, 1882, which is now designated as Land Office Record, Brewster County Rolled Sketch "B"; title of map being, "Map of Southern Part of Presidio County, Showing The Surveys Made for R. M. Gano & Sons and the State of Texas with the Adjoining and Connecting Surveys." The certification by John T. Gano is dated 1882.

Field notes for the Survey 295, originally designated Survey 296, were compiled by John T. Gano, are dated March 7, 1882, and were submitted to commissioner of the General Land Office for his examination and approval.

When on December 7, 1882, Commissioner Walsh advised E. G. Gleim, surveyor of Presidio County, that certain changes had been made in the numerical designations of his field notes, Survey 296 was changed to 295. Prior to the commissioner's letter to Gleim, the field notes which had been changed to 295 were endorsed "Correct on Map of Presidio County, December 6, 1882," and signed by Ernst von Rosenberg, examiner in the General Land Office.

The map designated as Rolled Sketch "B" of Brewster County reflects the changes mentioned in Commissioner Walsh's letter. Gano's Survey numbers were entered on the map in red ink. The last digit of the four sections referred to were changed in black ink. This is the map dated May 25, 1882. All subsequent maps of Brewster County, dated 1887 and 1896, in the General Land Office, reflect this correction.

The original copy of the Field Notes was executed in black ink, now faded. Across the last digit of the Survey Number heading 296, executed in black, appear two oblique slashes executed in red ink, with the figure five entered above. Likewise, similar changes were made in the metes and bounds description in the body of the field notes. Similar changes were made in Survey 295 so as to delegate it as Survey 296.

The same changes were made in red ink on the sketch drawn in the upper left-hand corner of the Field Notes. An endorsement, also in red ink, indicates that these changes were not effected until September 29, 1900, when they were signed by John J. Terrell, later a commissioner of the General Land Office.

The patent for Survey 295, issued on March 31, 1883, to Augustus Norton, assignee of the G. C. and S. F. Railroad, appears in Vol. 73, Patent No. 161 of the General Land Office Records. This patent is written in accordance with the description of the survey which embodied the corrections as mentioned in Commissioner Walsh's letter of December 7, 1882. Augustus Norton bought Certificate No. 3165 from G. C. and S. F. Railroad for $50.00.

The companion survey made under Certificate No. 3165, originally designated by John T. Gano as being Survey 296, and therefore being the even numbered survey, was set aside for the State of Texas.[1]

[1] Records of Surveys, General Land Office, Austin, Texas.

APPENDIX **C**

The Scott Furnace

"THE Scott furnace was twenty or more feet square, built of several thicknesses of brick, and about thirty-five to forty feet in height. The interior was constructed in such a manner that the ore would slide back and forth downward over large, flat tiles. In this way the ore moved by gravity practically all the way from the main hoist at the mine until it left the furnace and was transported farther down the canyon to the waste dump.

"Adjacent to the furnace, and on about the same level, were the condensers—eight or ten of them. They were somewhat smaller than the furnace, not quite as high, and were also made of thick brick walls. The condensers were connected with each other, near the tops, and to the furnace, by single iron pipes about 2.5 feet in diameter. Each condenser contained a brick partition extending from the ceiling downward to an archway at the floor, thus dividing it into two equal compartments. Beyond the last condenser was a smoke stack, 30 or 40 feet in height. Draft from the furnace, carrying smoke and quicksilver fumes, would pass out near the top of the furnace, into the top of the first condenser, thence downward and through the arch supporting the partition, thence upward into the pipe leading to the second condenser, and so on until reaching the smokestack. To insure sufficient draft a blower was installed in the pipe entering the stack.

"The reduction of quicksilver from the ore state to the liquid state is a simple but interesting process. In the first place, the furnace must be heated to a very high temperature before any ore is introduced. This takes two or three weeks of continuous firing. To prevent the escape of the fumes and consequent loss of quicksilver, and to keep the furnace workmen from becoming salivated, it is important that both furnace and condensers be made as nearly airtight as possible. Even then the peculiarly pungent and disagreeable odor of the fumes is noticeable as one approaches a quicksilver reduction plant.

Reprinted by permission from C. A. Hawley, "Life Along the Border," *Sul Ross State College Bulletin* 44 (September 1964): 44–45.

"When the furnace is in operation one workman, using a heavy hoe with a long iron handle, withdraws a certain quantity of the burned ore, or slag, from the bottom of the furnace. This permits the ore above to slide down over the flat tiles, thus creating space at the top just beneath the flat sliding iron doors. The man above seizes the level, and quickly opening the doors, dumps in a like quantity of ore at the top, closing the doors as quickly as possible. This operation is repeated every hour both day and night.

"Narrow platforms with railings are built around the furnace every six or eight feet, so that workmen may inspect the operations and make certain the ore does not become clogged in its movement downward. Peepholes fitted with removable iron plugs make such inspection possible. Long iron rods are used to loosen any stoppage of the ore.

"The interior of a quicksilver furnace in operation is cherry-red hot, ore and all. The temperature must be above 360° F. at all times. The ore does not melt, however, and appears much the same after passing through the furnace as it did before, except that it has lost the weight of the metal and its peculiar reddish color.

"The condensers serve but one purpose and that is to confine the smoke and fumes until the temperature lowers and the quicksilver condenses, much as dew on the grass is condensed from the moisture in the air. On account of the intense heat, no condensation takes place in the first condenser and but very little in the second. Heavy condensation takes place in the third and fourth condensers and rapidly diminishing amounts farther on. Little or none takes place after the eighth condenser. Complete condensation is, however, impossible. This is shown by the fact that the daily scrapings from a silver plate suspended in the smokestack always showed a trace of quicksilver, in the report made by the assayer.

"The floors of the condensers are slightly inclined to one side, where a small iron pipe about the size of a lead pencil passes through the wall." The reduced metal is removed from the condensers through these iron pipes directly into the iron flasks. It is then ready to be marketed.

APPENDIX D

Monthly Furnace Record, Chisos Mining Company

Date	Flasks Produced	Shipments	Flasks on Hand	Scott Furnace	Rotary Furnace	Retort	Cords of Wood	Operator
12/1919	184	353	173	95	89	—	31	Swanson
3/1920	435	360	340	330	105	—	31	Swanson
10/1920	179	352	112	166	13	—	35	Swanson
5/1923	117	—	487	—	—	—	30½	Dahlgren
8/1923	105	401	139	—	—	—	31	Dahlgren
9/1923	198	—	337	—	—	—	29	Dahlgren
2/1928	117	—	254	84	—	33	28¾	Dahlgren
7/1928	117	—	473	103	—	14	23½	Dahlgren
8/1928	183	362	294	183	—	—	21¾	Dahlgren
2/1930	141	1	356	141	—	—	22½	Dahlgren
8/1930	88	—	169	88	—	—	15¾	Dahlgren
12/1936	141	351	44	—	—	—	18½	Swanson
1/1937	51	—	95	51	—	—	6½	Swanson
2/1937	66	—	161	66	—	—	4¾	Dahlgren
3/1937	48	—	209	48	—	—	11	Dahlgren
4/1937	51	—	261	51	—	—	4	Dahlgren

Appendix D continued

Date	Flasks Produced	Shipments	Flasks on Hand	Scott Furnace	Rotary Furnace	Retort	Cords of Wood	Operator
5/1937	54	—	315	54	—	—	3½	Dahlgren
		(Ore bins empty)						
6/1937	39	—	354	39	—	—	5¾	Dahlgren
7/1937	73	344	83	73	—	—	13¾	Dahlgren
8/1937	47	—	130	—	—	—	5¼	Dahlgren
	(Cut tonnage—Ore bins empty)							
9/1937	26	—	158	—	—	—	6¼	Dahlgren
10/1937	65	—	223	—	—	—	10½	Dahlgren
	(Furnace closed down Oct. 19th. 7 P.M. Started fire in retort 4 P.M.)							
11/1937	259	345	138	—	—	—	22½	Dahlgren
	(Bought from Bob Speed 69 lbs. Quicksilver Nov. 19th. Through with clean up, total flasks from the clean up 203.)							
12/1937	36	—	177	—	—	—	—	Dahlgren
	(Bought one flask from Mr. Duncan. Retort closed down Dec. 20th. Total flasks from retort, 142. Total flasks from condensers and retort 345.)							
6/1940	56	51	17	—	—	—	10	Dahlgren
7/1940	14	16	0	—	—	—	36½	Dahlgren
1/1942	44	19	—	—	—	—	—	Dahlgren
9/1942	10	—	—	—	—	—	—	Swanson
	(Furnace closed down and Rotary closed down all month. Ten flasks quicksilver produced from soot made in August.)							

SOURCE: Chisos Mining Papers, Archives of the Texas State Library, Austin, Texas.

Bibliography

PRIMARY SOURCES

Correspondence

Charles Laurence Baker, Cardova, Illinois, November 16, 1964.
Walter K. Bailey, Cleveland, Ohio, May 24, 1968.
W. D. Burcham, Alpine, Texas, April 4, 1968.
James B. Burke, New York, New York, September 28, 1965.
J. E. Colcord, South Portland, Maine, September 15, 1965.
Ruth H. Fovargue, Painesville, Ohio, August 13, 1966.
Sonei Hohri, New York, April 15, 1968.
Edith Hopson, Alpine, Texas, March 25, 1968.
Stuart T. Penick, La Grange, Texas, November 22 and December 6, 1964.
Drury M. Phillips, Huntsville, Texas, January 17, 1966.
Sara Pugh, Alpine, Texas, December 29, 1965, and January 5, 1966.
Jeanette Dow Stephens, Shreveport, Louisiana, June 25, 1968.
Joseph C. Wolf, Chicago, Illinois, March 15, 1968.

Interviews

Mae Ament, Alpine, Texas, June 6, 1966.
Mr. and Mrs. Earl Anderau, Alpine, Texas, December 27, 1964.
C. T. Armstrong, San Antonio, Texas, August 24, 1966, April 3, 1968, and
    October 15, 1973.
Zola Avery, Dallas, Texas, November 28, 1973.
Mrs. G. E. Babb, Sanderson, Texas, December 27, 1964.
Jack Ball, San Antonio, Texas.
Mrs. Petra Benavídes, Study Butte, Texas, October 25, 1972.
Mrs. Elizabeth Bledsoe, Alpine, Texas, October 25, 1972.
Mr. and Mrs. William D. Burcham, Alpine, Texas, December 29, 1964.
Mrs. William D. Burcham, Alpine, Texas, October 24, 1972.
Father Calles, Alpine, Texas, October 27, 1972.
Robert L. Cartledge, Austin, Texas, various interviews from December,
    1964, to February, 1972.
Wayne Cartledge, Marfa, Texas, December 29, 1964.
Jim Casner, Alpine, Texas, June 6, 1966.

F. L. Dahlstrom, Austin, Texas, November 16, 1964, and October 17, 1965.

Hattie Grace Elliot, Alpine, Texas, December 29, 1964.

H. C. Hernandez, Alpine, Texas, October 23, 1972.

Elmo Johnson, Sonora, Texas, June 3, 1966.

Wilbur L. Matthews, San Antonio, Texas, August 24, 1966, and October 15, 1973.

Dr. Ross A. Maxwell, Austin, Texas, October 1, 1964, and January 8, 1965.

Hunter Metcalf, Marfa, Texas, June 6, 1966.

Dr. William W. Newcomb, Austin, Texas, April 17, 1965.

Laurance V. Phillips, Austin, Texas, December 14, 1964.

Dr. Donald E. Pohl, Austin, Texas, January 8, 1966.

Pablo Sandate, Alpine, Texas, October 24, 1972.

Harris Smith, Austin, Texas, July 7, 1965, January 21, 1966, and February 2, 1966.

W. D. Smithers, Alpine, Texas, June 7, 1966.

Mack Waters, Panther Junction, Big Bend National Park, October 24, 1972.

Mrs. Paz Valenzuela, Alpine, Texas, October 27, 1972.

Manuscript Collections

Annual Statement of School Funds, Brewster County. Records Division, Texas State Library, Austin, Texas.

Cartledge-Perry Chisos Mining Company Correspondence. In possession of Kenneth B. Ragsdale, Austin, Texas.

Cause No. 133, Equity. Howard E. Perry vs. W. M. Harmon, et al., Defendants. Federal Records Center, Fort Worth, Texas.

Cause No. 688, Watson-Anderson Grocery Company et al. vs. The Chisos Mining Company. Federal Records Center, Fort Worth, Texas.

Cause No. 1252, M. B. Whitlock et al. vs. Chisos Mining Company. Attorney's file, in possession of Wilbur L. Matthews, San Antonio, Texas.

Chisos Mining Company Accident Report. In possession of Peter Koch, Alpine, Texas.

Chisos Mining Company Papers. Archives of the Texas State Library, Austin, Texas.

Local History and Genealogy Collection. The Newberry Library, Chicago, Illinois.

National Recovery Administration. Record Group 9, National Archives, Washington, D.C.

Old Military Records. Record Group 94, National Archives, Washington, D.C.

Superintendent's Annual Report to the State Superintendent of Public Instruction. Records Division, Texas State Library, Austin, Texas.

Public Records

Bill of Sale Records. Office of the County Clerk, Brewster County, Alpine, Texas.

Deed Records. Office of the County Clerk, Brewster County, Alpine, Texas.

Election Records. Office of the County Clerk, Brewster County, Alpine, Texas.

Land Records. Office of the County and District Clerk, Brewster County, Alpine, Texas.

Minutes. Commissioners Court, Brewster County, Alpine, Texas.

Records of Incorporation. Office of the Secretary of State, Augusta, Maine.

Records of Incorporation. Office of the Secretary of State, Austin, Texas.

Records of Surveys. General Land Office, Austin, Texas.

School Records. Office of the County Superintendent, Brewster County, Alpine, Texas.

Tax Records. Office of the Tax Assessor-Collector, Brewster County, Alpine, Texas.

SECONDARY SOURCES

Articles and Periodicals

Blake, William P. "Cinnabar in Texas." In *Transactions of the American Institute of Mining Engineers*, XXV. New York: American Institute of Mining Engineers, 1896.

Burcham, William D. "Quicksilver in Terlingua, Texas." *Engineering and Mining Journal* 103 (January 13, 1917): 97.

Day, James M. "The Chisos Quicksilver Bonanza in the Big Bend of Texas." *The Southwestern Historical Quarterly* 64 (April 1961): 427–453.

Golby, Thomas. "The Story of a Quicksilver Mine." *Engineering and Mining Journal-Press* 118, no. 15 (October 1924): 579–580.

Hawley, C. A. "Life Along the Border." *Sul Ross State College Bulletin* 44 (September 1964): 7–84.

Hornaday, W. D. "The Cinnabar Deposits of Terlingua, Texas." *The Mining World* (December 17, 1910): 1133–34.

Johnson, J. Harlan. "A History of Mercury Mining in the Terlingua District of Texas." *The Mines Magazine* 36 (September 1946): 390–395; (October 1946): 445–448; 37 (March 1947): 28–38; (July 1947): 21–40.

Smithers, W. D. "Bandit Raids in the Big Bend Country." *Sul Ross State College Bulletin* 43 (September 1963): 75–105.

Utley, Robert M. "The Range Cattle Industry in the Big Bend of Texas." *The Southwestern Historical Quarterly* 69 (April 1966): 419–441.

Books

Allen, James B. *The Company Town in the American West.* Norman: University of Oklahoma Press, 1966.

Andreas, A. T. *History of Chicago.* 3 vols. Chicago: By the author, 1884–1886.

Bernstein, Marvin D. *The Mexican Mining Industry, 1890–1950: A Study of the Interaction of Politics, Economics, and Technology.* Albany: State University of New York, 1965.

Brobst, Donald A., and Walden P. Pratt, eds. *United States Mineral Resources, Geological Survey Professional Paper 820.* Washington, D.C.: Government Printing Office, 1973.

Broehl, Wayne G., Jr. *The Molly Maguires.* Cambridge: Harvard University Press, 1964.

Brown, John Briscoe. *Rescue From Chaos.* Oklahoma City: By the author, 1934.

Casey, Clifford B. *Mirages, Mysteries and Reality, Brewster County, Texas, The Big Bend of the Rio Grande.* Hereford, Texas: Pioneer Book Publishers, 1972.

Castañeda, Carlos E. *Our Catholic Heritage in Texas, 1519–1936.* 7 vols. Austin: Von Boeckmann-Jones, 1936–1958.

Clark, G. J., ed. *Memoir of Jeremiah Mason.* Reproduction of privately printed edition of 1873. Boston: Boston Law Book Company, 1917.

Cochran, Thomas C., and William Miller. *The Age of Enterprise: A Social History of Industrial America.* New York: Macmillan Company, 1942.

Corning, Leavitt, Jr. *Baronial Forts of the Big Bend.* San Antonio: Trinity University Press, 1967.

Dumble, E. T. *First Annual Report of the Geological Survey of Texas, 1889.* Austin: State Printing Office, 1890.

Emory, Major William H. *Report of the United States and Mexican Boundary Survey.* United States House of Representatives, 34th Cong., 1st Sess. Washington, D.C.: Cornelius Wendell, 1857.

Fergusson, Erna. *Our Southwest.* New York: Alfred A. Knopf, 1940.

Grodinsky, Julius. *Jay Gould: His Business Career, 1876–1892.* Philadelphia: University of Pennsylvania Press, 1957.

Haley, J. Evetts. *Jeff Milton: A Good Man With a Gun.* Norman: University of Oklahoma Press, 1948.

Heiman, Monica. *A Pioneer Geologist: Biography of Johan August Udden.* Kerrville, Texas: S. M. Udden, 1964.

Hofstadter, Richard. *The Age of Reform: From Bryan to F.D.R.* New York: Alfred A. Knopf, 1955.

Holden, William Curry. *The Espuela Land and Cattle Company: The Study of a Foreign-Owned Ranch in Texas.* Austin: Texas State Historical Association, 1970.

Jackson, W. Turrentine. *The Enterprising Scot: Investors in the American West after 1873.* Edinburgh: Edinburgh University Press, 1968.

Johnson, Hugh S. *The Blue Eagle From Egg to Earth.* Garden City: Doubleday, 1935.

Josephson, Matthew. *The Robber Barons.* New York: Harcourt, Brace & World, 1934.

Kirkland, Edward C. *Business in the Gilded Age.* Madison: University of Wisconsin Press, 1952.

Madison, Virginia. *The Big Bend Country of Texas.* Albuquerque: University of New Mexico Press, 1955.

————, and Hallie Stillwell. *How Come It's Called That?* Albuquerque: University of New Mexico Press, 1958.

Morgan, H. Wayne, ed. *The Gilded Age: A Reappraisal.* Syracuse: Syracuse University Press, 1963.

Navin, Thomas R. *The Whitin Machine Works Since 1831.* Cambridge: Harvard University Press, 1950.

Olmsted, Frederick Law. *Journey Through Texas: A Saddle-Trip on the Southwestern Frontier.* Edited by James Howard. Austin: Von Boeckmann-Jones, 1962.

Phillips, William B., and Benj. F. Hill. *The Terlingua Quicksilver Deposits, Brewster County, Bulletin Number 4.* Austin: The University of Texas Mineral Survey, 1902.

Roemer, Ferdinand. *Texas: With Particular Reference to German Immigration and the Physical Appearance of the Country; Described Through Personal Observation.* San Antonio: Standard Printing Company, 1935.

Schlesinger, Arthur M., Jr. *The Age of Roosevelt: The Coming of the New Deal.* Boston: Houghton Mifflin, 1959.

Schuette, C. N. *Quicksilver, Bureau of Mines Bulletin Number 335.* Washington, D. C.: U. S. Department of Commerce, 1931.

Sellards, E. H., and C. L. Baker. *Structural and Economic Geology. The Geology of Texas,* vol. 2. *University of Texas Bulletin No. 3401* (January 1, 1934).

Sheffy, Lester Fields. *The Francklyn Land & Cattle Company: A Panhandle Enterprise, 1882–1957.* Austin: University of Texas Press, 1963.

Shipman, Alice Jack (Mrs. O. L.). *Taming the Big Bend.* Austin: Von Boeckmann-Jones, 1926.

Southwest Reporter, 2nd Series. XCVI. Minneapolis: West Publishing Company, 1936.

Spence, Clark C. *British Investments in the American Mining Frontier.* Ithaca: Cornell University Press, 1958.

Streeruwitz, W. H. von. *Geology of Trans-Pecos Texas: Geological Survey of Texas, Preliminary Statement.* Austin: State Printing Office, 1890.

Texas House Journal, Regular Session, Forty-fourth Legislature, 1935. Austin: Von Boeckmann-Jones, 1935.

United States Department of the Interior. *Mineral Resources of the United States, 1903.* Washington, D.C.: Government Printing Office, 1903.

————. *Mineral Resources of the United States, 1905.* Washington, D.C.: Government Printing Office, 1905.

————. *Mineral Resources of the United States, 1917, Part I.* Washington, D.C.: Government Printing Office, 1917.

————. *Minerals Yearbook, 1935.* Washington, D.C.: Government Printing Office, 1935.

————. *Minerals Yearbook, Review of 1940.* Washington, D.C.: Government Printing Office, 1941.

Wecter, Dixon. *The Age of the Great Depression, 1929–1941.* New York: The Macmillian Company, 1928.

Wyllie, Irvin G. *The Self-Made Man in America: The Myth of Rags to Riches.* New York: The Free Press, 1954.

Yates, Robert G., and George A. Thompson. *Geology and Quicksilver Deposits of the Terlingua District.* Washington, D.C.: Government Printing Office, 1959.

Zevin, B. D., ed. *Nothing to Fear: The Selected Addresses of Franklin Delano Roosevelt, 1932–1945.* Cambridge: Houghton Mifflin, 1946.

Newspapers

Alpine *Avalanche*
Alpine *Times*
Chicago *Record*
Portland (Maine) *Evening Express*
San Angelo *Standard-Times*

Unpublished Material

Acker, Eva. "The Development of the Mineral Resources of Texas." Master's thesis, East Texas State Teachers College, 1939.

Cain, Alice Virginia. "A History of Brewster County, 1534–1934." Master's thesis, Sul Ross State Teachers College, 1935.

Walker, Katheryn B. "Quicksilver Mining in the Terlingua Area." Master's thesis, Sul Ross State Teachers College, 1957.

Index

DATE DUE